D1271110

HISTORY
of
MATHEMATICS:
STATES OF THE ART

Professor Christoph J. Scriba in his office at the Institut für Geschichte der Naturwissenschaften, Mathematik und Technik, Universtität Hamburg

HISTORY

of

MATHEMATICS:

STATES OF THE ART

Flores quadrivii—Studies in Honor of Christoph J. Scriba

Edited by

JOSEPH W. DAUBEN

The City University of New York
New York, New York

MENSO FOLKERTS

Institut für Geschichte der Naturwissenschaften
der Universität München, Museumsinsel
München, Germany

EBERHARD KNOBLOCH

Institut für Philosophie, Wissenschaftstheorie,
Wissenschafts- und Technikgeschichte
Technische Universität Berlin
Berlin, Germany

HANS WUSSING

Sächsische Akademie der Wissenschaften zu Leipzig
Leipzig, Germany

ACADEMIC PRESS

San Diego Boston New York
London Sydney Tokyo Toronto

Academic Press, Inc.
A Division of Harcourt Brace & Company
525 B Street, Suite 1900, San Diego, California 92101-4495

United Kingdom Edition published by
Academic Press Limited
24-28 Oval Road, London NW1 7DX

Library of Congress Cataloging-in-Publication Data

History of mathematics : states of the art / edited by Joseph W. Dauben [et al].
        p.   cm.
    Contributions in English, French, and German
    Includes index.
    ISBN 0-12-204055-4
    1. Mathematics—History.   I. Dauben, Joseph Warren, date.
QA21.H58   1995
510'.9—dc20
                                                                94-42042
                                                                CIP

PRINTED IN THE UNITED STATES OF AMERICA
96   97   98   99   00   01   QW   9   8   7   6   5   4   3   2   1

Contents

Contributors   vii
Preface: Christoph J. Scriba—65 Years   ix
Vorwort: Christoph J. Scriba —65 Jahre   xiii
Bibliographie: Christoph J. Scriba   xvii

Art

1. The Mathematical Treatment of Anamorphoses from Piero della
   Francesca to Niceron   3
   Kirsti Andersen

2. Mozart 18, Beethoven 32: Hidden Shadows of Integers in Classical
   Music   29
   I. Grattan-Guinness

3. La style mathématique de Dürer et sa conception de la géométrie   49
   Jeanne Peiffer

People and Events

4. Einige Nachträge zur Biographie von Karl Weierstraß   65
   Kurt-R. Biermann
   Gert Schubring

5. Mathematics at the University of Toronto: Abraham Robinson in
   Canada (1951–1957)   93
   Joseph W. Dauben

6. N. N. Luzin and the Affair of the 'National Fascist Center' 137
   Sergei S. Demidov
   Charles E. Ford

7. Johannes Praetorius (1537–1616)—ein bedeutender Mathematiker
   und Astronom des 16. Jahrhunderts   149
   Menso Folkerts

The Transmission and Evolution of Ideas: Various Traditions

8. Einiges über die Handschrift Leiden 399/1 und die arabisch-
   lateinische Übersetzung von Gerhard von Cremona   173
   Hubert L. L. Busard

9. The *Book of Assumptions* by Thābit ibn Qurra (836–901)   207
   Yvonne Dold-Samplonius

10. Partielle Differentiation im Briefwechsel Eulers mit Niklaus I
    Bernoulli—eine Miszelle   223
    Emil A. Fellmann

11. Zur Rezeption der arabischen Astronomie im 15. und 16. Jahrhundert   237
    Eberhard Knobloch

12. Die Rückführung des allgemeinen auf den Sonderfall—Eine
    Neubetrachtung des Grenzwertsatzes für binomiale Verteilungen von
    Abraham de Moivre   263
    Ivo Schneider

The Evolution of Notation

13. Zur Geschichte der negativen Zahlen   279
    Helmuth Gericke

14. Begriffs- und Zeichenkonzeptionen in der Mathematik des lateinischen
    Mittelalters und der Renaissance   307
    Wolfgang Kaunzner

15. Die Rolle Arnold Sommerfelds bei der Diskussion um die
    Vektorrechnung, dargestellt anhand der Quellen im Nachlaß des
    Mathematikers Rudolf Mehmke   319
    Karin Reich

Links with Physics and Commerce

16. Experimental Physics at the University of Leuven during the 18th
    Century   345
    Paul Bockstaele

17. Zur Gründungsgeschichte der Polytechnischen Gesellschaft zu Leipzig,
    1825–1827   363
    Hans Wußing

Index   377

Contributors

Kirsti Andersen
History of Science Department
University of Aarhus
DK-8000 Aarhus, Denmark

Kurt-R. Biermann
Lindenberger Weg 22
D-13125 Berlin, Germany

Paul Bockstaele
Graetboslaan 9
B-3050 Oud Heverlee, Belgium

Hubert L. L. Busard
Herungerstraat 123
NL-5911 AK Venlo, The Netherlands

Joseph W. Dauben
Department of History
Herbert H. Lehman College
of the City University of New York
Bronx, New York 10468

Sergei S. Demidov
Institute for History of Science and Technology
103012 Moscow, Russia

Yvonne Dold-Samplonius
Mathematisches Institut
University of Heidelberg
D-69120 Heidelberg, Germany

Emil A. Fellmann
Euler-Archiv
Arnold Böcklinstrasse 37
CH-4051 Basel, Switzerland

Menso Folkerts
Institut für Geschichte der Naturwissenschaften
der Universität München, Museumsinsel
D-80306 München, Germany

Charles E. Ford
Department of Mathematics and Computer Science
Saint Louis University
Saint Louis, Missouri 63103

Helmuth Gericke
Sonnenbergstraße 31
D-79117 Freiburg, Germany

Ivor Grattan-Guinness
Middlesex University at Enfield
Middlesex EN3 4SF, England

Wolfgang Kaunzner
Zoller Straße 9
D-93053 Regensburg, Germany

Eberhard Knobloch
Institut für Philosophie, Wissenschaftstheorie,
Wissenschafts- und Technikgeschichte
Technische Universität Berlin
D-10587 Berlin, Germany

Jeanne Peiffer
Laboratoire d'histoire des sciences et des techniques
F-75013 Paris, France

Karin Reich
Institut für Geschichte der Naturwissenschaften,
Mathematik und Technik
Universität Hamburg
D-20146 Hamburg, Germany

Ivo Schneider
Institut für Geschichte der Naturwissenschaften
der Universität München, Museumsinsel
D-80306 München, Germany

Gert Schubring
Institut für Didaktik der Mathematik
Universität Bielefeld
D-33615 Bielefeld, Germany

Hans Wußing
Sächsische Akademie der Wissenschaften zu Leipzig
D-04107 Leipzig, Germany

Preface
Christoph J. Scriba—65 Years

On October 6, 1994, the internationally renowned German historian of science, Christoph J. Scriba, celebrated his 65th birthday. Professor Scriba, son of Lutheran minister Hans Scriba and his wife Walberta (née Becker), was born in Darmstadt. He studied physics, mathematics, and philosophy in Marburg and Gießen, where he received his doctorate in 1957 for a dissertation on "James Gregory's Early Works on Infinitesimal Calculus." From 1957 until 1962 he taught at various universities in North America, including the University of Kentucky (Lexington), the University of Massachusetts (Amherst), and the University of Toronto (Canada). Supported by a stipend from the *Deutsche Forschungsgemeinschaft,* he spent the following two years at Oxford, studying the mathematical papers of John Wallis.

In 1964 he returned to Germany, where he accepted a position at the University of Hamburg as a lecturer and assistant (1965) at the Institute for History of Science then headed by Bernhard Sticker (later the Institute was renamed "Institute for History of Science, Mathematics and Technology"). In 1966 he presented his *Habilitationsschrift* devoted to "Studies on the Mathematics of John Wallis," and in 1968 was promoted to the rank of *Universitätsdozent.* A year later, in 1969, he was appointed to a newly established position as Full Professor of the History of Exact Sciences and Technology at the Technical University in Berlin. In 1975 he returned to Hamburg where, as the successor of Bernhard Sticker as Professor of History of Science, Professor Scriba has been engaged ever since in teaching and research.

The focus of Professor Scriba's research has been the history of mathematics. But he has also made contributions to the history of science and technology as well. The bibliography of his works (see below) gives further information about this. Here above all mention should be made of his publications on mathematics in the 17th century (Wallis, Leibniz, Newton), on C. F. Gauss, and on the history of number theory from Fermat to Jacobi. He has also had a special interest in the treatment of questions concerning how to use the history of mathematics in mathematics education, in methodological questions concerning the historiography of mathematics, and in connections between mathematics and teaching, as well as mathematics and music.

Thanks to his scientific accomplishments, early in his career Professor

Scriba began to receive both national and international recognition, reflected primarily in membership and positions of leaderhip in numerous academies and institutions. These include:

1967 (corresponding), 1971 (effective) membership in the *Académie internationale d'histoire des sciences* (Paris), which he served as Vice-President (1981–1985).

1972, member of the *Deutsche Akademie der Naturforscher Leopoldina* (Halle), which he also served as a Senator (1982–1992).

1972–1979, expert consultant on History of Science and Technology for the *Deutsche Forschungsgemeinschaft.*

1976–1979, Chairman of the *Deutsche Gesellschaft für Geschichte der Medizin, Naturwissenschaft und Technik.*

1977–1985, President, then Vice-President of the *Nationalkomitee der Bundesrepublik Deutschland* of the International Union of the History and Philosophy of Science (IUHPS).

1977–1985, Chairman of the International Commission on the History of Mathematics, a joint Commission of the International Mathematical Union and the IUHPS.

1989, Chairman of the Organizing Committee of the XVIIIth International Congress of History of Science (August 1–9, Hamburg and Munich). As Chairman of the "Administrative Group Responsible for the XVIIIth ICHS," he played an essential part in planning this Congress.

1991, Vice-Administrative Director of the Graduate College devoted to "Transmission of Greek and Byzantine Texts—History of Science—Humanistic Research and Neo Latin," University of Hamburg, in which the historians of science at the University of Hamburg are active as faculty members.

1991, Corresponding member of the Hausdorff-Kommission of the *Rheinisch-Westfälische Akademie* (Düsseldorf).

1991, Foreign member of the Royal Belgian Academy of Sciences, Letters and Fine Arts (Brussels).

Professor Scriba has also served influentially as a member of the Executive Committee of the *Deutsches Museum* (1976–1992), of the administrative boards of the *Wilhelm Blaschke Foundation* (since 1980) and the *Hans Schimank Memorial Foundation* (since 1982), and is also a member of the Advisory Board of the *Gottfried Wilhelm Leibniz Society* (since 1981).

He has also served (or continues to serve) as Editor or Associate Editor of numerous scientific journals and series, including *Arbor Scientiarum* (1971–1986), *Archive for History of Exact Sciences* (1973–1987), *Archives internationales d'histoire des sciences* (since 1974), *Historia Mathematica* (1974–1976), *History of Science* (1973–1987), *Humanismus und Technik* (1969–1979), *Isis* (1971–1975), *Studia Leibnitiana* (since 1984), *Sudhoffs Archiv* (1969–1979), *Ganita Bhāratī* (since 1983), and *Acta historica Leopoldina* (since 1990).

In his unobtrusive and yet successful way, Professor Scriba has played an essential role in the national as well as the international development of the history of science and technology during the last twenty-five years. Here the numerous meetings he has organized, sometimes by himself, sometimes with the help of colleagues, on history of mathematics at the *Mathematisches Forschungsinstitut* in Oberwolfach/Walke have been of special importance. Similarly, he has made a significant contribution to the continued professional advance of the history of mathematics through his teaching and the graduate students with whom he has worked. Among the dissertations he has directed are those by E. Knobloch (1972), H. J. Zacher (1972), H. Mehrtens (1977), A. Djafari Naini (1981), O. Blumtritt (1984), G. Graßhoff (1985), B. Elsner (1987), D. Herbert (1991), and C. Ketelsen (1994). He was also involved with the *Habilitations* of K. Mauel (1971), M. Folkerts (1973), E. Knobloch (1975), J. Fischer (1986), and C. Meinel (1987).

Perhaps no better recognition could summarize the high esteem of his colleagues for the many contributions he has made to the international development of the history of mathematics than the award he received in 1993 from the International Commission on History of Mathematics at the XIXth International Congress for History of Science held in Zaragoza, Spain (August, 1993). There Professor Scriba was awarded the second Kenneth O. May medal for outstanding contributions to the history of mathematics, in recognition of a scholarly career that has successfully combined both scholarship of the highest quality with service to the profession, contributing substantially to the growth and prominence of the history of mathematics internationally. On the occasion of his 65th birthday, it is a special pleasure to wish him continued success and productivity in all of his endeavors.

Menso Folkerts
Eberhard Knobloch

Earlier appreciations of Christoph J. Scriba's accomplishments:

Folkerts, M. and E. Knobloch. 1989. Christoph J. Scriba—60 Jahre. *Historia Mathematica* **16**, 207–212 (with bibliography through 1988).

Weyer, J. 1990. Christoph J. Scriba 60 Jahre. *Nachrichten aus dem Institut für Geschichte der Naturwissenschaften, Mathematik und Technik* **20**, 22–23.

Vorwort
Christoph J. Scriba—65 Jahre

Am 6. Oktober 1994 wurde der international renommierte deutsche Wissenschaftshistoriker Christoph J. Scriba 65 Jahre alt. Scriba wurde als Sohn des Pfarrers Hans Scriba und seiner Ehefrau Walberta, geb. Becker, in Darmstadt geboren. Er studierte Physik, Mathematik und Philosophie in Marburg und Gießen, wo er 1957 mit einer Arbeit über "James Gregory's frühe Schriften zur Infinitesimalrechnung" den Doktorgrad erwarb. In den Jahren 1957 bis 1962 unterrichtete er an verschiedenen nordamerikanischen (Lexington, Kentucky, und Amherst, Massachusetts) und kanadischen Universitäten (Toronto, Ontario). Ein Stipendium der Deutschen Forschungsgemeinschaft ermöglichte ihm, in den beiden folgenden Jahren den mathematischen Nachlaß von John Wallis zu untersuchen.

1964 kehrte er nach Deutschland zurück und wurde Lehrbeauftragter bzw. Wissenschaftlicher Assistent (1965) an dem damals von Bernhard Sticker geleiteten Hamburger "Institut für Geschichte der Naturwissenschaften." 1966 habilitierte er sich mit "Studien zur Mathematik des John Wallis," wurde 1968 Universitätsdozent und 1969 auf die neugeschaffene ordentliche Professur für Geschichte der exakten Wissenschaften und der Technik an der Technischen Universität Berlin berufen. 1975 wurde er als Nachfolger von Bernhard Sticker ordentlicher Professor für Geschichte der Naturwissenschaften an der Universität Hamburg, wo er seitdem in Lehre und Forschung tätig ist. Das Institut wurde später in "Institut für Geschichte der Naturwissenschaften, Mathematik und Technik" umbenannt.

Der Schwerpunkt seiner wissenschaftlichen Arbeiten liegt auf dem Gebiet der Mathematikgeschichte. Er hat sich jedoch auch mit Beiträgen zur Geschichte der Naturwissenschaften und der Technik hervorgetan. Die Bibliographie seiner Schriften (siehe unten) gibt darüber Auskunft. Hier sind vor allem seine Veröffentlichungen zur Mathematik im 17. Jahrhundert (Wallis, Leibniz, Newton), zu C. F. Gauss, zur Geschichte der Zahlentheorie von Fermat bis Jacobi zu nennen. Sein besonderes Interesse galt und gilt der Behandlung mathematikgeschichtlicher Fragen innerhalb der Didaktik der Mathematik, methodologischen Fragen der Historiographie der Mathematik, den Beziehungen zwischen Mathematik und Technik bzw. Mathematik und Musik.

Scriba gewann durch seine wissenschaftlichen Leistungen schon früh nationales und internationales Ansehen, wovon Mitgliedschaften und leitende

Funktionen in zahlreichen Akademien und Institutionen zeugen. Hier seien die folgenden aufgeführt: 1967 wurde er korrespondierendes, 1971 effektives Mitglied der Académie internationale d'histoire des sciences (Paris), von 1981–1985 einer ihrer Vizepräsidenten. 1972 wurde er Mitglied der.Deutschen Akademie der Naturforscher Leopoldina (Halle), 1982–1992 einer ihrer Senatoren. Er war Fachgutachter der Deutschen Forschungsgemeinschaft für Geschichte der Naturwissenschaften und der Technik (1972–1979). Er war Vorsitzender der Deutschen Gesellschaft für Geschichte der Medizin, Naturwissenschaft und Technik (1976–1979), Präsident, später Vizepräsident des Nationalkomitees der Bundesrepublik Deutschland in der Internationalen Union für Geschichte und Philosophie der Naturwissenschaften (IUHPS) (1977–1985), Chairman der *International Commission on the History of Mathematics* (ICHM) (1977–1985). Vom 1. bis 9. August 1989 fand in Hamburg und München der 18. Internationale Kongreß für Wissenschaftsgeschichte statt. Als Vorsitzender des Organisationskomitees und als Vorsitzender des "Vereins zur Durchführung des 18. Internationalen Kongresses für Geschichte der Wissenschaften e.V." war er in maßgeblicher Weise an der Planung und Durchführung dieses Kongresses beteiligt. Er wurde stellvetretender Sprecher des Graduiertenkollegs "Griechische und byzantinische Textüberlieferung—Wissenschaftsgeschichte—Humanismusforschung und Neulatein," das am 1. April 1991 an der Universität Hamburg unter wesentlicher Beteiligung der Hamburger Wissenschaftshistoriker eingerichtet wurde. In demselben Jahr wurde er zum auswärtigen Mitglied in die Hausdorff-Kommission der Rheinisch-Westfälischen Akademie in Düsseldorf berufen und am 9. Oktober 1991 zum auswärtigen Mitglied der Koninklijke Academie voor Wetenschappen, Letteren en Schone Kunsten van België in Brüssel gewählt.

Er wirkte bzw. wirkt maßgeblich im Verwaltungsausschuß des Deutschen Museums mit (1976–1992), in den Vorständen der Wilhelm-Blaschke-Stiftung (seit 1980) und der Hans Schimank-Gedächtnis-Stiftung (seit 1982), im Beirat der Gottfried-Wilhelm-Leibniz-Gesellschaft (seit 1981).

Scriba war oder ist noch bei zahlreichen wissenschaftshistorischen Zeitschriften und Reihen als (Mit-) Herausgeber tätig: *Arbor Scientiarum* (1971–1986), *Archive for History of Exact Sciences* **11–37** (1973–1987), *Archives internationales d'histoire des sciences* seit **24** (1974), *Historia Mathematica* **1–3** (1974–1976), *History of Science* **11–25** (1973–1987), *Humanismus und Technik* **53–63** (1969–1979), *Isis* **62–66** (1971–1975), *Studia Leibnitiana*, seit **14** (1984), *Sudhoffs Archiv* **53–63** (1969–1979), *Gaṇita Bhāratī*, seit **5** (1983), und *Acta historica Leopoldina* (seit 1990).

In seiner unauffälligen und doch so erfolgreichen Art zu wirken hat er am nationalen wie internationalen Aufschwung der Wissenschafts- und Technikgeschichte während der vergangenen fünfundzwanzig Jahre wesentlichen Anteil. Dazu zählen die von ihm seit Jahrzehnten teils allein, teils mit Kollegen organisierten Tagungen zur Geschichte der Mathematik im Mathematischen Forschungsinstitut Oberwolfach/Walke. Besonders erwäh-nenswert sind auch seine Verdienste um das Heranbilden des wissenschaftlichen Nachwuchses für diese Disziplin. Er betreute u.a. die Dissertationen von E. Knobloch (1972), H. J. Zacher (1972), H. Mehrtens (1977), A. Djafari Naini (1981), O. Blumtritt (1984), G. Graßhoff (1985), B. Elsner (1987), D. Herbert (1991) und C. Ketelsen (1994). Er wirkte bei den Habilitationen von K. Mauel (1971), M. Folkerts (1973), E. Knobloch (1975), J. Fischer (1986) und C. Meinel (1987) mit.

Wie hoch seine Kollegen seinen Beitrag zur internationalen Entwicklung der Mathematikgeschichte einschätzen, zeigt sich u.a. an der Auszeichnung, die ihm die ICHM auf dem 19. Internationalen Kongreß für Geschichte der Natur-wissenschaften in Zaragoza in August 1993 zuerkannte. Ihm wurde die zweite Kenneth-O.-May-Medaille für außerordentliche Verdienste um die Geschichte der Mathematik verliehen. Mit dieser Auszeichnung wurden nicht nur seine bedeutenden wissenschaftlichen Arbeiten gewürdigt, sondern auch seine Aktivitäten innerhalb des Faches, die wesentlich dazu beigetragen haben, daß die Geschichte der Mathematik international an Umfang und Bedeutung gewon-nen hat. Die Glückwünsche zu seinem 65. Geburtstag verbinden wir mit der Hoffnung, daß er weiterhin so erfolgreich wirken wird.

Menso Folkerts
Eberhard Knobloch

Bisherige Würdigungen:

Folkerts, M., und Knobloch, E. 1989. Christoph J. Scriba—60 Jahre. *Historia Mathematica* **16**, 207–212 (mit Bibliographie bis 1988).

Weyer, J. 1990. Christoph J. Scriba 60 Jahre. *Nachrichten aus dem Institut für Geschichte der Naturwissenschaften, Mathematik und Technik* **20**, 22–23.

Bibliographie: Christoph J. Scriba

1957. *James Gregorys frühe Schriften zur Infinitesimalrechnung* (Mitteilungen aus dem Mathematischen Seminar Gießen, Heft 55). Gießen: Mathematisches Seminar.

1961. Zur Lösung des 2. Debeauneschen Problems durch Descartes. Ein Abschnitt aus der Frühgeschichte der inversen Tangentenaufgaben. *Archive for History of Exact Sciences* 2, 406–419.

1963. Bemerkungen zu einem zahlentheoretischen Problem von Roberval. *Nova Acta Leopoldina, Neue Folge* 27, Nr. 167, 339–349.

1964. The inverse method of tangents: A dialogue between Leibniz and Newton (1675–1677). *Archive for History of Exact Sciences* 2, 113–137.

1964. Mercator's Kinckhuysen-Translation in the Bodleian Library at Oxford. *The British Journal for the History of Science* 2, 45–58.

1964–1965. Wallis and Harriot. Centaurus 10, 248–257.

1965. Die Tagungen zur Geschichte der Mathematik im Mathematischen Forschungsinstitut Oberwolfach/Schwarzwald. *Nachrichtenblatt der Deutschen Gesellschaft für Geschichte der Medizin, Naturwissenschaft und Technik e.V.*, Nr. 26, 63–67.

1966. John Wallis's *Treatise of angular sections* and Thâbit ibn Qurra's generalization of the Pythagorean theorem. *Isis* 57, 56–66.

1966. *Studien zur Mathematik des John Wallis (1616–1703). Winkelteilungen, Kombinationslehre und zahlentheoretische Probleme. Im Anhang: Die Bücher und Handschriften von Wallis.* Wiesbaden: Steiner (Boethius. Texte und Abhandlungen zur Geschichte der exakten Wissenschaften, Bd. 6).

1966–1976. Artikel über Geschichte der Mathematik und Biographien von Mathematikern. In *Brockhaus-Enzyklopädie*, 20 Bde. Wiesbaden: Brockhaus.

1967. Über Aufgaben und Probleme mathematikhistorischer Forschung. In *Beiträge zur Geschichte der Wissenschaft und der Technik* (Veröffentlichung der Deutschen Gesellschaft für Geschichte der Medizin, Naturwissenschaft und Technik, 9: Beiträge zur Methodik der Wissenschaftsgeschichte), W. Baron, Hrsg., Wiesbaden: Steiner, 54–80.

1967. A tentative index of the correspondence of John Wallis, F. R. S. *Notes and Records of the Royal Society of London* 22, 58–93.

1968. Geschichte der Mathematik. In *Überblicke Mathematik. 1*, D. Laugwitz, Hrsg. (BI-Hochschultaschenbücher, Nr. 161/161a), Mannheim: Bibliographisches Institut, 9–33.

1968. *The concept of number. A chapter in the history of mathematics, with applications of interest to teachers.* (BI-Hochschulskripten, Nr. 825/825a). Mannheim: Bibliographisches Institut.

1968. Wie läuft Wasser aus einem Gefäß? Eine mathematisch-physikalische Aufzeichnung von John Wallis aus dem Jahr 1667. *Sudhoffs Archiv* **52**, 193–210.

1968. Zur Entstehung der Royal Society. *Sudhoffs Archiv* **52**, 269–271.

1968. Das Problem des Prinzen Ruprecht von der Pfalz. *Praxis der Mathematik* **10**, 241–246.

1969. Ewald Fettweis †. *Nachrichtenblatt der Deutschen Gesellschaft für Geschichte der Medizin, Naturwissenschaft und Technik e.V.* **19**, 17–18.

1969. Neue Dokumente zur Entstehungsgeschichte des Prioritätsstreites zwischen Leibniz und Newton um die Erfindung der Infinitesimalrechnung. In *Studia Leibnitiana—Supplemente II: Mathematik, Naturwissenschaften*, Wiesbaden: Steiner, 69–78.

1969. Eine mathematische Festvorlesung vor 300 Jahren. *Janus* **56**, 182–190.

1970. 36 Biographien von Mathematikern. In *Große Naturwissenschaftler*, A. Meyer-Abich und F. Krafft, Hrsgg. (Fischer-Bücherei, Nr. 6010). Frankfurt/Main: Fischer.

1970. Zur Entwicklung der additiven Zahlentheorie von Fermat bis Jacobi. *Jahresbericht der Deutschen Mathematiker-Vereinigung* **72**, 122–142.

1970. The autobiography of John Wallis, F. R. S. *Notes and Records of the Royal Society of London* **25**, 17–46.

1970. Geschichtsschreibung der Mathematik. *Gießener Universitätsblätter* **2**, 44–51.

1970–1978. 9 Biographien: Blaschke, Wilhelm Johann Eugen (Vol. 2, 1970, 191–192); Borchardt, Carl Wilhelm (Vol. 2, 1970, 298–299); Crelle, August Leopold (Vol. 3, 1971, 466–467); Jacobi, Carl Gustav Jacob (Vol. 7, 1973, 50–55); Lambert, Johann Heinrich (Vol. 7, 1973, 595–600); Reidemeister, Kurt Werner Friedrich (Vol. 11, 1975, 362–363); Wallis, John (Vol. 14, 1976, 146–155); Wieleitner, Heinrich (Vol. 14, 1976, 336–337); Grassmann, Hermann Günther (Vol. 15, 1978, 192–199; with W. Burau). In *Dictionary of Scientific Biography*, C. Gillispie, Ed. New York: Scribner's.

1971. The French edition of Newton's "Principia" (Translation of the Marquise du Chatelet): 1759 or 1756? In *Actes du XIIe Congrès International d'Histoire des Sciences, Paris 1968.* Tome III B: *Science et philosophie, XVIIe et XVIIIe siècles*, Paris: Albert Blanchard, 117–119.

1971. Geschichte der Mathematik im Spiegel der Zeit. Zugleich eine Würdigung des Schaffens von J. E. Hofmann. *Mitteilungen aus dem Mathematischen Seminar Gießen* **90**, 2–24. (Ebenso: *Verzeichnis der Schriften, Vorlesungen und Vorträge von Joseph Ehrenfried Hofmann*, 51–73).

1971. Siegfried Heller (1. Dezember 1876–9. Juni 1970). *Jahresbericht der Deutschen Mathematiker-Vereinigung* **73**, 1–5.

1971. Zur Entwicklung und Verbreitung der Algebra im 17. Jahrhundert. *Mededelingen uit het Seminarie voor Geschiedenis van de Wiskunde en de Natuurwetenschappen aan de Katholieke Universiteit te Leuven (Communications from the Seminar in the History of Sciences at the University of Louvain)* **4**, 13–22.

1971–1979. Artikel über Geschichte der Mathematik und Technik sowie Biographien von Mathematikern. In *Meyers Enzyklopädisches Lexikon*, 25 Bde. Mannheim: Bibliographischer Verlag.

1972. Einführung in den Nachdruck: John Wallis. *Opera Mathematica*, Bd. I, Hildesheim/New York: Olms, V–XIII.

1972. (mit B. Sticker) Walter Baron. 1904–1971. *Isis* **63**, 384–387.

1972. Die Geschichte der Naturwissenschaften im Spiegel des XIII. Internationalen Kongresses in Moskau. *TUB (= Technische Universität Berlin)* **6**, 535–546.

1973. Joseph Ehrenfried Hofmann † (7. März 1900–7. Mai 1973). *Nachrichtenblatt der Deutschen Gesellschaft für Geschichte der Medizin, Naturwissenschaft und Technik e.V.* **23**, 86–88.

1973. Artikel "Number." In *Dictionary of the History of Ideas*, P. P. Wiener, Ed., Vol. 3, New York: Scribner's, 399–407.

1974. John Pell's English edition of J. H. Rahn's *Teutsche Algebra*. In *For Dirk Struik*, R. S. Cohen et al., Eds., Boston Studies in the Philosophy of Science, Vol. **15**, Dordrecht/Boston: Reidel, 261–274.

1974. Die Behandlung mathematikgeschichtlicher Probleme im Unterricht. In *Beiträge zum Mathematikunterricht*, Vorträge der 8. Bundestagung für Didaktik der Mathematik vom 12. bis 15. März 1974 in Berlin, E. Knobloch, H. Meschkowski, und H. Schütz, Hrsgg. Hannover: Schroedel, 43–54.

1975. Chronology of J. E. Hofmann, bibliographic note, and supplementary bibliography of his publications. *Historia Mathematica* **2**, 147–152.

1975. Die Bedeutung der Forschungen von Professor J. E. Hofmann für die Geschichte der Mathematik. In *XIVth International Congress of the History of Science, Tokyo and Kyoto, Japan, 19–27 August 1974*, Proc. No. 2, Tokyo: Science Council of Japan, 154–157.

1975. The place and function of a "historical introduction" in the curriculum for mathematics students. *Historia Mathematica* **2**, 327–331.

1975. (Katalog, mit Kurt Mauel und Helmut Lindner). *Leonardo da Vinci. Eine Ausstellung von Modellen nach seinen Entwürfen in der Technischen Universität Berlin*, Berlin: Universitätsbibliothek der Technischen Universität (Holländische Ausgabe: Twente/Delft 1977).

1976. (mit M. Folkerts und H. Mehrtens) Bibliographie zur Geschichte der Mathematik. *Nachrichten aus dem Institut für Geschichte der Naturwissenschaften Hamburg* **6**, 8–23.

1976. Naturwissenschaft als geistiger Prozeß. In *Bernhard Sticker, Erfahrung und*

Erkenntnis. Vorträge und Aufsätze zur Geschichte der naturwissenschaftlichen Denkweisen, 1943–1973, Hildesheim: Gerstenberg, 1–7.

1976. Mathematische Theorie und mathematische Erfahrung. *Sudhoffs Archiv* **60**, 338–353.

1977. Zum 200. Geburtstag von Carl Friedrich Gauß. *Beiträge zum Mathematikunterricht*, Vorträge auf der 11. Bundestagung für Didaktik der Mathematik von 8.3. bis 11.3.1977 in Hamburg, Hannover: Schroedel, 253–261.

1977. Carl Friedrich Gauß in der Wissenschaftsgeschichte. *Abhandlungen der Braunschweigischen Wissenschaftlichen Gesellschaft* **27**, 39–56. (Festschrift der Braunschweigischen Wissenschaftlichen Gesellschaft und der Technischen Universität Carolo Wilhelmina zu Braunschweig zur 200. Wiederkehr des Geburtstages von Carl Friedrich Gauß. Göttingen 1977.)

1977. (Herausgabe von) Hofmann, J. E., *Register zu G. W. Leibniz: Mathematische Schriften und Der Briefwechsel mit Mathematikern, hrsg. von C. I. Gerhardt* (Olms Paperbacks **49**), Hildesheim/New York: Olms.

1977. Nachruf auf Prof. Dr. Hans Kangro. *uni hh (Hamburg)* **8** (Nr. 4) (Berichte, Meinungen aus der Universität Hamburg), 25.

1977. Bernhard Sticker † (2. August 1906–30. August 1977). *Studia Leibnitiana* **9**, 159–167.

1978. (mit Hans-Werner Schütt) Bernhard Sticker, 1906–1977. *Archives Internationales d'Histoire des Sciences* **28**, 102–105.

1978. Hans Kangro, 1916–1977. *Archives Internationales d'Histoire des Sciences* **28**, 105–107.

1978. Hans Kangro. June 12, 1916–September 15, 1977. *Isis* **69**, 255–256.

1978. A memorial tribute to Kenneth O. May. *Historia Mathematica* **5**, 8–9.

1978. Hans Schimank 90 Jahre. *Nachrichtenblatt der Deutschen Gesellschaft für Geschichte der Medizin, Naturwissenschaft und Technik e.V.* **28**, 9–11.

1979. Geschichte der Naturwissenschaften als neue Disziplin. Zur Frühgeschichte der Jahresversammlungen in Deutschland und der Internationalen Kongresse. In *Disciplinae Novae. Zur Entstehung neuer Denk- und Arbeitsrichtungen in der Naturwissenschaft. — Festschrift zum 90. Geburtstag von Hans Schimank* (= Veröffentlichung der Joachim Jungius-Gesellschaft der Wissenschaften Hamburg, Nr. **36**), Göttingen: Vandenhoeck & Ruprecht, 9–24.

1979. Heinrich Hermelink †. *Nachrichtenblatt der Deutschen Gesellschaft für Geschichte der Medizin, Naturwissenschaft und Technik e.V.* **29**, 15–16.

1979. Eine zahlentheoretische Aufgabe Bhaskaras II und die sogenannte Pellsche Gleichung. *Der Mathematikunterricht* **25**, 93–101.

1979. Viggo Brun (1885–1978). *uni hh (Hamburg)* **10** (Nr. 4), 35.

1979. Heinrich Hermelink in memoriam. *Historia Mathematica* **6**, 233–235.

1980. Viggo Brun in memoriam. *Historia Mathematica* **7**, 1–6.

1980. Selbstverständnis und Öffentlichkeitsverständnis der Mathematik. In *Exakte Wissenschaften im Wandel. Vier Vorträge zur Chemie, Physik und Mathematik in der Neuzeit. Von Hans Schimank und Christoph J. Scriba*

(= Beiträge zur Geschichte der Wissenschaft und der Technik, Heft **16**), Wiesbaden: Steiner, 36–53.

1981. Von Pascals Dreieck zu Eulers Gamma-Funktion. Zur Entwicklung der Methodik der Interpolation. In *Mathematical Perspectives. Essays on Mathematics and its Historical Development*, J. W. Dauben, Ed. (Festschrift für Kurt-R. Biermann), New York: Academic Press, 221–235.

1981. Hans Schimank, 1888–1979. *Archives Internationales d'Histoire des Sciences* **31**, 200–202.

1981. (Herausgabe von) Schimank, Hans. *Rückschau auf neun Jahrzehnte. Ein Interview mit Harald von Troschke*. Hamburg: Joachim Jungius-Gesellschaft der Wissenschaften.

1981. Wie kommt "Napoleons Satz" zu seinem Namen? *Historia Mathematica* **8**, 458–459.

1981. Hans Schimank (17. März 1888–25. August 1979). *Berichte zur Wissenschaftsgeschichte* **4**, 149–153.

1981. Adolf Pawlowitsch Juschkewitsch 75 Jahre. *Nachrichtenblatt der Deutschen Gesellschaft für Geschichte der Medizin, Naturwissenschaft und Technik e.V.* **31**, 132–134.

1981. Der XVI. Internationale Kongreß für Geschichte der Wissenschaften, Bukarest, 26. August bis 3. September 1981. *Nachrichtenblatt der Deutschen Gesellschaft für Geschichte der Medizin, Naturwissenschaft und Technik e.V.* **31**, 138–142.

1982. Friedrich Engel (1861–1941) / Mathematiker. In *Giessener Gelehrte in der ersten Hälfte des 20.Jahrhunderts*, H. G. Gundel, P. Moraw, und V. Press, Hrsgg. (Lebensbilder aus Hessen **2**. Veröffentlichungen der Historischen Kommission für Hessen 35, 2), Marburg, 212–223.

1982. (Mit A. Kleinert) Bericht über den XVI. Internationalen Kongreß für Geschichte der Wissenschaften in Bukarest (26. August bis 3. September 1981). *Berichte zur Wissenschaftsgeschichte* **5**, 231–236.

1983. Die Rolle der Geschichte der Mathematik in der Ausbildung von Schülern und Lehrern. *Jahresbericht der Deutschen Mathematiker-Vereinigung* **85**, 113–128.

1983. Gregory's converging double sequence. A new look at the controversy between Huygens and Gregory over the "analytical" quadrature of the circle. *Historia Mathematica* **10**, 274–285.

1983. Carl Ludwig Siegel, 1896–1981. *Archives Internationales d'Histoire des Sciences* **33**, 127–129.

1983. Friedrich Klemm, 1904–1983. *Archives Internationales d'Histoire des Sciences* **33**, 328–330.

1984. Eulers zahlentheoretische Studien im Lichte seines wissenschaftlichen Briefwechsels. In *Zum Werk Leonhard Eulers. Vorträge des Euler-Kolloquiums im Mai 1983 in Berlin*, E. Knobloch, I. S. Louhivaara, und J. Winkler, Hrsgg. Basel: Birkhäuser, 67–94. (Ebenfalls *Humanismus und*

Technik **26** (1983), 78–105.)

1984. *Zur Geschichte der Bestimmung rationaler Punkte auf elliptischen Kurven. Das Problem von Behā-Eddin ͨAmūlī* (Berichte aus den Sitzungen der Joachim–Jungius-Gesellschaft der Wissenschaften e. V. Hamburg, **1**, Heft 6). Hamburg.

1984. Abriß der Geschichte der Analytischen Geometrie und Linearen Algebra. In K. Endl, *Analytische Geometrie und Lineare Algebra*, Kap. 12, Gießen: Würfel-Verlag, 206–217, 219.

1984. Hans Schimank. In *Hans Schimank — Eine Bibliographie seiner Veröffentlichungen, bearbeitet von Pia Köppel. Ergänzt um eine Biographie*, A. Kleinert und C. J. Scriba, Hrsgg. (Beiträge zur Geschichte der Wissenschaft und der Technik, **19**), Stuttgart: Steiner, 9–28.

1985. Zur Konstruktion des regelmäßigen Neunecks in der islamischen Welt. In *Mathemata. Festschrift für Helmuth Gericke*, M. Folkerts und U. Lindgren, Hrsgg. (Boethius. Texte und Abhandlungen zur Geschichte der exakten Wissenschaften **12**), Stuttgart: Steiner, 87–94.

1985. Eric Gray Forbes †. *Nachrichtenblatt der Deutschen Gesellschaft für Geschichte der Medizin, Naturwissenschaft und Technik e.V.* **35**, 9–11.

1985. Dreißig Jahre Tagungen zur Geschichte der Mathematik im Mathematischen Forschungsinstitut Oberwolfach. *Berichte zur Wissenschaftsgeschichte* **8**, 47–49.

1985. Nachlese zum Euler-Jahr. *Suhoffs Archiv* **69**, 91–95.

1985. Die Entwicklung der Wissenschaften. Mathematik. In *Panorama der Fridericianischen Zeit. Friedrich der Große und seine Epoche. Ein Handbuch*, J. Ziechmann, Hrsg., Bremen, 93–96.

1985. Zur Erinnerung an Viggo Brun. Zum 100. Geburtstag am 13. Okt. 1985. *Mitteilungen der Mathematischen Gesellschaft in Hamburg* **11**, 271–290.

1985. Thirty Years of the "History of Mathematics" at Oberwolfach. In remembrance of J. E. Hofmann (1900–1973). *Historia Mathematica* **12**, 369–371.

1985. Die mathematischen Wissenschaften im mittelalterlichen Bildungskanon der Sieben Freien Künste. *Acta historica Leopoldina* **16**, 25–54.

1985. Kurt Vogel, 1888–1985. *Archives Internationales d'Histoire des Sciences* **35**, 418–420.

1986. 41 Biographien von Mathematikern. In *Große Naturwissenschaftler. Biographisches Lexikon*, F. Krafft, Hrsg., 2. Aufl. Düsseldorf: VDI Verlag.

1987. On the so-called "classical problems" in the history of mathematics. In *History in Mathematics Education*, I. Grattan-Guinness, Ed. (Cahiers d'Histoire et de Philosophie des Sciences, Nouvelle Série **21**), Paris: Belin, 73–99.

1987. Uses of history in teaching number theory. In *History in Mathematics Education*, I. Grattan-Guinness, Ed., Paris: Belin, 102–115.

1987. Auf der Suche nach neuen Wegen. Die Selbstdarstellung der Leopoldina und der Royal Society in London in ihrer Korrespondenz der ersten Jahre (1664–1669). In *Salve Academicum. Festschrift der Stadt Schweinfurt anläßlich des 300. Jahrestages der Privilegierung der Deutschen Akademie der Naturforscher Leopoldina durch Kaiser Leopold I. vom 7. August 1687* (Veröffentlichungen des Stadtarchivs Schweinfurt, **1**), Schweinfurt, 69–85.

1988. Welche Kreismonde sind elementar quadrierbar? Die 2400jährige Geschichte eines Problems bis zur endgültigen Lösung in den Jahren 1933/1947. *Mitteilungen der Mathematischen Gesellschaft in Hamburg* **11**, 517–539.

1989. Hans Schimank. In *Zum Gedenken an Hans Schimank (1888–1979). Festkolloquium, verbunden mit der Verleihung des Schimank-Preises, aus Anlaß seines 100. Geburtstages am 9. Mai im Geomatikum der Universität* (Hamburger Universitätsreden Nr. **48**), Hamburg, 9–17.

1989. (Mit F. Krafft.) "Abstracts." *XVIIIth International Congress of History of Science*, General Theme: "Science and Political Order—Wissenschaft und Staat," 1st–9th August 1989, Hamburg/Munich. Hamburg: IUHPS/DHS.

1989. (Mit M. Folkerts, K. Reich.) Das Schriftenverzeichnis von Ewald Fettweis (1881–1967) samt einer Würdigung von Olindo Falsirol. *Historia Mathematica* **16**, 360–372.

1990. Matematik og musik. *Nordisk Matematisk Tidskrift* **38**, 3–17 (Dänische Übersetzung von J. Lützen).

1990. (Herausgabe von) Joseph Ehrenfried Hofmann, *Ausgewählte Schriften*. 2 Bde. Hildesheim: Olms.

1990. (Herausgabe von, mit W. Schmidt.) *Frauen in den exakten Naturwissenschaften. Festkolloquium zum 100. Geburtstag von Frau Dr. Margarethe Schimank (1890–1983)* (Beiträge zur Geschichte der Wissenschaft und der Technik, Heft **21**). Stuttgart: Steiner.

1991. (Mit B. Maurer.) Technik und Mathematik. In *Technik und Kultur*. Bd. 3. *Technik und Wissenschaft*, A. Hermann und C. Schönbeck, Hrsgg., Düsseldorf: VDI Verlag, 31–76.

1991. Die Kunst-Rechnungsliebende Societät in Hamburg (gegr. 1690) und die Deutsche Akademie der Naturforscher Leopoldina (gegr. 1652): Gemeinsamkeiten und Unterschiede bei der Förderung der Wissenschaften seit dem 17. Jahrhundert. *Mitteilungen der Mathematischen Gesellschaft in Hamburg* **12**, 629–661. (Festschrift zum 300jährigen Bestehen der Gesellschaft, Dritter Teil).

1992. Zum historischen Verhältnis von Mathematik und Musik. In *Braunschweigische Wissenschaftliche Gesellschaft, Jahrbuch 1990* (Göttingen 1991), 115–152.

1992. In memoriam Clas-Olof Selenius (1922–1991). Historia Mathematica **19**, 325–327.

1992. Einige Bemerkungen zu antiken Konstruktionen. In *Amphora, Festschrift für Hans Wußing zu seinem 65. Geburtstag*, S. S. Demidov, M. Folkerts, D. E. Rowe, und C. J. Scriba, Hrsgg., Basel: Birkhäuser, 677–692.

1993. (Herausgabe von, mit F. Krafft.) *XVIIIth International Congress of History of Science Hamburg–Munich, 1st–9th August 1989, Final report*

(Sudhoffs Archiv, Beiheft **30**), Stuttgart: Steiner.

1993. The beginnings of the International Congresses of the History of Science. *XVIIIth International Congress of History of Science, Hamburg–Munich, 1st–9th August, 1989*, Final Report (= *Sudhoffs Archiv*, Beiheft **30**), 3–10.

1993. Historiographie der Mathematik als Wissenschaft, Kunst — und Macht? *Archives Internationales d'Histoire des Sciences* **42**, 10–26.

1993. Zur Aufgabe 86 des Byzantinischen Rechenbuchs Cod. Vindob. Phil. Gr. 65. In *Vestigia Mathematica. Studies in Medieval and Early Modern Mathematics in Honour of H. L. L. Busard*, M. Folkerts and J. P. Hogendijk, Eds., Amsterdam and Atlanta: Editions Rodopi, 309–314.

1993. Adolf Pawlowitsch Juschkewitsch (15.7.1906–17.7.1993). *Nachrichtenblatt der Deutschen Gesellschaft für Geschichte der Medizin, Naturwissenschaft und Technik, e.V.* **43**, 148–161.

1994. (Mit anderen.) Der XIX. Internationale Kongreß für Geschichte der Naturwissenschaften in Zaragoza, 22.–29. August 1993. *Berichte zur Wissenschaftsgeschichte* **17**, 45–59.

1994. Die International Commission on the History of Mathematics. *Berichte zur Wissenschaftsgeschichte* **17**, 59–60.

1994. International Academy of History of Science. *Berichte zur Wissenschaftsgeschichte* **17**, 132–133.

1994. Zur Geschichte des Hundert-Vögel-Problems — Wege durch Jahrhunderte und von Kontinent zu Kontinent. *MU — Mathematikunterricht* **40**, Heft 3, 34–41.

Art

The Mathematical Treatment of Anamorphoses from Piero della Francesca to Niceron

Kirsti Andersen

History of Science Department, University of Aarhus, DK-8000 Aarhus, Denmark

Mathematically, anamorphic and perspective representations are closely related. Historically, the connection is less obvious. Apart from Piero della Francesca, no writers on anamorphoses applied their knowledge on perspective constructions before the 17th century. At virtually the same time as anamorphic constructions became a part of perspective, a new problem of anamorphoses reached Europe: how to construct cylindrical mirror anamorphoses. This problem was given an exact mathematical solution by Vaulezard. For practical use, Vaulezard's solution was too complicated and was superseded by a much easier method whose mathematical foundation was never explained.

1. INTRODUCTION

Professor Scriba is known for the active interest he takes in the research of his colleagues and for making them aware of special publications. Several years ago when I started my investigations of the history of perspective he gave me a book on anamorphoses. This has inspired me to use the occasion of this *Festschrift* to present the mathematical development of the subject from the Renaissance to the mid-17th century.

Some aspects of the fascinating history of anamorphoses have already been treated in books [Elffers, Leeman, and Schuyt 1975, Baltrušaitis 1977, Meyere and Weijma 1989, Kemp 1990]. Thus many of the artifacts preserved have been reproduced and part of the old literature dealing with anamorphoses has been surveyed. Little attention, however, has been given to how the authors dealt with the mathematical problems involved in producing anamorphoses, and I therefore intend to consider this question here.

3

2. VARIOUS TYPES OF ANAMORPHOSES

The beginning of the article on anamorphosis in the *Oxford Companion to Art* states:

> The word first appears in the 17th century and refers to a drawing or painting that is so executed as to give a distorted image of the object represented but which, if viewed from a certain point or reflected in a curved mirror, shows the object in true proportion, the purpose being to mystify or amuse. [Osborne 1970, 43]

There is no definition of an anamorphosis that is so precise that it can be decided exactly what the concept covers; thus it is not easy to delimitate the concept from other concepts as for instance a *trompe l'oeil*.

In this chapter I mainly deal with a category of anamorphoses for which the following is characteristic: Each of them has to be seen from one fixed eye point and for each a plane exists so that its intersection with the light rays from points in the anamorphosis to the eye shows a "usual" composition. For some anamorphoses the light rays have to be reflected in a mirror before they reach the eye point; this type I call *mirror anamorphoses*. In the case in which no mirror is involved the anamorphosis obeys the rule of perspective and has, like a perspective image, been produced by a central projection; I call this type *perspectival anamorphoses*. The placement of the eye point is decisive for whether a picture appears as a perspective image or a perspectival anamorphosis—as we shall see later.

The two kinds of anamorphoses, just described, do not exhaust the field of what is usually treated in the literature on the history of the topic. It is common to include some distorted pictures that are not perspectival and that are constructed in a very simple way—I return to these in Section 5. Moreover, some authors also include a special representation that was used in connection with objects placed in high positions. The idea of this representation was not to create a "hidden" picture, but to solve a visual problem. Before dealing with the anamorphoses that are the subject of this chapter, I very briefly discuss this problem and its history.

To explain the problem I refer to Fig. 1 which shows an eye point, O, and a profile of a wall containing three equal line segments, AB, CD, and EF. It is obvious from the figure that segment AB, nearest to the eye, is seen under a larger visual angle than CD and a much larger angle than the high-up EF. According to one of Euclid's optical postulates, this implies that AB appears larger than CD and much larger than EF or equivalently

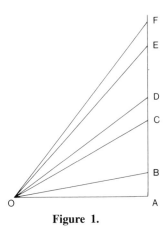

Figure 1.

that an object with a given height appears much smaller when it is placed high above the eye than when it is placed at the level of the eye.

Actually, the change of visual angles is not unique for elevated objects as can be realized by considering Fig. 1 as a horizontal section—or as any other section. Historically, however, the diminishing of the visual angle with an increasing distance from the eye has been closely connected to the problem of deciding the size of elevated objects. This problem was taken up in antiquity, an example being that in his *Sophist* (235–236) Plato seems to be referring to a representation in which objects placed in high positions are made larger to compensate for the smaller visual angles. The question of visual angles was also treated in connection with the perspective problem of depicting a row of equidistant equal cylindrical columns parallel to the picture plane. The latter problem actually existed in two versions; one in which only the diameter of the column is considered and one in which the visible parts of the cylinders are taken into account (for a very thorough discussion of the history of the column problem see [Frangenberg 1992]).

From the Renaissance and onward the problems of depicting elevated objects and columns were often discussed in tracts on perspective. The authors were in doubt whether they should follow the rules of perspective or represent equal line segments so that they appear equal. Many—and in particular artisans—chose the second solution, whereas several mathematicians argued for the first solution. For elevated objects the solution with equal visual angles led to the special representation mentioned earlier, and it has become well known through Albrecht Dürer's illustra-

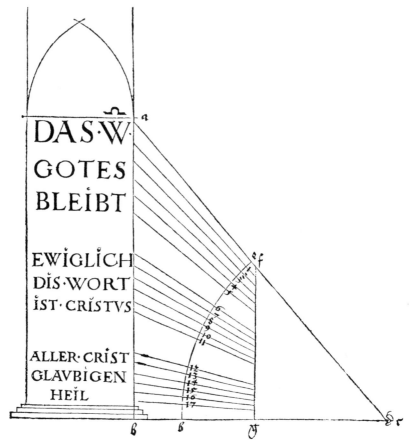

Fig. 2. Dürer's illustration of how to decide the sizes of letters on a column so that visual angles are constant over intervals. Albrecht Dürer, *Underweysung der Messung, mit dem Zirckel und Richtscheyt.* Nürnberg 1525.

tion (Fig. 2). Special names such as *negative perspective* or *decelerated perspective* have been given to this representation [Carter 1970, 847, Baltrušaitis 1977, 6]. It might be considered a part of the history of perspective, because it was treated in connection with perspective, but its relation to anamorphoses is so subtle that I leave it out of the following considerations.

The histories of the perspectival anamorphoses and the mirror anamorphoses run partly parallel; however, rather than following a chronological order I treat the two types separately.

3. EARLY ANAMORPHOSES

The oldest descriptions of the anamorphic art give the impression that an anamorphosis was a popular piece of decoration on walls and furniture during the 16th century, especially in Italy and Germany. Indeed, the idea of a hidden picture matches the Mannerism's fascination with symbols reflecting a magic or mysterious cosmos. With changing fashions many artifacts were lost, so only a little part of the produced anamorphoses exists today.

Among the preserved and more elaborated works are a number composed by Dürer's pupil Erhard Schön; particularly known is his *Vexierbild* from the mid-1530s, hiding four portraits (Fig. 3). Around the same time Hans Holbein constructed an anamorphosis that has become very famous and that was incorporated in his *Ambassadors* (Fig. 4). The painting itself has a usual perspective, but one of its many symbolic objects—the skull—is anamorphic. In the group of Schön's and Holbein's contemporary German authors I am aware of no one dealing with the question of how anamorphoses were constructed. The information about the technique applied in early anamorphoses is to be found in the Italian literature on perspective.

Several of the early reports on anamorphoses deal with a very simple technique for producing distortion—this is presented in section 5. The very first known description of the construction of anamorphoses concerns, however, three sophisticated examples of perspectival anamorphoses. The description occurs at the end of *De Prospectiva Pingendi*, which Piero della Francesca wrote in the second half of the 15th century. In his first example Piero gave instructions for making a drawing on a table so that the drawing seen from a given point gives the visual illusion of a bowl standing on the table. The outcome of Piero's construction is shown in Fig. 5; if this diagram is placed horizontally and seen from the eye point it does look like a standing bowl—thus it could also be called a *trompe l'oeil*. Because, seen from positions other than the eye point, the drawing is distorted, it is also an anamorphosis and it is perspectival, because Piero constructed it by making a central projection of a bowl from the eye point upon the table.

One might wonder how Piero's bowl and other perspectival anamorphoses differ from usual perspective compositions, which are also produced by central projections. I shall return to this question in Section 4, but already here it can be remarked that the choice of the relative positions of the eye point, the object to be projected, and the picture plane is essential for the anamorphic effect.

Fig. 3. Anamorphic portraits of Charles V, Ferdinand I, Pope Paul II and Francis I. Woodcut by Erhard Schön, c. 1535. Below is seen the corrected portrait of Ferdinand I. Reproduced from Baltrušaitis [1977].

Fig. 4. Hans Holbein, *The Ambassadors*, 1533, National Gallery, London. The painting is 206 × 209 cm; this means that, if it is hung with its ground line very near to the floor, the eyes of the ambassadors are on a level with the eyes of the spectator. Moreover, if the painting is placed to the left of a door at a distance that is a bit less than the width of the painting, a spectator passing the door and turning his head will see the skull from its eye point [Baltrušaitis 1977, 104–105].

Piero proceeded with an example in which a sphere instead of a bowl is considered and then closed his book with a construction of an anamorphic drawing on a ceiling. This anamorphosis suggests that a ring is hanging from the ceiling (for a description of the technical details in Piero's construction see Andersen [1992]).

Although more than 400 years passed before *De Prospectiva Pingendi* was printed, its content was known and used by the succeeding generations of Italian perspectivists. However, nobody took up Piero's work on anamorphoses, wherefore it fell into complete oblivion. Even in the modern literature on anamorphoses, one seldom sees a reference to Piero's impressive examples, whereas a couple of Leonardo da Vinci's anamorphic drawings have become famous.

Fig. 5. Piero's anamorphic bowl redrawn, some invisible circle arcs are left out. The drawing should be kept horizontally. Its eye point lies at a distance from the drawing that is almost equal to the height of the anamorphic bowl; moreover, the projection of the eye point upon the plane of the drawing falls at the line of symmetry of the drawing at a distance from its lower part that is approximately equal to $1\frac{1}{2}$ times the height of the anamorphic bowl.

4. PERSPECTIVAL ANAMORPHOSES PRODUCED BY GRIDS

Piero's way of creating anamorphoses by making a perspective projection of an entire object is exceptional in the history of anamorphoses. His successors did not deal with a construction of an anamorphic object, but with the problem of how a given drawing can be changed into an anamorphosis. The drawing was commonly assumed to be situated in a vertical plane perpendicular to a wall that functioned as the picture plane of the anamorphosis. The means to solve this problem was to equip the drawing with a grid of squares and then to construct the anamorphic image of the grid. The latter served as a system of coordinates in which the original drawing could be redrawn.

When the grid upon the drawing is projected upon the wall with the help of a central projection—as illustrated in Fig. 6—there is mathematically

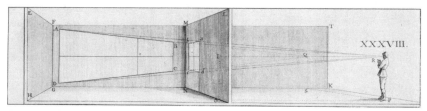

Fig. 6. Niceron's illustration of how to project a grid upon a wall [Niceron 1646].

no difference between the construction of the anamorphic grid and one of the fundamental problems in perspective: to construct in a vertical plane the perspective image of a floor tiled with squares. The early writers on anamorphoses, however, did not seem to realize this parallel, because none of them applied their knowledge of perspective when they constructed anamorphoses.

It is an advantage to clear up the mathematics connected to the grid problem before giving an account of how this problem was solved by various authors. Therefore I discuss the mathematics in this section and present the history in the next. Let us assume that an eye point, O, and a grid of squares situated in a vertical plane, α, are given and that the grid has to be projected from O upon a wall perpendicular to α. We first consider the images of the horizontal lines of the grid (Fig. 7); the vanishing point of this set of parallel lines is the orthogonal projection, F,

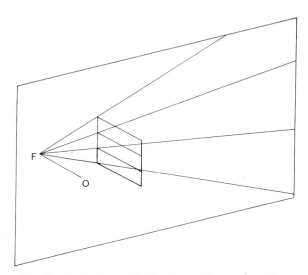

Fig. 7. Projection of the horizontal lines of a grid.

of *O* upon the wall. This means that the images are lines radiating out from *F*. It can be remarked that if we wanted to project the images of the horizontal lines back to the original plane we would have an example of a result that engaged Girard Desargues when he was working on perspective: a pencil of lines is depicted in a set of parallel lines when the line joining the vertex of the pencil and the eye is parallel to the picture plane [Andersen 1991, 74–75].

Returning to the problem of projecting the grid of squares upon the wall we look at its vertical lines. Because these are parallel to the wall, i.e., the picture plane, their images are also vertical lines. To decide the positions of the vertical lines on the wall various techniques can be applied; one of the most convenient procedures corresponds to the distance point method used for perspective constructions. The idea of this solution—which I refer to as the *diagonal method*—is to construct the image of a diagonal in the grid such as the line *SR* (Fig. 8). This line intersects the wall in the point *S*, and its vanishing point is the point *D* in which the line through *O* parallel to *SR* intersects the wall; because ∠*DOF* = 45° the point *D* can be characterized as the point on the vertical line through *F* for which *DF* = *FO*. According to one of the most fundamental results in the theory of perspective the line *DS* is the image of *SR*. The points in which *DS*

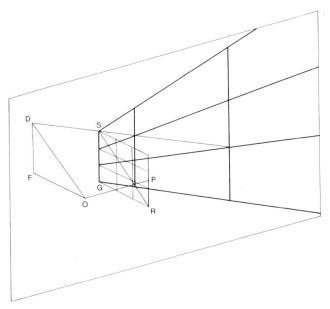

Fig. 8.   Determination of the image of a diagonal in the grid.

intersects the images of the horizontal lines of the grid therefore determine the positions of the images of the vertical lines.

The construction of the projection of the grid of squares enables us to
study the relationship between the anamorphic effect and the placements
of the object to be depicted as an anamorphosis and of the eye point. If
the object that is projected upon the wall is one that, seen in the grid,
looks to have natural dimensions, then these are largely distorted on the
wall. For instance, the distances between the images of the equidistant
vertical lines increase considerably with increasing distance to α (Fig. 8).
In fact, the vertical line through the orthogonal projection, P, of O upon
α is depicted in the line at infinity. Moreover, vertical lines to the right of
P are depicted as lines that lie "behind" O. Thus the distance OF should
be chosen so that it is larger than the width of the drawing from which the
anamorphosis is constructed. This consideration also shows that, if one
requires that the drawing itself should be a perspective composition with
eye point O, then it is necessary that its principal vanishing point—which
is the point P—falls at least a bit outside the drawing.

In the anamorphic picture, F is the principal vanishing point, and it is
situated outside the image of the grid. The position of F is determined by
OP; the latter has to be chosen to be so large that the drawing in α lies
inside the maximal angle, which from O can be covered in one glance—this
is often set to 90°. If OP is very large, the projection upon the wall almost
becomes a parallel projection. Although not presented as such, parallel
projections were used to construct anamorphoses during the 16th and 17th
centuries. Under a parallel projection equidistant vertical lines again
become equidistant vertical lines with a changed distance (except in one
case). The change in distance does give rise to a distortion, but contrary to
what is the case with perspectival anamorphoses this distortion cannot be
eliminated by looking at the image from the eye point.

5. THE TREATMENT OF ANAMORPHOSES IN THE LITERATURE ON PERSPECTIVE AFTER PIERO

The drawing reproduced in Fig. 9 shows that already in the mid-16th
century a diagonal method was applied for producing an anamorphosis. A
knowledge of such a method is, however, not reflected in the contemporary literature despite the fact that several treatises on perspective included a section on anamorphoses. Daniele Barbaro, for instance, devoted
a couple of pages to "the beautiful and secret part of perspective" in his

Fig. 9. The diagonal method applied for an anamorphic drawing by the Master H. R., Nürnberg, c. 1540. Reproduced from Baltrušaitis [1977].

Pratica della perspettiva [Barbaro 1569, pp. 159–161]. His description almost gives the impression that he also wanted to keep the technique of this part —the anamorphoses—a secret. Thus he gave a very blurred description of a geometrical construction and a vague indication of how an anamorphosis can be obtained by using sunlight for projecting a drawing upon a surface. Barbaro's second solution was later criticized by Niceron; using other words, Niceron pointed out that, if sunlight is used, then a parallel instead of a central projection, required for anamorphoses, is performed [Niceron 1638, 55–66]. Another practical way of producing anamorphoses was touched upon by Paolo Giovanni Lomazzo in his tract on perspective [Lomazzo 1584]. Lomazzo suggested the use of threads rather than light rays [Baltrušaitis 1977, 32]; it is unclear whether Lomazzo had the idea that the threads should radiate from one point so that a central projection was performed.

A presentation of how to construct an anamorphosis clearer than those given by Barbaro and Lomazzo is found in the comments that the Italian mathematician Egnazio Danti added to his edition of Vignola's work on perspective [Vignola 1583]. Danti's anamorphosis was, however, a very simple one, in which a grid of squares is depicted as a grid of rectangles (Fig. 10)—corresponding to a parallel projection. It is difficult to decide whether Danti avoided the perspectival anamorphoses because they were

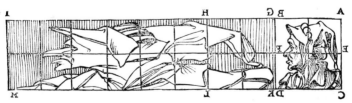

Fig. 10. Danti's illustration of an anamorphosis. The letters are mirrored because as an extra sophistication Danti placed his anamorphosis in a box with a mirror [Vignola 1583].

unknown to him or because he found them mathematically too challenging. The latter could be the case, because when Danti edited Vignola's work the rules of perspective constructions were still a kind of mystery for artists as well as mathematicians.

At the end of the 16th century another Italian mathematician, Guidobaldo del Monte, managed to make the rules of perspective transparent by creating a theory of perspective based on the concept of a vanishing point. Guidobaldo's theory was so comprehensive that it was able to solve the mathematical problems connected with the construction of anamorphoses, but it was not immediately applied to these problems. Neither Guidobaldo himself nor Simon Stevin, who polished Guidobaldo's theory, took up the theme of anamorphoses, whereas Samuel Marolois—one of the mathematicians who had grasped the basic principles of the theory—kept to Danti's simple forms of anamorphoses (Fig. 11).

The first 17th century author on perspective who applied the diagonal method for constructing anamorphoses and related it to perspective seems to have been Salomon de Caus [Bessot 1991, 307–311]. In his *La Perspective* (1612), De Caus gave a description of how to construct an anamorphic grid [De Caus 1612, Chapters 28 and 29]. De Caus' description does, however, belong to the category of instructions that can be understood only by persons who know the procedure beforehand.

A quarter of a century later an understandable description of the construction of an anamorphic grid appeared. It was written by the French Minim Jean-François Niceron, who considered the optical laws a kind of natural magic and was so fascinated by the wonders of anamorphoses that he decided to compose a work on them. He realized how closely anamorphoses and perspective were related and opened his book—called *La*

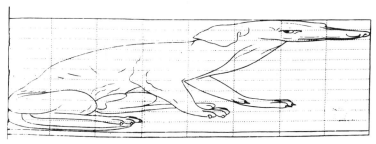

Fig. 11. Simple anamorphosis from Samuel Marolois, *La Perspective Contenant la Theorie et la Pratique.* Den Haag, 1614.

Perspective Curieuse—with a chapter on perspective in which he presented the distance point construction. He proceeded by describing the diagonal construction of the anamorphic grid as an application of the distance point construction in a way that is similar to my exposition in Section 4 [Niceron 1638, 52]. Niceron's construction thereafter became standard in books dealing with anamorphoses.

Niceron not only dealt theoretically with anamorphoses, but also was active in producing them [Baltrušaitis 1977, 50]. Moreover, as a professor of mathematics at the Minim convent in Rome, the S. Trinità dei Monti, he inspired Emmanuel Maignan to design a huge anamorphosis. Seen from the eye point, this anamorphosis shows a portrait of the founder of the Minim Order, S. Francesco di Paola, whereas seen from other positions it illustrates Francesco's dramatic crossing of the Strait of Messina. The Minims have left the Trinità dei Monti but Maignan's creation can still be admired there.

Liking to play with optical effects, Niceron also described how an anamorphic effect can be obtained by projecting a picture upon the surface of a cone or a pyramid. This form of anamorphoses was later nicely illustrated by Jean Dubreuil (Fig. 12). Finally Niceron dealt with mirror anamorphoses as we shall see in the next section.

Fig. 12. Conic mirror anamorphosis from Jean Dubreuil, *La Perspective Pratique ... Troisiesme ... Partie*. Paris, 1649.

6. VAULEZARD'S MASTERING OF THE MIRROR ANAMORPHOSES

Before the perspectival anamorphoses were mathematically demystified, a new challenge reached Europe, namely cylindrical mirror anamorphoses. These are pictures that, placed in a horizontal plane and seen in a vertical cylindrical mirror, give the impression of a "natural" composition. The art of making these anamorphoses had been cultivated in China for some time (Fig. 13), and it was presumably the Chinese pictures that inspired Europeans to take up this new form of representation in the beginning of the 17th century [Baltrušaitis 1977, 159–169]. In China the construction of the cylindrical anamorphoses seems to have been based on observation and experience. European drawers, however, looked for construction rules.

Among those who searched for rules was a group that happened to be students of the French mathematician Jean-Louis Vaulezard, known as a translator of some of François Viète's works. The students asked Vaulezard

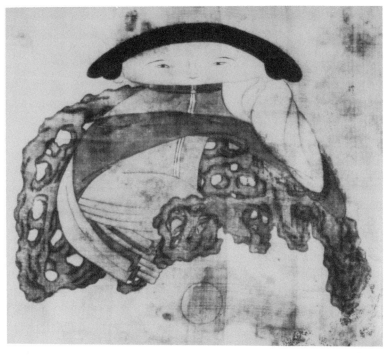

Fig. 13. Chinese cylindrical mirror anamorphosis from about 1600. Reproduced from Baltrušaitis [1977].

for help, and because he could find no literature on the subject he made his own investigations [Vaulezard 1630, *Advertissement*]. Vaulezard was apparently fascinated by this new mathematical problem, and he found an exact solution, which he published in *Perspective Cilindrique et Conique* (1630) together with all the necessary mathematics and optics; the theory he presented either as axioms or as proven theorems. Taking over the grid construction from the usual anamorphoses, he dealt with the problem of how to draw a grid that appears as a grid of squares when it is reflected in a mirror. Formulated more precisely, Vaulezard's problem is as follows (Fig. 14). A horizontal plane, γ, and an eye point, O, whose orthogonal projection upon γ is F, are given; moreover a cylindrical mirror standing on γ is given by the radius and the center X of its circular base, and finally a grid of squares is given in that tangent plane to the cylinder that is perpendicular to FX. It is required to construct curves in γ so that reflected in the mirror—seen from O—they appear as the given grid.

Vaulezard solved this problem in several rather pedagogical steps. In accounting for his solution I make some short cuts, introduce other terms than he used, and draw different diagrams, but essentially I follow his exposition.

Because the eye does not distinguish between points on the same visual ray, it is not essential for the construction that the given grid lies in the mentioned tangent plane; any grid defined by the same visual rays and situated in a plane parallel to the tangent plane can be used as well. Among these grids Vaulezard chose one that is conveniently placed, namely in the plane defining the part of the cylinder that is visible from O. When FA and FB are the tangents to the circular base of the cylinder touching it in A and B (Fig. 15), the vertical plane through AB determines the visible part of the cylinder; hence AB is also the maximal base of a reflected grid situated in this plane. Vaulezard let the grid be given by a number of division points, D_i, of AB—to avoid too many lines I have limited this number to three.

If the reflected light has to appear as coming from a given point, G, in the grid (Fig. 16), the light has to be reflected at the point R on the cylinder that lies on the visual ray OG; I call R the point of reflection corresponding to G. One step in Vaulezard's solution was to find the points of reflection corresponding to the vertical lines of the grid—by him called *montantes* [Vaulezard 1630, 20]. For a given vertical line like GD in Fig. 16, the points of reflection lie on the intersection of the surface of the cylinder and the plane OGD. This intersection is a line whose position can be determined by looking at the configuration in the ground plane γ (Fig.

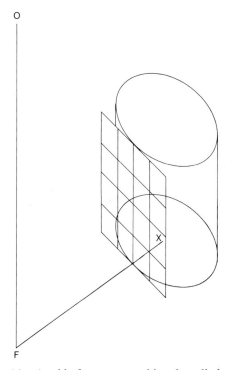

Fig. 14.   A grid of squares touching the cylinder.

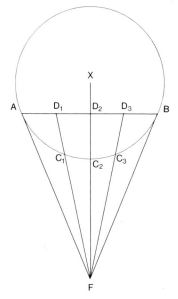

Figure 15.

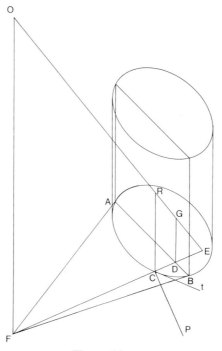

Figure 16.

15); because the D_i's determine the position of the vertical lines of the grid, the positions of the corresponding lines are determined by the points C_i in which the lines FD_i cut the base of the cylinder.

Vaulezard called the horizontal lines of the grid the *transversales*; copying him I term them "transversals." The curve of reflection—that is, the locus of the points of reflection—corresponding to a transversal is similarly the intersection of the surface of the cylinder and the plane determined by O and the transversal. Apart from the case in which the latter plane is horizontal, this intersection is an ellipse. Vaulezard did not mention these reflecting ellipses; looking at his drawing, reproduced in Fig. 17, one could even get the impression that he assumed that they were parallel circles. He was, however, well aware—as we shall see later—that the points of reflection corresponding to the points on a tranversal have different distances to γ.

Another step in Vaulezard's construction was the following: given a point, R, on the cylinder determine the point P in the ground plane γ so

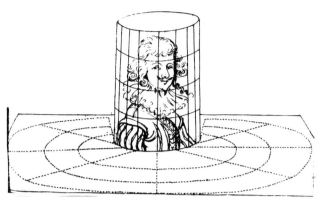

Fig. 17. Illustration from the title page of Vaulezard's *Perspective Cilindrique et Conique*. Paris, 1630.

that the light passing from P to O via the mirror is reflected at R (Fig. 16). Let C be the orthogonal projection of R upon γ, and let t be the tangent to the circle at C; then according to the law of reflection P lies on the line l, which is characterized by making the same angle with t as FC, F being as earlier the orthogonal projection of O upon γ. Before the question of P's position on l is settled, it is adequate to notice that the determination of l shows that all points on l are reflected in points situated on the vertical line RC. Using this result Vaulezard solved the first part of his general problem: Determine the curve in γ that appears as a given vertical line of the grid [Vaulezard 1630, 23]. When the given vertical is GD (Fig. 16), his solution was to find the point of intersection, C, of the circular base and FD and then construct the line l in the just described way.

Returning to the problem of the position of P on l, we let E be the point of intersection of OR and γ, and then the law of reflection tells us that

$$CP = CE. \qquad (1)$$

When R is the point of reflection corresponding to the point G on the grid, P is the point in γ that appears as G. Thus, when G is given, all that is needed to construct P is the position of l and the distance CE. Vaulezard used this observation to solve the second part of his general problem: Determine the curve in γ that appears as a given transversal [Vaulezard 1630, 23–25]. He characterized the curve by describing how it can be constructed pointwise. To determine the points on the curve that

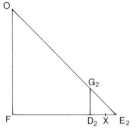

Fig. 18. Let G_2 be the point that lies on a given transversal as well as in the plane determined by O, F, and X. The point of intersection, E_2, of OG_2 and FX determines the position of the line in which the plane through O and the given transversal cuts the ground plane γ.

appears as the transversal through G, Vaulezard first considered the plane through this transversal and the eye point O. This plane cuts γ in a straight line that is parallel to AB and whose distance to AB can be determined in a vertical profile through O and the center X of the base circle (Fig. 18). Having determined the position of this line, Vaulezard continued his construction in γ. For each division point, D_i of the base of the grid (Fig. 19), he found the distances $C_i E_i$, required by Eq. (1), and

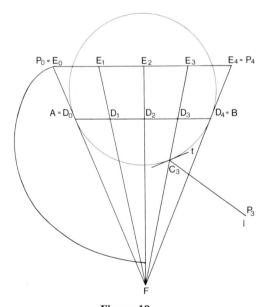

Figure 19.

then cut off $C_i P_i = C_i E_i$ on the line l_i, which makes the same angle with the tangent at C_i as FC_i.

The resulting curve is part of an oval that is symmetric around the line FX—the left part of the curve is shown in Fig. 19. Vaulezard did not identify this curve, which is no wonder because it is not one of the then known curves, and, writing his tract before the emergence of analytical geometry he did not search for its equation. To find this equation, which is quite a calculation, I have had help from Eisso Atzema, who used the computer program Maple. Thanks to Atzema and Henk Bos, I can tell that the equation of the curve is a polynomial of degree 6 with coefficients that contain many terms.

Having finished his presentation of the construction of the coordinate curves in the ground plane, Vaulezard remarked that another method for constructing the coordinate curves was in use. This method—whose origin is unclear—consists of letting the lines that appear as the vertical lines of the grid be lines radiating from the center X of the base circle of the cylinder and letting the curves that appear as transversals be concentric circles (cf. Fig. 21). Vaulezard did not at all approve of this method [Vaulezard 1630, 28]. Instead he offered a couple of other constructions that approximated his own exact solution. He also dealt with the situation in which the reflection takes place on the inner side of a mirror that is a part of cylindrical surface.

Finally, Vaulezard treated anamorphoses that should be seen reflected on the surface of a right cone. This sounds complicated, but it is not really, because Vaulezard chose to place the eye point on the axis of the cone above the apex. He furthermore gave up the usual practice of constructing curves that appear as a grid of squares and instead introduced what corresponds to polar coordinates. Thus he asked for the curves that appear as a system of lines radiating from the center F of the base circle and circles concentric with this (Fig. 20).

Vaulezard's *Perspective Cilindrique et Conique* gives the impression that he enjoyed to solve problems connected to mirror anamorphoses. He did, for instance also include the problem of determining the curve that in a cylindrical mirror appears as a given line that cuts the cylinder [Vaulezard 1630, 42–44]. His lectures on the material might have been so enthusiastic that the students who had asked him for a solution found some excitement in his construction. But, as a textbook to be read on its own, his work is too complicated for practitioners.

In his approach to mirror anamorphoses, Niceron was very much influenced by Vaulezard. In the first edition of *La Perspective Curieuse*—from

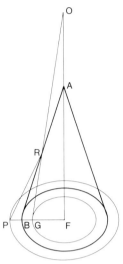

Fig. 20. Vaulezard's treatment of conic mirror anamorphoses. The eye point O is given on the axis of a given conic mirror with apex A. The orthogonal projection of O upon the ground plane is hence the center, F, of the base circle of the cone. Let G be a point that lies in the ground plane, let OG intersect the mirror in R, and let P be the point in the ground plane that reflected in the mirror appears as G. The law of reflection implies that P lies on the line FG and that P's position on FG can be determined by using the fact that the angles ORA and PRB are equal. Moreover, it can be concluded that the circle with center F passing through P appears as the concentric circle passing through G.

1638—Niceron limited his acknowledgment of Vaulezard's work to stating that as far as he knew Vaulezard was the first to write *doctement* (scholarly) on this subject [Niceron 1638, 87]. The 1663 edition, which appeared posthumously contains the following more praising phrase:

> Vaulezard . . . a fort bien ecrit sur ce sujet, & . . . est l'un des grands Analystes & des sçavans Geometres d'auiourd'huy. [Niceron 1663, 161]

This might be a sign that Niceron later realized that he owed more to Vaulezard than he had expressed in his first edition. There is, however, also the possibility that the change was made not by Niceron, but his editor Gilles Personne Roberval. Anyway, Niceron copied Vaulezard's treatment of conic mirror anamorphoses [Niceron 1638, 94]. He also copied Vaulezard's method of constructing a grid for cylindrical mirrors. As his first solution to this problem, however, Niceron presented the one Vaulezard had criticized (Fig. 21), that is, the one in which the curves that

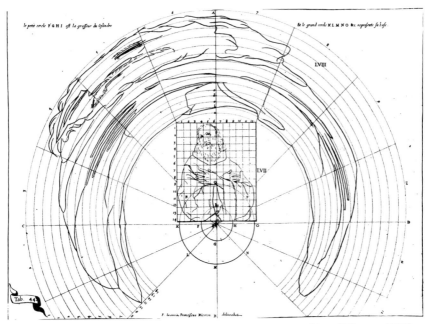

Fig. 21. Niceron's solution of the cylindrical mirror problem [Niceron 1638].

should appear as transversals are drawn as concentric circles. Niceron prescribed that the radii of the circles should increase with the factor $\frac{21}{20}$ [Niceron 1638, 85], but it is unclear how he came to this ratio.

After Niceron the concentric circle solution became standard in books on anamorphoses. In this century the method has even been rediscovered, independently of earlier treatments, by the Swedish artist Hans Hamngren, who created several mirror anamorphoses, one of which is reproduced in Fig. 22 [Hamngren 1981].

7. CONCLUDING REMARKS

The history of the anamorphic art is an example of how practical problems give rise to mathematical problems. It is also an example of how difficult it is to bridge the gap between practitioners' and mathematicians' traditions.

In the case of perspectival anamorphoses, the practitioners applied a correct solution for a longer period before any mathematicians—at least in writing—engaged themselves with this solution. In the case of cylindri-

Fig. 22. Hans Hamngren *The Pump*, 1988. Reproduced from Hans Hamngren, *Anamorfoser*, Carlsson, Stockholm, 1992.

cal mirror anamorphoses, there existed a practical and a mathematical solution. The practical solution is the one of unknown origin that Vaulezard criticized; whereas the mathematical one was his own exact solution. The latter was too complicated for practical use and did not have any influence on how artisans constructed anamorphoses (apart from, perhaps, Vaulezard's students). The mathematician Niceron presented both methods; he recommended the easy method without giving any scientific argument that could show that this method was reasonable.

The history of anamorphoses does not end with Niceron's *La Perspective Curieuse*; throughout the rest of the 17th century mirror anamorphoses kept their magical attraction. Besides the cylindrical and the simple conic anamorphoses described by Vaulezard and Niceron, conical mirror anamorphoses that should be seen from the side emerged (Fig. 23). My familiarity with the later literature is limited, but I do have the impression that after Vaulezard and Niceron no new attempts were made to solve mathematical problems concerning anamorphoses.

Fig. 23. Conic mirror anamorphosis from Mario Bettini, *Apiaria Universae Philosophiae Mathematicae*, Bologna, 1642.

REFERENCES

Andersen, K. 1991. Desargues' method of perspective. *Centaurus* **34**, 44–91.

Andersen, K. 1992. Perspective and the plan and elevation technique, in particular in the work by Piero della Francesca. In *Amphora. Festschrift für Hans Wussing*, S. S. Demidov, M. Folkerts, D. E. Rowe and C. J. Scriba, Eds., pp. 1–23, Basel: Birkhäuser.

Baltrušaitis, J. 1977. *Anamorphic Art*. Cambridge, UK: Chadwyck-Healey.

Barbaro, D. 1569. *La pratica della perspettiva*. Venezia. Facsimile Sala Bolognese: Arnaldo Forni, 1980.

Bessot, D. 1991. Salomon de Caus (c. 1576–1626): Archaïque ou précurseur. In *Destin de l'art, desseins de la science. Actes du colloque A.D.E.R.H.E.M. Université de Caen, 24–29 octobre 1986*, D. Bessot, Y. Hellegouarc'h and J.-P. Le Goff, Eds.

Carter, B. A. R. 1970. Perspective. In *Oxford Companion to Art*, H. Osborne, Ed. Oxford: Clarendon.

Caus, S. de. 1612. *La perspective avec la raison des ombres et miroirs*. London.

Elffers, J., Leeman, F. and Schuyt, M. 1975. *Anamorphosen. Ein Spiel mit der Wahrnehmung, dem Schein und der Wirklichkeit*. Köln: DuMont. Taschenbuchausgabe Köln: DuMont, 1981.

Frangenberg, T. 1992. The angle of vision: Problems of perspectival representation in the fifteenth and sixteenth centuries. *Renaissance Studies*, **6**, 1–45.

Hamngren, H. 1981. My anamorphoses: Types that produce three kinds of images in cylindrical mirrors. *Leonardo* **14**, 198–201.

Kemp, M. 1990. *The Science of Art*. New Haven, CT: Yale University Press.

Lomazzo, G. P. 1584. Libro quinto della prospettiva in *Trattato dell'arte della pittura, scoltura, et architettura*. Milano. Several later editions. Facsimile of the 1584 edition Hildesheim 1968. English translation edited by Richard Haydock, Oxford 1598.

Meyere, J. de and Weijma H. 1989. *Anamorfosen. Kunst met een omweg*. Bloemendaal: Aramith.

Niceron, J.-F. 1638. *La perspective curieuse, ou magie artificiele des effets merveilleux . . . .* Paris. Later editions Paris 1652, Paris 1663, and Paris 1679. Edited in Latin with the title *Thaumaturgus opticus*, Paris 1646.

Osborne, H., Ed. 1970. *Oxford Companion to Art*. Oxford: Clarendon.

Piero della Francesca. ~ 1470s. *De prospectiva pingendi*, manuscript. First printing appeared together with a German translation in *Petrus Pictor Burgensis: De prospectiva pingendi*, C. Winterberg, Ed. Strasbourg 1899. Second printing *De prospectiva pingendi*, G. Nicco Fasola, Ed., 2 vols., Firenze 1942; second edition Firenze: G. C. Sansoni, 1974.

Vaulezard, J.-L. 1630. *Perspective cilindrique et conique ou traite des apparences vuës par le moyen des miroirs . . . .* Paris.

Vignola, J. B. da. 1583. *Le due regole della prospettiva*, E. Danti, Ed. Roma. Several later editions. Facsimile of the 1583 edition Bologna: Casa di Risparmio di Vignola, 1974.

Mozart 18, Beethoven 32: Hidden Shadows of Integers in Classical Music

I. Grattan-Guinness

Middlesex University at Enfield, Middlesex EN3 4SF, England

This paper explores the connections between mathematics and music, especially in the works of Mozart and Beethoven. Attention is focused primarily on *Die Zauberflöte* (1791), with additional examples from his last three symphonies of 1788, as well as several examples of numerology in works by Beethoven. A comparative analysis of these two composers is also given, noting especially some important differences. Reasons that musicologists have not treated these subjects more seriously are suggested, and a number of other composers are listed whose works might well be of numerological interest. In closing, the limits and scope of considering numerology in music are discussed.

1. MATHEMATICS IN MUSIC

Links between mathematics and music have long been known, and they take several different forms: tuning and temperament, acoustics, influences on notation, mathematical structures in musical forms, and so on. The link discussed here is very ancient and may well belong to the origins of each science: numerology in music, in which mystical properties assigned by a given culture to certain integers were used to guide the composition and even performance of that culture's music.

In Western music the case of J. S. Bach is particularly intriguing. Numerology is evident in, for example, his chorale "Das alte Jahr vergangen ist" (BWV 614), which contains 365 notes. But he also practiced gematria, the sister discipline in which integers are assigned in order to the letters of the alphabet of a language and a word or phrase takes the integer given by the sum of those corresponding to its letters. In the 24-letter alphabet of his day (which is explained in Note 4), BACH = 14 and JSBACH = 41, integers that guided works such as the "Crab canon," so called for its palindromic structure. Bach's interest in these practices belongs to a rich tradition of numerology in German cabbalism, which itself was an offshoot of the vogue for *Rechenbücher* on practical arithmetic [1].

This paper treats two unknown numerologists and one gematricist. Wolfgang Amadeus Mozart (1756–1791), the subject of the next five sections, practiced both disciplines; the chief example, the opera *Die Zauberflöte* (1791), is followed by the last three symphonies of 1788, and then a general appraisal of his case is offered. Sections 7–9 treat the numerology of Ludwig van Beethoven (1770–1827) in a similar way. Section 10 points out some principal differences between the two composers.

The two concluding sections are reflective. Reasons are sought for the deliberate and even anxiously pursued ignorance of these topics among musicologists; then by contrast, aspects of numerological research are described, and other known or candidate composers are mentioned. When general remarks are made, especially in those two sections, only numerology is mentioned, but the comments usually hold for gematria also.

2. MOZART IN HIS MASONIC TRADITION

For both Mozart and Beethoven, Christianity and Freemasonry provide the basic doctrines from which the numerology sprung. Let us start with the latter, which is not well known in general, because it is supposed to be a secret movement. Yet an astonishing amount has been written about it and at times even by its practitioners. In the case of Mason Mozart, the literature is very specific, because the Viennese lodge group that he joined in December 1784 had recently started its own *Journal für Freimaurer*, under the chief stimulus of the mineralogist Ignaz Born (1742–1791). But Freemasonry in general was then running foul of monarchies and governments, especially when the French Revolution of 1789 made them chary of all international movements. The situation in the Habsburg Empire was particularly serious; for the Austrian Enlightenment was deeply informed by Freemasonry, and Mozart knew many of its leading Viennese figures [Till 1992].

The crisis peaked in 1790, when Holy Roman Emperor Leopold II planned to ban Freemasonry entirely. In response Mozart and his librettist Emmanuel Schikaneder (for his role, see Section 4) prepared *Die Zauberflöte*, an opera that extolled Masonic ideals and virtues. The apparently silly story is in fact a brilliant account of how the virtues of Freemasonry, as apotheosized in the High Priest Sarastro (Saros, the Sun), conquer the dark ignorance of the Königin der Nacht (Moon, the silver reflected light), especially in the training of Tamino ("animo", spirit; TAM, Tugend, Arbeit, and Menschheitsziele, the principal triad of Masonic virtues) to pass the Masonic trials. Chailley [1967] has explained in detail how this scenario is worked out and has also given most of the principal integers; Dalchow, Duda, and Kerner [1966] contains other essential insights. However, neither text explores the consequences of the numerology; this is analyzed in Grattan-Guinness [1992], from which the examples below are taken.

The line in Freemasonry to which Mozart belonged took Egyptian lore (as then understood) as the main origin of civilized knowledge (hence the opera is set in an Egypt-like land), and Egyptian arithmetic affected some of the choices made of integers. The principal one is 3, with the triads of Masonic virtue (one was just mentioned); it also conveys a masculine connotation, and so does 6. Femininity is encompassed in 5 (hence the five-note panpipe of bird-watcher Papageno) and 8, whereas 7 and 18 carry holy ramifications.

The number 18 is especially prominent, due to the property (known to the Egyptians) of the perimeter of a 6×3 rectangle equalling its area. For the Masons other attractive features were available: not only

$$6 \times 3 = 6 + 3 + 6 + 3 \text{ but also } 18 \times 2 = 36 = 6^2 = 1 + 2 + 3 \cdots + 8,$$
(1)

that is, the triangular integer to 8. For a Mason 18 also governed time, in that the day started in the evening, at 6 o'clock (1800 hr).

Roles were assigned to 42 and 55, also significant in Egyptian lore via the so-called "pyramidal" or "triangular" integers:

$$42 = 2 \times 21, \quad \text{where } 21 = 1 + 2 + 3 \cdots + 6;$$

and
(2)

$$55 = 1 + 2 + 3 + \cdots 10.$$

The other integers are 2 and 10; and 4 carried various important connotations (for example, the number of trials that Tamino had to undertake), above all in the triad 3:4:5. This triad of integers had a special significance for Masons, as "the Masons' square." For the cathedral builders of the Middle Ages, among whom numerology gained great prominence, this right-angled triangle was a principal tool, for it was used to construct not only perpendiculars but also rectangles in the proportions 2:1 ("ad quadratum") and 3:1 ("ad triangulum") [Lund 1921]. Among their successors, the Masons, it was worn as an ornament on their costume, in the form of a suspended "L" of the shorter sides; a portrait exists of Mozart and his fellow Masons so attired when gathered in their "temple" [Robbins Landon 1991].

3. A FEW EXAMPLES FROM *DIE ZAUBERFLÖTE*

Every aspect of *Die Zauberflöte* is controlled by the numerological system just outlined. The music runs on it from beginning to end (literally: from 5,3,5 in the Overture before the 36-note subject opens the three-voiced fugue to 18,4,5,4,5 in the final bars). Among other facets are the numberings of both the "Auftritte" (AF) into which music and dialogues are divided and also the musical items (MI); the names of the principal characters (not only Sarastro, Königin, and Tamino, mentioned above, but also Tamino's helper Pamina: "anima" there again); the three-syllable names of these principal characters, as opposed to the four-syllable names of lower orders such as Papageno, Papagena, and Monostasos; the 18 singing parts (counting the chorus as 1); and the scene changes and set designs, which reflect many aspects of Freemasonry [Curl 1991]. Even the day of the first performance was carefully chosen: 30 7ber (September 30th), the end of the 3rd 4ter of the year, but 1 8ber (October 1st) for Masons such as Mozart, because the performance started after 1800 hr. However, to the Masons the year was not 1791 but 5791, dating from the supposed origins of Egyptian knowledge; hence there is no role for $1 + 7 + 9 + 1 = 18$.

I shall give a few musical examples concerning the principal protagonists. The most beautiful case of numerology occurs in the cadenza of the revenge aria of the Königin (AF 8 in Act 2, MI 14). Figure 1 shows her line, in which the feminine integers, 5 and 8, are deployed in a wonderful way:

$$5, 8, 5, 8, 5, 8, 8, 8. \tag{3}$$

Here are eight groups of notes comprising three 5s and five 8s, with three pairs at the beginning and three singles at the end; the sum of the notes is 55 . . . . The 5 (not counting the accidentals) is a leitmotiv in the work; for example, it appears twice in the 36 notes of the overture theme. Note that for good measure (as it were), the cadenza is completed by a $5 + 3$.

Fig. 1. Cadenza for Königin's revenge aria.

The other examples concerns the High Priest Sarastro, Saros the Sun God. Ignaz Born died while the opera was being composed, and maybe Sarastro was conceived in honor to him. At all events, his entire role is controlled by 18 in the most astonishing fashion. He first appears in the Finale of Act 1 at AF18, on a "triumph cart" pulled by six lions (the lion symbolized the sun)—that is, as 6 × 3. In the opera he has 18 entries and sings in 180 bars (6 × 30 as well as 10 × 18, I suspect). But the greatest detail comes in his solos.

Each of Sarastro's two arias has 18-note melodies. Figure 2 shows the first, "O Isis und Osiris" (AF1 of Act 2, MI 10), in vocal score, which accurately reflects the orchestral original. Note how the orchestration preserves the 18 scheme by giving only one note to some *but not all* tied quavers, which therefore are to be counted as 1 (for example, "dem" in bar 10 but not "−ris" in bar 7). Then six 9-sequences follow, then two 9s, and two 6s, and finally three 9s on "nehmet sie". The four-part chorus repeats the last line of each verse (over 4 bars), giving 10 lines of text in all. With 4 bars for orchestra at the beginning and 3 at the end, the number of bars in the item is 55.

Sarastro's other aria, "In diesen heiligen Hallen" (AF12 of Act 2, MI 15) also has an 18-note melody (if the tied notes on "Hal-" and "die" are counted as 1 each); this phrase is repeated, and some later passages are on or close to 18 notes. The text comprises two verses, each of six lines and sung over 24 bars: the words starting each fifth line are sung three times, and once again the last phrase is sung three times at the end. Counting the opening upbeat in the usual way, the number of bars is 54; *designed* as 2 × 24 plus 6 for the orchestra, but *conceived* as 18 × 3.

In addition to these two arias, Sarastro's priests—18 of them, we are told at the head of Act 2—open scene-change 5. Each priests carrying a pyramid, the sing to the Egyptian gods "O Isis und Osiris" on their own [AF of Act 2, MI 18 (sic)]. Their six lines of text take 18 + 18 bars, with 4 bars at the end to allow the last words to be sung three times. Allowing for upbeats, the orchestra has 6 other bars to itself, giving 42 in all.

Fig. 2.   The first verse of Sarastro's "heiligen Hallen."

4.   MOZART AS GEMATRICIST

In section 1 I mentioned that Mozart also deployed gematria in this opera. This is the discovery of Irmen [1991], who conducted a most important study on Mozart's Masonic career (for example, he has used the secret Masonic files in Vienna to transcribe the ceremony of Mozart's induction in December 1784) [2]. Gematria is "visible" chiefly in the integers corresponding to the names of persons involved in preparing the opera and in words and phrases in the text. It is also manifest in the numbers of notes assigned to the instruments in parts of the score *in the autograph* [Köhler 1979]; a degree of arbitrariness was available to Mozart in the places at which he chose to stop writing out repeated notes and use "&c." or some such sign. Thus all printed editions are useless for this analysis.

The most interesting example comes at the start of the first scene, when the story is marvellously encapsulated in this encounter between the frightened Tamino and the Königin's three silver-clad ladies, who (in their veiled ignorance) kill Freemasonry by chopping up into three pieces its symbol, the serpent. The 16 $\frac{1}{2}$ bars of orchestral introduction encode a title page for the opera, in the normal form of words then used following the 24-letter code of the German alphabet:

$$a = 1, \qquad b = 2, \ldots, i = j = 9, \ldots, u = v = 20 \ldots, z = 24. \qquad (4)$$

first violins	129	Zauberfloete
second violins + violas	56 + 90	Compositeur
Basses	87	Mozart
All strings	362	Neues Singspiel in zwey Aufzuegen
Oboes	69	[Uninterpreted] [Musik]
Bassoons	71	Poeten
Bassoons and horns	71 + 60	Ludwig Giesecke [Gedichte]
Clarinets [3] + tympani	79	Emmanuel
All winds	200	Schikaneder der Juengere
All instruments	562	Wolfgang Amadé Mozart, Kapellmeister in wirklichen k.k. Diensten

Note that in this reading both Giesecke and Schikaneder are named as librettists; the question of which *one* was responsible has been a matter of controversy. The entries in square brackets are my suggestions.

5. MOZART'S LAST THREE SYMPHONIES

One test of a new theory is to try it out in a context from which it was not created and to see how it fares. In this case I have been able to come up with a very satisfactory answer to this well-known mystery in Mozart's career: why did he write three symphonies in the summer of 1788 with no commission in hand, when he was short of money? (The fact that they turned out to be his last symphonies is, of course, not involved here.) When one recognizes that his passion for Freemasonry included numerology, the solution comes readily: he wrote this trio of works in order to fulfil a major commitment to Masonic ideals. Here are some main features, summarized from Grattan-Guinness [1994c].

At the time, Mozart was in his 33rd year of life. Although 33 does not feature in *Die Zauberflöte* (unless the 16 $\frac{1}{2}$ bars just mentioned are $\frac{33}{2}$), it is a most important Masonic integer. Christianity was probably the source: in the orthodox interpretation of the Bible it was the supposed length in years of the life of Jesus; in the Apocryphal tradition it served as 3 × 11, where the digit string 1 1 symbolized their belief that Jesus was a rabbi married to Mary Magdelene [Thiering 1992].

The three keys of the symphonies are Masonic; Eb (three flats), the most favored Masonic key, to be used also as the main key of *Die Zauberflöte*; C, the so-called "white" key; and, most interestingly of all, g, a very special key in Mozart. "G" was an important symbol in Freemasonry, although (to outsiders like me, anyway) its purpose is unclear. It may have carried three (?) meanings: "Gott"; "Gnosis," a term in the Apocryphal tradition for knowledge, especially when involving elistist insight; and "gnomon," the triangular device long used to determine the time and latitude, and represented by the Mason's square.

The symphony in g is a miracle of 18s, for most of its main themes are so based. The opening subject is a particularly instructive case. It appears to be a melody of 4 × 10 notes, suitable for this Symphony No. 40, a very important Christian integer; but in fact Mozart never assigned numbers to any of his works (the symphonies were numbered by publisher Breitkopf & Härtel in the early 19th century). In the manuscript Mozart followed his practice, used elsewhere (such as with Sarastro), of tying pairs of notes to count as one, here, D–Bb and C–A of the first and third clauses, with the repeated Cs and Bs of the other two also to be so understood (Fig. 3). Thus we have a quartet not of 10s but of 9s—indeed, the important 36 of (1).

The other first subjects also manifest 18s; in the three 6s of quavers in the Andante, before rounding off with an upbeat and a rhythmic 6; as 18 + 18 + 9 in the minuet; and as (7 + 11) + (7 + 11) of the concluding Finale. In addition, the second subjects of the second and fourth movements use 18s: respectively, 4 + 4 + 4 + 6 (including a pair of tied quavers as 1) and as 8 + 10.

The other two symphonies show numerology at work, although not as intensively. For example, the Eb symphony announces its Masonism at once, in the threefold statement of the opening flourish, and the first part

Fig. 3. Mozart's Symphony in g (1788), the first subject of the first movement.

Fig. 4. Mozart's Symphony in C (1788), the first subjects of the first and fourth movements.

of the second subject is again an 18 (6 + 6 + 3 + 3), as is the first subject of the second movement (6 + 12).

The third and final symphony, the "Jupiter" (not Mozart's name) in C, announces the importance of 3 very explicitly in the first part of the first subject (Fig. 4a). The number 4 and its multiples are featured frequently, especially in the motive C D F E in the first part of the first subject of the last movement (Fig. 4b). This theme goes back to Haydn's Symphony No. 13 (1763) and to Mozart's Missa Brevis K 192 (1774). Here it will have signified "Cre-do, cre-do"; in addition, the four notes shape in print the cross of Saint Andrew, which has a symbolic meaning in Freemasonry derived from its own Scottish origins [Irmen 1991, 279].

6. MOTIVES AND ORIGINS IN MOZART

Although Mozart joined the Masons at the end of December 1784, numerology begins to appear in works from around 1782. Baron van Swieten might have been a crucial acquaintance, made at that time: an aristocratic Masonic intellectual of a kind characteristic of that society and a supporter of Mozart, especially with his rich library, which brought to Mozart access to the music of Handel and Bach. The quartet K452 of 1784 for piano and four wind instruments may be a key initial work in numerology (I leave note counting as an exercise); not only was the combination of instruments original, but also it included especially wind instruments, which the Masons extolled above all others. This is why there is so such wonderful wind music in Mozart, and an opera by him with the somewhat surprising title "*Die Zauberflöte.*"

Outside *Die Zauberflöte* the most evident example of numerology is the presence of 18-note melodies. There are many of them, including some of Mozart's most famous ones. For example, sing for yourself the opening of "Eine kleine Nachtmusik" (1787), and note the 9 + 9—and also the title (Mozart's own, although not new in music), redolent of the name of the Königin der Nacht still to come and of the secrecy of Freemasonry in general.

When one is oriented to the use of integers in this way, some of the evidence is very obvious. For example, in the manuscript of *Die Zauberflöte* [Köhler 1979], Mozart counted up the numbers of bars of every musical item, even the long finales (which, doubtless following (2), are written out on respectively, 42 and 55 pages of manuscript paper...). He also shows two examples in letters to his wife Constanze. In 1789 "drucke ich 1095060437082 mal: hier kannst du dich im aus-sprechen über," where the hint and the lack of commas in the string of digits clearly invite the decoding

$$10 + 9 + 50 + 60 + 43 + 70 + 82 = 324 = 18^2, \tag{5}$$

the power (square) of holy love. Again, on June 6 (6/6) 1791, while working on *Die Zauberflöte*, "es fliegen 2999 und ein $\frac{1}{2}$ bissel von mir," which yields

$$29 + 99 = 128 = 2^7, \tag{6}$$

the seven-sacred strength of love between man and woman ("and a $\frac{1}{2}$" for the baby whom she was carrying) [Mozart, Vol. 4, 84, 135]. The distinctions made here between an integer and a digit string are very clear evidence of Mozart's *conscious* intent—just what one would expect of a numerologist.

7. BEETHOVEN'S NUMBERINGS

Mozart is a relatively easy case to analyze numerologically, not only for the wealth of examples that he provided but also because of the detailed information available on the Masonic movement to which he belonged. I now take a less straightforward case, another great composer who lived in Vienna in the generation after Mozart's death and therefore fell under similar social and religious influences.

In Beethoven's numerology there are some of the same numbers as those used by Mozart, such as 3, 6, and 18; but the quartet 27, 30, 32, and

33 is also prominent. He also used this quartet:

$$3 + 37 = 40, \qquad 3 \times 37 = 111 \quad \text{(the Trinity number)}. \qquad (7)$$

Again the origins are partly Christian and partly Masonic, but I have not determined the significance of 32 to him; maybe it was as 2^5, for the hands and fingers of a man whose own instrument was the piano. There is a possibility from gematria, for in the German alphabet (1) LVB gives 32; however, I do not think that Beethoven practiced gematria, and so would not have noticed this property.

Like Mozart, although to a lesser extent, Beethoven used these integers to plot out the numbers of notes in a melody and more often of bars in (a section of) a piece. But the most striking use, quite absent in Mozart, is in his numbering of works of a given kind and in his placing works at "nice" opus numbers. The word "his" here is worth stressing, for *he chose the numbers himself*, and did so from the start of his publishing career—quite contrary to the contemporary habit of publishers (if anyone) assigning them. His works as listed by Opus-number order show significant variation from the chronological order of completion and also a measure of divergence from that of publication.

8. SOME EXAMPLES FROM BEETHOVEN'S OUTPUT

The most intense period of gematriac concern occurred in the early 1820s, when Beethoven was in his early fifties. I shall take two examples from his output; they are summarized in Grattan-Guinness [1994b], where many others are given.

Beethoven wrote 32 numbered piano sonatas. The last 3 were done together, in a consecutive run of Operas 109–111; and the last integer [see (7)] does not disappoint us. In three flats, it is run on 2, 3, and 32 in a remarkable way. In the first movement, 18 bars of introduction are followed by the first subject, stated in a dual manner: its first phrase has $3 + 3$ notes, stated twice, to be followed by $5 + 5$ notes, again stated twice —giving a total of 32 notes. Moreover, the theme is expressed in octaves; that is, with the two hands together (Fig. 5). The second and last movement is an Arietta on a theme over 4×8 bars, as are its first three variations; however, the last one is much freer. Why? The answer is well known: interaction with Beethoven's next major work for piano, already started in 1819 but only finished after (or with) Opus 111. And that work is the "33 variations" (to quote from its title) on a theme in C by Diabelli,

Fig. 5. Beethoven Piano Sonata no. 32, Opus 111, the first subject of the first movement.

the 33rd major work for piano after the 32 sonatas. I am sure that at this stage of composition Beethoven *decided* that the total of variations was to be 33. The last five were written contemporaneously with the completion of the last movement of Opus 111, which sounds rather like another variation on the theme [4]. Of this quintet No. 32 is the most interesting numerologically: performed in E♭ (that is, three flats, unusual for the work), lasting 5 × 32 bars, and with a first subject of 32 notes.

This pair of works also shows well his policy of Opus numbering. Although they were completed very close together, the 33rd work is 9 away from the 32nd, as Opus 120 (3 × 40 = 4 × 30, I suspect), not only another nice integer but also the *first* such after 111 ... . Again, as with Mozart's symphonies (Section 5), the fact that these piano works were to be the last of their kind is not our concern; however, it is worth noting that Beethoven did not add any more during his exceptionally creative final years, presumably having found a good position at which to pause.

During these years Beethoven also worked on his major religious work, the "Missa Solemnis" (as I think he wanted it to be known), completing it in 1823. He assigned to it the Opus number 123, which, as the string 1–2–3, pronounces Christian threeness in a most obvious way. His deliberate choice of number is confirmed by the fact that when he sent it to his publisher, Schott, he left empty the boring number 122 [Beethoven 1985, 39–40].

Threeness occurs all though the work in both small- and large-scale cases. The latter is evident in the tripartite design of several of the five movements, in major parts of them, or in both. Small cases start right at the beginning, for his opening theme started with three rising notes stated three times, and moreover in bars 6, 8, and 10, twice the Pythagorean ratios of the Mason's square described in Section 3.

I shall describe in detail the "Credo," the third movement and the heart of the work: "I believe," the only first-person declaration in the text. It is divided into three parts, of which the first and last are two of the three

Fig. 6. Beethoven's sentiment at the head of the manuscript of "Missa solemnis."

fugues in Bb—or B (for "Beethoven"?), in the German system of lettering. The first is an allegro of 123 bars, no less, and no more. Divided carefully into 30-bar blocks, it also exhibits a remarkable addition at bar 30: the repetition "et, et" invites one to *add on* the rest of the phrase, which ends at bar 33. The second statement of "Credo" starts at bar 37 [see (7)], and the other changes occur at bars 60 (the start of "Deum de Deo"), 90 (for "Qui propter"), and 120 (the end of the "descendit" part), when three orchestral bars complete the movement. The other fugue (a double one) makes up the third part and is itself divided into three subparts by 33s: the first commences at bar 306 and lasts 66 bars, when there occurs a strange transition bar (372) incorporating a tempo change; then 66 more cover "et vitam," before the concluding "Amen" subpart of 33 bars is launched by the soloists at bar 439 (with a final bar for the orchestra).

Beethoven threw his numerology into this life-work piece and not only into the music. At the head of the "Kyrie" he wrote a sentiment that is not normally included in editions but is frequently quoted—*and almost always transcribed incorrectly*. Figure 6 shows his hand [Beethoven 1966]:

Von Herzen—Möge es wieder—zu Herzen gehn!

The form had been carefully chosen, especially with the correct but somewhat colloquial spelling of the last word. There are three clauses, with dashes stressing the divisions; 2 + 3 + 3 words, 3 + 5 + 4 syllables (note the Pythagorean triad), and 9 + 12 + 12 letters, and the last sum is 33.

9. BEETHOVEN AS MASON AND AS ARITHMETICIAN

It is not clear whether Beethoven was a Mason; but the question is in fact *not* crucial for my analysis, because he clearly went along with Masonic sentiments without having been a member—just like his Deism, in fact, which he affirmed without going to church very often.

More important is the question of Beethoven's capacities for effecting arithmetical calculations, for he was notoriously bad at the subject. Musicologists take this fact as sufficient reason to discount the possibility of numerology in him; but mathematical educators (to whom the question properly belongs) recognize that the issue has to be considered in several respects. All that is required of numerologist Beethoven is the mental capacity to do addition: multiplication does not arise, nor scribal competence (which indeed is sometimes wanting in his conversation-book sums). A revealing anecdote tells that on his deathbed he attempted to do multiplication [Thayer 1964, 942]. The implication that he failed is hardly surprising; far more suggestive is the fact that he *tried* to multiply in the first place. It is well known that people sometimes return to childhood practices when dying; presumably he was reverting to early and life-long concerns with arithmetic.

10. COMMENTS AND COMPARISONS

Clearly the mathematics involved in numerology is quite elementary; thus no mathematical ability is presumed in any composer who practiced it. Hence Beethoven's difficulties with arithmetic do not refute the interpretation, as we saw. By contrast, Mozart was fond of numbers—in 1770 (his 15th year) he even signed a letter to his sister Nannerl "Freund des Zahlenhausens" [Mozart, Vol. 1, 311]—but we do not need hypotheses asserting mathematical gifts to understand his later numerological career.

The fact of numerology's occurrence is indubitable for both composers, and indeed is not surprising when one recalls the history of numerology in (pre-) classical music and its place in the Freemasonry of the Austrian Enlightenment. The most extraordinary feature is that nearly two centuries passed before anybody properly realized its presence for either composer (not at all in Beethoven's case). Some methodological problems, which are discussed in the next section, arise here.

11. NUMEROLOGY IN MUSIC: ITS ACTUAL LIMITS

First, numerological doctrines were (and are) almost always held *silently*, whether they occurred in music, religion, architecture, painting, or whatever; only in rare cases did the numerologist leave written testimony about the details or even the practice of numerology. Neither Mozart nor

Beethoven is known to have done so, and to expect to find such sources shows ignorance of the activity. (The structure of *Die Zauberflöte* is so rich that Mozart and Schikaneder/Gieseke may have kept an aide-mémoire at the time of its preparation; but, even if so, it would have been destroyed soon after completion.) Thus the scholar has to resort to contextual understanding in order to be aware of the possibility of the occurrence of numerology and to observation of the kinds of places where it is likely to occur—numbers of notes in a melody, Opus numbers, or whatever. But such research techniques run quite counter to normal criteria for historical evidence.

Second, there is the tyranny of professionalization within disciplines, especially strong in modern academic life. On the principle of leaving war to the generals, only a musicologist can produce pukka musicology; hence if a mere historian of mathematics comes up with such a suggestion, then it cannot be taken seriously. Given the well-known presence of numerology in the music of earlier periods (noted in the next section), I expected historians of classical music to be aware of at least numerology's basic principles and also of its importance to its adherents; but instead I have been confronted with unawareness even down to the level of not knowing the difference between an integer and a digit string (123 versus 1–2–3)—ignorance that Mozart and Beethoven did not exhibit.

I find my analysis dismissed also on the grounds of some vague (and certainly uninformed) appeal to statistical coincidence. But if you want to think that, for example, Sarastro sings 18-note melodies and all the rest because of some very strong law of strong numbers, then that really is your credulous (and ahistorical) hostility. Further, note that some relatively large integers are involved; one does not constantly confront 18, 27, 30, 32, 33, 123, ...

12. NUMEROLOGY IN MUSIC: ITS POSSIBLE SCOPE

Learning about numerology (and gematria) is not too difficult (for a summary history, see Grattan-Guinness [1994a]), but the problems involved have to be understood. These doctrines use only the simplest mathematics, but their history is very long, going back to ancient times; the oldest seems to be Jewish gematria, way back in B.C. territory [Scholem 1971]. Music was possibly not the prime motivation; it seems to have informed more deeply religious principles and practices of a culture and determined features of its architecture and art.

Contrary to the view of the instant nonexperts, the associations between integer and property or symbol are *not* arbitrary; but they may be surprising or unexpected, with connotations local to the originating culture—and very likely not often described in print anyway. The real difficulty is quite another thing, namely, that the *plurality* of doctrines that have evolved over the centuries entail that an integer can have different interpretations in different traditions ([Hopper 1938], a masterly survey). For example, 9 in Hinduism has connotations quite distinct from 9 in Dante; this example is pertinent to Mozart (and his librettist da Ponte), in that Dante's use of 9 and 10 surely guided the Commendatore's music in the Finale of Act 2 of *Don Giovanni* [Grattan-Guinness 1992, 231].

In addition, whatever numerological doctrine was being adopted, it was *important* to its followers, not just a "Zahlenspielerei" (to quote a note to me from the editor of the *Mozart Jahrbuch*, apparently the leading organ for Mozart studies...). In Mozart's case, as a typical example, it affected composition of both internal aspects (such as the numbers of notes in a melody or of bars in a piece) and external ones (such as Masonic or Christian ideals or the dates of first performance).

When alerted to the practices and means of numerological study, one perceives a long stream of thought in the history of (not only) Western music; and a few cases have already been explored to some extent. In the Western tradition, the place of numerology has been well recognized for many figures of the middle ages (for example, Obrecht [van Crevel 1959] and de Vitry [Bent and Howlett 199?]), when it penetrated deeply into all aspects of cultural thought. In our century some cases have been recorded, especially Debussy [Howat 1983], Bartók [Lendvai 1971], Schönberg in his serialist phase [Sterne 1982, 1993], and Berg [Carner 1983, 121–142]—usually without Christian or Masonic connections.

But the classical and romantic periods have largely been omitted from this kind of analysis. Yet it is easy to think of other candidates. An obvious choice is Haydn, whom Mozart talked into joining the Masons in 1785 [Irmen 1991, 120–125]; [5] *Die Schöpfung* (with libretto by van Swieten) is known to carry many numerological and Masonic connotations, and doubtless so do other works.

Among romantic composers, prime "targets" include Schumann and Tchaikovsky, for they are known to have placed codes of other kinds in their music. The chromatic tradition from Chopin to Liszt and Wagner, and maybe on to followers such as Busoni and Reger, is also likely to be a fruitful area. Modern candidates surely include Messiaen, Tippett, and Stockhausen, and there is a clear and lovely use of Fibonacci numbers in the structure of the 34 (sic) common chords used in Britten's *Billy Budd* (1951) to represent the fateful conversation between Budd and Captain Vere [6].

The message of this chapter is that a serious lacuna exists in the understanding of the work of at least two of the greatest composers and probably of several others. But for reasons discussed in the previous section, this important aspect of musical composition will continue to evade the minds of commentators for a few more centuries. Perhaps the historians of mathematics can take up numerology and gematria in music, as part of their concern with the links between mathematics and music.

NOTES

[1] Tatlow [1991] rather revises our understanding of Bach's numerology by suggesting that the assignment of numbers may not necessarily have been made by alphabetical order. For other examples of Bach's numerology in the (30) Goldberg variations, see Mellers [1980, 262–289].

[2] Irmen's book first appeared in 1988 and is, of course, patronized by musicologists. I am chary of certain claims; in particular, on p. 362 he interprets gematriacally the 143rd item in Mozart's thematic catalogue when the numbering is not in Mozart's hand. However, he definitely refutes my original belief that Mozart did not deploy gematria.

[3] Irmen misreads Mozart's "Clarinette" for "Trombe" ("trumpets"). Both parts were deleted on the manuscript at a later stage.

[4] The analysis of the manuscripts pertaining to the Diabelli variations made in Kinderman [1987] closely corroborates this conjecture on the chronology of composition of the variations. It also seems that some of the earlier ones were written late. Of particular interest among them is No. 22, which uses a Mozart theme and is structured not upon the normal 16 bars but on 18 . . . .

[5] H. C. Robbins Landon has kindly informed me of the recent discovery of a previously unknown Masonic membership for Haydn in Pressburg.

[6] In the original (1951) version of Britten's opera, these 34 chords complete the third of its four acts. In general, Fibonacci numbers cause some historical complications for the numerologist. They certainly drove Bartok and aspects of Debussy, but otherwise their place seems to me to be exaggerated, especially for the classical era, when they were of little interest even to mathematicians. For example Peter [1983] analyzes intervals, rhythms, keys, and so on of *Die Zauberflöte* in their terms (see especially pp. 298–300 on Sarastro's aria "O Isis und Osiris"). Similarly, Rutter [1977, 21–28] claims that they run the first movement of Beethoven's Fifth Symphony (but on its undoubtedly intentional 5s, see Grattan-Guinness [1994, Section 2.7]).

REFERENCES

Beethoven, L. van. 1966. *Missa Solemnis Op. 123 Kyrie. Facsimile.* Tutzing: Schneider.

—, 1985. *Der Briefwechsel mit dem Verlag Schott.* Munich: Henle.

Bent, M., and Howlett, D. 199?. [Book in preparation on de Vitry.]

Carner, M. 1983. *Alban Berg...*, 2nd ed. London: Duckworth.

Chailley, J. 1967. *The Magic Flute. Masonic Opera...* (H. Weinstock; Trans). London: Gollancz 1972. (French original: *'La flute enchantée' opéra maçonnique.* Paris: Laffont, 1968).

Curl, J. S. 1991. *The Art and Architecture of Freemasonry.* London: Batsford.

Dalchow, J., Duda, G., and Kerner, K. 1966. *W. A. Mozart. Die Dokumentation seines Todes.* Pähl: Bebenburg.

Grattan-Guinness, I. 1992. Counting the notes: numerology in the works of Mozart, especially *Die Zauberflöte. Annals of Science* **49**, 201–232.

—, 1994a. Numerology and gematria, In (Ed.), *Companion Encyclopedia of the History and Philosophy of the Mathematical Sciences*, 2 vols., pp. 1585–1592. London: Routledge.

—, 1994b. Some numerological features of Beethoven's output. *Annals of Science* **51**, 103–135.

—, 1994c. Why did Mozart write three symphonies in the summer of 1788? *The Music Review* **53**, 1–6.

Hopper, V. F. 1938. *Medieval Number Symbolism.* New York: Cooper Square. [Reprint 1969.]

Howat, R. 1983. *Debussy in Proportion: A Musical Analysis.* Cambridge, UK: Cambridge University Press.

Irmen, H.-J. 1991. *Mozart Mitglied geheimer Gesellschaften*, 2nd ed. Zülpich: Prisca. [1st ed. 1988.]

Kinderman, W. 1987. *Beethoven's Diabelli Variations.* Oxford: Clarendon Press.

Köhler, K. H., Ed. 1979 *Mozart. Die Zauberflöte, Faksimile der autographen Partitur* (Documenta Musicologica, Ser. 2, Vol. 7. Kassel: Bärenreiter). [With an accompanying booklet *Beiheft* which alone carries copyright information.]

Lendvai, E. 1971. *Béla Bartók. An Analysis of His Music.* London: Kahn and Everill.

Lund, J. L. M. 1921. *Ad Quadratum. A Study of the Geometric Bases of Classic and Medieval Religious Architecture*, London: Batsford.

Mellers, W. 1980. *Bach and the Dance of Death.* London: Faber & Faber.

Mozart, W. A. 1962–1975. *Briefe und Aufzeichnungen*, 7 vols. Kassel: Bärenreiter.

Peter, C. 1983. *Die Sprache der Musik in Mozarts Die Zauberflöte...* Stuttgart: Freies Geistesleben.

Robbins Landon, H. C. 1991. *Mozart and the Masons...*, 2nd ed. London: Thames and Hudson.

Rutter, J. 1977. *Elements of Music*, Course A241, Unit 15. Milton Keynes, Open University Press.

Scholem, G. 1971. Gematria. In *Encyclopaedia Judaica*, Vol. 4, cols. 369–374. Jerusalem: Keter.

Sterne, C. C. 1982–1983. Pythagoras and Pierrot: an approach to Schoenberg's use of numerology in the construction of "Pierrot lunaire." *Perspectives on New Music* **21**, 506–534.

—, 1993. *Arnold Schoenberg. The Composer as Numerologist*. Lewiston, PA: Queenstown/Lampeter: Edwin Mullen Press.

Tatlow, R. 1991. *Bach and the Riddle of the Number Alphabet*. Cambridge, UK: Cambridge University Press.

Thayer, 1964. *Thayer's Life of Beethoven*, E. Forbes, Ed. Princeton, NJ: Princeton University Press.

Thiering, B. 1992. *Jesus the Man*. London: Doubleday. [Corgi paperback 1993.]

Till, N. 1992. *Mozart and the Enlightenment*. London: Faber & Faber.

van Crevel, M. 1959. Secret structure. In J. Obrecht, *Opera omnia*, Vol. 1, Pt. 6. xvii–xxv. Amsterdam: Alsbach.

Le style mathématique de Dürer et sa conception de la géométrie

Jeanne Peiffer

Laboratoire d'histoire des sciences et des techniques (UPR 21) du CNRS, Paris, France

My work on Albrecht Dürer (1471–1528), which will appear as a book with my French translation of the painter's *opus geometricum, Underweysung der messung* (1525), owes a great deal to Professor C. J. Scriba, who has always been generous in providing bibliographic information, and with whom I had the pleasure of organizing a seminar on a related subject at the Institut für Geschichte der Naturwissenschaften, Mathematik und Technik of Hamburg University in the summer of 1989. I take this opportunity to express my sincere thanks to him. In the present paper, I will characterize in a very synthetic form Dürer's mathematical style. I show first the heterogeneity of his sources (the practical geometry of handicrafts, classical mathematics, and Italian artists' manuals), then I describe, with the help of examples, the highly personal use the artist makes of the various sources of information at his disposal. Very often he proceeds from the abstract to the concrete, but he also applies techniques from the crafts to abstract mathematical objects. Practical utility is one of his constant preoccupations, but sometimes he is simply absorbed in the tremendous pleasure of just doing mathematics. His geometry is constructive and certainly not demonstrative. The art of measuring (*Kunst der Messung*) is nothing else than that of construction, in fact the point by point construction of geometric forms. Dürer's style is certainly in conformity with his conception of geometry.

1. BREVE REVUE BIBLIOGRAPHIQUE ET BUT DE L'ARTICLE

Depuis un bon siècle, le contenu mathématique de l'*Underweysung der messung* (1525), manuel de géométrie rédigé par le grand artiste allemand Albrecht Dürer (1471–1528) à l'adresse des peintres, artistes et artisans, est relativement bien connu, grâce surtout à l'étude fondamentale que

Hermann Staigmüller lui a consacrée en 1891. Leonardo Olschki s'est intéressé aux sources classiques de l'ouvrage et a analysé, à travers quelques exemples tirés de l'*Underweysung*, la position de Dürer par rapport aux Anciens [Olschki 1919, I, 414–459]. Durant la seconde guerre mondiale, Max Steck a consacré ses recherches à la philosophie sousjacente aux mathématiques de Dürer, devenu le représentant emblématique du génie allemand, et il a publié un volume trop nourri d'idéologie pour constituer encore une référence [Steck 1948]. Plus récemment, Marshall Clagett a étudié l'héritage spécifiquement archimédien dans l'*Underweysung* et a abouti à des conclusions intéressantes [Clagett 1978, 3.III].

Grâce à la monumentale publication, par Hans Rupprich, du *Nachlaß* de Dürer [Rupprich 1956–1969], il est aisé de comparer le texte imprimé en 1525 à des états antérieurs et à des notes diverses que l'artiste a accumulés pendant de longues années. Joseph E. Hofmann est le premier à avoir entrepris ce travail. Dans une série d'articles [Hofmann 1971a, b, c], rédigés à l'occasion du 500ᵉ anniversaire de la naissance de Dürer—dont un [Hofmann 1971c] dédié à Ludwig Bieberbach de sinistre mémoire, éditeur sous le régime fasciste de *Deutsche Mathematik*—Hofmann signale pour chacune des parties étudiées les pièces du *Nachlaß* qui s'y réfèrent. Cette confrontation laisse entrevoir l'enrichissement que constituerait pour les spécialistes de Dürer la publication d'une édition critique de l'*Underweysung*, que j'appelle de tous mes voeux. Plus récemment, Walter L. Strauss a publié une traduction américaine de l'*Underweysung* [Strauss 1977].

Dürers schriftlicher Nachlaß permet d'ores et déjà d'étudier, sur quelques points particuliers, le style de travail de Dürer, sa façon de recueillir les matériaux, de les transformer et de les plier à ses exigences de peintre. Dans cet article, je caractériserai très synthétiquement les démarches de Dürer géomètre. Après avoir montré la grande hétérogénéité de ses sources, je décrirai, à l'aide d'exemples, l'utilisation très personnelle que l'artiste en fait.

2. L'HETEROGENEITE DES SOURCES DE DÜRER

Lorsqu'à son retour d'Italie, en 1507, Dürer conçut le projet de rédiger un manuel encyclopédique à l'usage des apprentis peintres [Rupprich 1956–1969, I, 7], il se mit aussitôt à la recherche d'information dans les domaines de l'optique, de la perspective, de la géométrie, de l'architecture et de l'anthropométrie. L'*Underweysung der messung*, dont il décida la publication en 1523 et qui parut après un délai assez bref en 1525, est le résultat de près de vingt ans de recherche et d'étude. On peut distinguer [Staigmüller 1891, Strauss 1977] trois grands groupes de sources: la géométrie des métiers; les corpus médiévaux et renaissants de la géométrie ancienne (euclidienne et archimédienne notamment); et les manuels des artistes italiens du quattrocento. Dürer pouvait, à Nuremberg même, avoir accès à des collections de livres parmi les plus riches du monde germanique, dont celle de Regiomontanus-Walther, qui comprenait de prestigieux manuscrits anciens, tout particulièrement dans le domaine scientifique. Ne citons à titre d'exemple qu'une copie de la traduction latine de Jacobus Cremonensis du corpus archimédien, qui fut à la base de l'editio princeps (Bâle 1544). Willibald Pirckheimer, ami intime de Dürer ayant fait ses études en Italie du Nord, possédait un manuscrit du *De pictura* (vers 1435) d'Alberti, un exemplaire de la *Summa de arithmetica, geometria, proportioni e proportionalita* (1494) de Luca Pacioli et un autre du *De sculptura* (1504) de Pomponio Gaurico.

2.1. La géométrie des métiers

Traditionnellement on a présenté Dürer pétri de connaissances géométriques héritées des métiers du moyen âge et acquises par le peintre dans les ateliers de Nuremberg, celui de son père orfèvre tout d'abord, puis celui du peintre Michael Wolgemut. Ainsi, Hermann Staigmüller s'exclame: "Nein, sicher nicht Euklid, sondern die Zunfttradition der mittelalterlichen Bauhütte ist die erste und wichtigste Quelle der mathematischen Kenntnisse Dürers" [Staigmüller 1891, 50]. Transmises oralement, les connaissances géométriques des métiers échappent cependant en grande partie, vu l'état lacunaire de la documentation, à notre analyse. Lon R. Shelby a reconstruit de façon convaincante celles des tailleurs de pierre médiévaux et a forgé l'expression très adéquate de "constructive geometry" pour les caractériser [Shelby 1972, 409]. Les problèmes techniques furent résolus grâce à la construction et à la manipulation de formes géométriques simples, comme le triangle, le carré, les polygones, le cercle, etc. Les calculs complexes qu'aurait pu impliquer la résolution de ces problèmes, furent évités grâce à des procédures exécutées par les artisans, pas à pas, avec leurs outils et instruments. Les quelques traités pratiques d'architecture gothique parus à la fin du 15e siècle, comme ceux de Matthäus Roritzer [Geldner 1965], Hans Schmuttermayer [Schmuttermayer 1881] et Lorenz Lechler [Recht 1989, 281] témoignent de ces méthodes.

Que Dürer en soit jusqu'à un certain point tributaire, il n'y a aucun
doute. N'utilise-t-il pas, dans l'*Underweysung*, un certain nombre d'expressions appartenant au vocabulaire corporatif du moyen âge, comme
"Fischblase" (vessie de poisson) pour les configurations obtenues par
l'intersection de deux arcs de cercle et "Vierung" (quadrilatère ou quad-
rangle) pour le carré ou parfois le rectangle. Par la forme d'expression
choisie, l'*Underweysung* s'apparente aussi aux traités pratiques ci-dessus
cités. Elle est souvent prescriptive: "Si tu veux obtenir tel résultat, procède
de la façon suivante." Pour certaines de ses parties, elle a le caractère
constructif mis en avant par Shelby.

Dürer reprend, dans son traité, toutes les constructions, sauf une, du
petit livre (10 feuillets) de Roritzer (vers 1487/1488) censé rassembler les
quelques procédés mathématiques en usage dans les ateliers et intitulé
Geometria deutsch [Geldner 1965, 56–60]: construction d'un angle droit,
détermination du centre d'un arc de cercle, construction (approchée) du
pentagone et de l'heptagone réguliers, …Dans certains cas, Dürer va
cependant plus loin en distinguant entre constructions approchées ("me-
chanice") et constructions exactes ("demonstrative").

Il lui est aussi arrivé de recueillir directement ses informations auprès
des artisans. Nous en avons un témoignage explicite dans un document du
Nachlaß [Rupprich 1956–1969, III, 336–337, Tab. 69, No. 235], où Dürer
rapporte comment les journaliers ("degliche erbetter") ont coutume de
construire un ennéagone régulier à l'aide de trois vessies de poisson. Après
quelques modifications grammaticales ou de style, comme le passage de
l'indicatif à l'impératif, le paragraphe a été inclus dans l'*Underweysung*
[Dürer 1525, II, Fig. 18].

2.2. *Les mathématiques classiques*

Notre connaissance des corpus médiévaux et renaissants des textes euclidi-
ens et archimédiens ayant beaucoup progressé ces dernières décennies, il
est devenu plus aisé de répondre à la question de l'héritage ancien dans un
livre aussi singulier que l'*Underweysung* de Dürer. Il faut partir du constat
que, formé plus dans les ateliers et les voyages que dans les écoles, Dürer
n'est ni latiniste ni helléniste, ce qui constitue un handicap sérieux pour
accéder aux connaissances mathématiques de l'antiquité. Il y remédie en
ayant recours à ce que nous appellerions des conseillers scientifiques
(Willibald Pirckheimer, Johannes Werner, Johann Tschertte, etc.) et en
développant un style de travail collectif. Les feuilles de notes publiées par

Rupprich en portent les traces. Ainsi nous sont parvenues quelques pages contenant des extraits en allemand de l'*Optique* du Pseudo-Euclide, qui suivait les *Eléments* d'Euclide dans l'édition de Bartolomeo Zamberti, que Dürer avait acquise en 1507 à Venise [Zamberti 1505]. Le manuscrit est autographe à l'exception d'un théorème, qui est de la main de Pirckheimer. Rupprich en a suggéré l'interprétation suivante [Rupprich 1971, 91]: Dürer a écrit sous la dictée de Pirckheimer, qui traduisait oralement le texte latin en allemand. Pour une raison ou une autre—on imagine les deux hommes attablés dans l'atelier de Dürer être dérangés par un apprenti venu demander quelque chose au maître, qui s'absente alors pour quelques minutes—Pirckheimer prit la plume pour noter le théorème en question, puis il la céda de nouveau à Dürer.

Le problème de la duplication du cube, auquel Dürer semble accorder une place importante dans son livre, fournit un autre exemple d'une telle collaboration. Dürer présente trois solutions différentes (celles attribuées par Eutocius à Sporus, Platon et Héron) et même une démonstration pour l'une d'entre elles. En fait, ce sont des copies—au moins pour deux d'entre elles—d'un texte préliminaire [Rupprich 1956–1969, III, 354–367] préparé par un de ses conseillers non encore identifié (Johannes Werner très probablement). Marshall Clagett a su trouver la source de ce texte: c'est le commentaire d'Eutocius au Traité de la sphère et du cylindre d'Archimède, extrait d'un manuscrit médiéval, le *De arte mensurandi* de Jean de Murs [Clagett 1978, 3.III, 1163–1170] et traduit en allemand.

Or, Dürer et son conseiller avaient, à Nuremberg, le choix entre plusieurs versions, du texte grec à la paraphrase de Werner [Werner 1522]. C'est un fragment latin provenant de la tradition archimédienne du moyen âge qu'ils ont finalement retenu. Ce choix situe Dürer bien loin des préoccupations des humanistes, soucieux de rétablir les textes anciens dans leur authenticité et dont l'historiographie moderne a tendance à le rapprocher pour en faire même une figure emblématique [Wuttke 1978, 1980].

Même si, sans aide extérieure, Dürer était incapable de comprendre les textes classiques à la lettre, il a su en général s'en approprier le substrat mathématique, ce dont témoigne Joachim Camerarius dans la courte biographie qui précède sa traduction latine des deux premiers livres de *Vier Bücher von menschlicher Proportion* [Dürer 1528]: "Litterarum quidem studia non attigerat, sed quae illis tamen traduntur, maxime naturalium et mathematicarum rerum scientiae, fere didicerat" [Rupprich 1956–1969, I, 307]. L'*Underweysung* n'est pas une compilation d'extraits de textes d'origines diverses. C'est rare que Dürer se contente de reprendre pure-ment et simplement des matériaux préparés par autrui. L'exemple ci-dessus est le seul que je connaisse. En général, il sélectionne et exerce son esprit critique et créatif.

2.3. L'influence italienne

Erwin Panofsky a mis en évidence le fort impact de l'Italie renaissante sur Dürer et son œuvre [Panofsky 1915, 1987]. D'abord le programme, qu'il conçoit à son retour de Venise, s'inspire de la démarche des théoriciens italiens du quattrocento, comme Alberti. Donner à la peinture une plus grande exactitude en la fondant sur les lois de la géométrie, était considéré comme un moyen d'en faire un art libéral et d'améliorer ainsi le statut social des peintres. Puis, Dürer présente, dans l'*Underweysung*, le premier exposé au Nord des Alpes des règles de la perspective linéaire telle qu'elle a été développée en Italie. De nombreux rapprochements entre textes et figures de Dürer et de Piero della Francesca suggèrent que Dürer devait connaître le *De prospectiva pingendi*, pourtant encore sous forme manuscrite [Nicco Fasola 1942, Le Goff 1991]. Il devait également disposer de certains écrits de Leonardo da Vinci. Finalement, il est redevable aux Italiens, de bon nombre d'idées et de constructions en architecture et en mathématiques. Sans que nous connaissions précisément les voies qu'a empruntées la transmission, l'influence sur Dürer du *Divina proportione* de Luca Pacioli est indéniable en ce qui concerne le tracé géométrique des lettres de l'alphabet ainsi que les polyèdres platoniciens et archimédiens.

Dürer inclut dans l'*Underweysung* la description de 7 corps archimédiens, puis de 9 dans l'édition posthume [Dürer 1525, IV, Figs. 35–41, 1538, IV, Figs. 43–43a], sur les 13 dont Pappus avait donné la liste dans sa *Collection* [Hultsch 1965, I, 352–354]. Dans ce cas, il est difficile d'établir ses sources. Certes, la *Collection* circulait sous forme manuscrite, mais aucun exemplaire n'en a pu être localisé à Nuremberg. Quant aux traités de Piero [Mancini 1916] et de Luca Pacioli [Pacioli 1494, 1509], le traitement qu'y subissent les polyèdres semi-réguliers est bien différent de celui mis en œuvre par Dürer. Piero décrit 5 corps obtenus par troncature des angles des 5 corps platoniciens. Pacioli déduit de chacun de ces derniers le corps tronqué et le corps surélevé (pas forcément archimédien) obtenu en ajoutant une pyramide sur chacune des faces, puis représente en perspective les corps d'abord massifs, puis évidés, sur des figures admirablement exécutées par Léonard de Vinci. Rien de tel chez Dürer, qui se contente de donner les développements des solides sur une surface plane, après les avoir succinctement caractérisés d'un point de vue géométrique. Ce qui lui importe, c'est de pouvoir facilement les construire au moyen de carton, de colle et de ciseaux. Face à une telle attitude, on tend à chercher la source d'inspiration de Dürer, non pas dans les traités, mais dans la fabrique. Celle des fameux modèles des 5 corps platoniciens et des corps dérivés, dont il est à plusieurs reprises question dans Pacioli [Pacioli 1494, II, fol.68v, 1509, I, fol.28v] et dont plusieurs collections ont dû exister. Dürer n'aurait-il pu voir l'une d'elles à Venise? Une rencontre avec Pacioli dès 1494–95 n'est d'ailleurs pas à exclure. Le principe simple sur lequel repose la confection de ces modèles se serait assurément gravé dans la mémoire d'un homme aussi rompu à la reconnaissance des formes, pour pouvoir recréer librement son ensemble de solides archimédiens, dont deux d'ailleurs ne se trouvent ni chez Piero, ni chez Luca et sont l'œuvre de Dürer.

3.　LE MODE D'UTILISATION DES SOURCES

Dans certains cas, très rares, nous disposons de chaînes de manuscrits allant de notes de lecture (en allemand, donc déjà traduites) au texte définitivement retenu par Dürer pour l'impression, en passant par des états intermédiaires. Ces séquences permettent de suivre les transformations successives que Dürer fait subir au texte initial, qu'il retravaille sans cesse. Souvent des éléments très hétérogènes interagissent pour donner des résultats intéressants et parfois même nouveaux. Dürer sait faire feu de tout bois.

3.1.　La démarche de l'abstrait vers le concret

Sa démarche, figeant les définitions d'objets mathématiques (tels la spirale d'Archimède ou les coniques d'Apollonius) dans des opérations concrètes de construction graphique, va de l'abstrait vers le concret. Le substrat géométrique est incorporé dans une certaine matérialité. Le traitement de la spirale dans l'*Underweysung* [Dürer 1525, I, Fig. 7] (Fig. 1) est très parlant à cet égard: Dürer part de la définition d'Archimède dont il réalise un modèle matériel. C'est le lieu des points qui parcourent un rayon avec une vitesse constante, tandis que le rayon tourne lui-même avec une vitesse angulaire constante. Dürer fixe un point a au centre et il trace un cercle de rayon ab. Il divise cette circonférence en 12 parties égales, numérotées de 1 à 12, dans le sens inverse des aiguilles d'une montre. L'extrémité b est désignée par 12. Il fabrique ensuite une règle de longueur ab qu'il divise en 24 segments égaux (numérotés de 1 à 23). Il

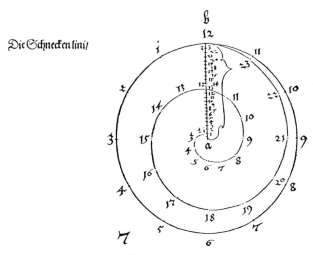

Die Schnecken lini/

Figure 1.

déplace cette règle, l'extrémité *a* restant toujours en *a*, l'autre extrémité
étant successivement positionnée sur les points 1, 2,…, 11 du cercle.
Lorsque la règle joint *a* à 1, Dürer repère sur la règle le point 1; ce sera le
premier point de la spirale. Puis il place l'extrémité de la règle en 2 et
repère le point correspondant sur la règle…. Après le premier tour, il en
commence un second: il joint *a* et 1 sur le cercle et repère le point 13 de la
règle. Il obtient ainsi 24 points de la spirale qu'il n'a qu'à relier par une
ligne courbe. [Dürer 1525, I, Fig. 7]

De nombreux éléments caractéristiques des procédés dürériens sont
présents dans cet exemple: la construction est effectuée point par point.
Un nombre fini de points sont choisis et pour chacun d'eux Dürer répète,
pas à pas, une même opération élémentaire facile à exécuter à l'aide de la
règle et du compas. Il opère sur des points matérialisés sur des règles, des
bois et des pierres. Il pose des points sur des droites et les déplace (comme
une boule sur un boulier).

3.2. Application des méthodes artisanales à des objets mathématiques abstraits

Une des contributions les plus originales de Dürer à la géométrie repose
sur la technique de la double projection, qui plonge ses racines dans les
pratiques graphiques des architectes et des tailleurs de pierre, et qu'il
suppose connue. Par cette méthode [Andersen 1992, Sakarovitch 1989,

Taton 1951], l'objet trimensionnel est représenté dans un plan par deux projections orthogonales, l'une sur un plan horizontal et l'autre sur un plan vertical, que l'on fait coïncider dans le plan de représentation en effectuant une rotation autour de la droite commune aux deux plans de projection. Un point de l'espace est alors représenté par deux points, situés sur une même droite: la ligne de rappel. Dürer cite cette méthode parmi les préliminaires indispensables à la lecture de ses *Vier Bücher von menschlicher Proportion*. Il exige que son lecteur ait bien compris "wie alle ding in grund gelegt/und auffgezogen sollen werden/wie dann die kunstlichen Steinmetzen in teglichem geprauch habenn/dann on das wirdet er mein underrichtung nit volkommenlich vernemen mögen" [Dürer 1528, fol. Aijv], c'est-à-dire comment tracer, à la façon des tailleurs de pierre, plan et élévation de toute chose. Ce qu'il y a de plus remarquable dans l'usage que fait Dürer de cette technique, c'est d'abord la conception claire de la correspondance, point par point, entre les deux projections [Taton 1951, 52], ensuite sa mise en œuvre dans la construction d'objets mathématiques abstraits, comme les sections coniques [Dürer 1525, I, Figs. 34–38]. Dürer en obtient ainsi le tracé d'une façon hautement originale, dont Gaspard Monge codifiera la méthode, à la fin du 18e siècle, dans sa géométrie descriptive [Taton 1951]. Concevant, dans la partie consacrée à l'architecture, une colonne torse, Dürer en vient à considérer sa surface comme enveloppe des sphères de rayon constant et ayant leur centre sur une courbe plane [Dürer 1525, III, Fig. 10].

3.3. Souci de l'utilité pratique

On peut dire que l'*Underweysung* est surtout une géométrie appliquée à des problèmes auxquels doit quotidiennement faire face un artisan, qu'il soit orfèvre, tailleur de pierre, charpentier ou peintre. Très souvent, Dürer insiste lourdement sur l'utilité pratique des formes construites. Ainsi les spirales doivent pouvoir servir au dessin des volutes des chapiteaux ioniques [Dürer 1525, I, Fig. 6], d'une crosse d'évêque ou des bosses de feuillage garnissant les pinacles gothiques. Il remarque [Dürer 1525, I, Fig. 17] que les tailleurs de pierre utilisent l'hélice pour les escaliers en colimaçon. Etc., etc.

Dans l'exemple de la duplication du cube [Dürer 1525, IV, Figs. 44–51], la visée de Dürer est apparente et très claire. Son apport personnel, après avoir repris le texte d'une version médiévale d'Eutocius, consiste dans les applications qu'il en propose. Conscient de l'ancienneté du problème, Dürer rend hommage à Platon pour avoir su indiquer une solution et sauver ainsi la ville d'Athènes de la peste—Dürer dévie ici du récit habituel qui situe les événements à Delos, lieu de l'oracle d'Apollon—puis il se propose de divulguer la solution tenue au secret par les érudits "weyl nun solichs ein ser nutze kunst ist und allen werckleuten dient." Et pourquoi cet art est-il utile? Je cite Dürer lui-même: "dann auß diser kunst kan man puxen und kloken giessen die sich vergrössen und dupliren wie man wil/und doch alweg ir rechte proporcion/auch ir gewicht behalten" [Dürer 1525, IV, Fig. 44]. Il peut servir dans l'armurerie et dans la fonte des cloches. Dürer ne s'arrête pas à la duplication du cube, mais enseigne aussi le triplement et le quadruplement du cube et envisage même la multiplication par quelque facteur n entier. Il constitue alors une série de 4 cubes, dont le deuxième est le double du premier, le troisième le triple et le quatrième le quadruple, puis il indique un moyen d'en déduire rapidement d'autres séries de cubes conservant entre eux la même proportion 1:2:3:4. Le poids, pour un même métal, étant proportionnel au volume, Dürer peut ainsi obtenir une série de boulets de canon, dont le poids augmente continûment d'une livre. Cet exemple illustre l'aisance avec laquelle Dürer sait plier un problème classique aux exigences de la production en série d'un armement standardisé.

3.4. Géométrie et art de la mesure

Dans toute l'*Underweysung*, Dürer n'utilise qu'une seule fois le terme de géométrie. C'est au tout début de son ouvrage, lorsqu'il se réfère à Euclide: "Der aller scharff sinnigst Euclides/hat den grundt der Geometria zusamen gesetzt …". Le titre que Dürer a donné à son ouvrage mentionne la mesure ("Messung") et traduit sans aucun doute ce que Dürer a voulu faire, non une géométrie euclidienne, démonstrative, mais un enseignement de la mesure. Par là il n'entend pas du tout un calcul d'aires ou de volumes, il n'y en a pas un seul dans son livre, mais la construction réglée de formes géométriques. Dürer met à la disposition des artistes et artisans un réservoir de formes correctement construites, dans lequel ils pourront puiser, en sélectionner certaines et les faire varier. Dürer incite souvent son lecteur à ne pas suivre passivement ses instructions, mais à les dépasser en inventant des variantes. Lui-même ne suit pas son propre enseignement. Alors qu'il insiste constamment sur l'utilité pratique des constructions décrites, il ne saisit pas l'occasion de les appliquer dans ses *Vier Bücher von menschlicher Proportion*. Cette attitude, qui peut nous paraître un peu paradoxale, est conforme à la conception de la géométrie que Dürer expose dans ce qu'il est convenu d'appeler l'excursus esthétique du Livre III de ce traité [Dürer 1528, III, fol.T$_{i-iv}$]. Dürer est conscient de l'impuissance de la géométrie à décrire les formes vivantes et naturelles. Le corps humain, ne serait-ce que dans ses contours, ne peut être décrit par les droites et cercles de la géométrie euclidienne. Il faut une géométrie plus fine qu'il entrevoit éventuellement avec Luca Pacioli dans l'approximation de corps complexes par des solides non réguliers à un nombre croissant de côtés, obtenus par des troncatures successives des angles des polyèdres platoniciens. Mais une géométrie puissante et certaine est un idéal d'essence divine, inaccessible à l'homme. Le savoir géométrique, dont celui-ci peut disposer est nécessairement limité. Il ne faut pas pourtant y renoncer. Dürer, avide de savoir, rejette cette idée avec véhémence: "Etwas kunen ist gut. Dan dardurch werd wir destmer vergleicht der pildnus gottes, der alle ding kan" [Rupprich 1956−1969, 2, 106]. Même si notre intelligence est trop limitée pour parvenir à la vérité, nous pouvons l'aiguiser par l'étude et l'exercice. L'art de la mesure est alors une espèce de propédeutique pour le peintre. Il dote l'oeil d'un juste sens des proportions et exerce la main, dispensant l'artiste de toujours tout mesurer. L'apprentissage de cet art, conférant au peintre un bon jugement et une assurance intérieure, le libère en fin de compte.

REFERENCES

Andersen, K. 1992. Perspective and the plan and elevation technique, in particular in the work by Piero della Francesca. In *Amphora. Festschrift für Hans Wussing zu seinem 65. Geburtstag*, S. Demidov, M. Folkerts, D. E. Rowe, and C. J. Scriba, Eds. Basel: Birkhäuser.

Clagett, M. 1978. *Archimedes in the Middle Ages*. Vol. 3. *The Fate of the Medieval Archimedes 1300−1565* [= *Memoirs of the American Philosophical Society* **125**].

Dürer, A. 1525. *Underweysung der messung mit dem zirckel und richtscheyt in Linien ebnen unnd gantzen corporen, durch Albrecht Dürer zusamen getzogen und zu nutz aller kunstliebhabenden mit zu gehörigen figuren in truck gebracht.* Nürnberg.

—, 1528. *Hierinn sind begriffen vier bücher von menschlicher Proportion durch Albrechten Dürer von Nürnberg erfunden und beschriben zu nutz allen denen so zu diser kunst lieb tragen.* Nürnberg: Hieronymus [Andreae] Formschneider [Faksimile-Neudruck der Originalausgabe Nördlingen: Verlag Dr. Alfons Uhl 1980].

—, 1538. *Underweysung der Messung, mit dem Zirckel und richtscheyt, in Linien Ebnen und gantzen Corporen, durch Albrecht Dürer zusamen gezogen, und durch in selbs* (*als er noch auff erden war*) *an vil orten gebessert*, ...Nürnberg: Hieronymus [Andreae] Formschneider.

Geldner, F., Ed. 1965. *Matthäus Roritzer—Das Büchlein von der fialen Gerechtigkeit und die Geometria deutsch.* Wiesbaden: Guido Pressler.

Hofmann, J. E. 1971a. Dürers Verhältnis zur Mathematik. In *Albrecht Dürers Umwelt. Festschrift zum 50. Geburtstag Albrecht Dürers am 21. Mai 1971*, 132–151. Nürnberg: Selbstverlag des Vereins für Geschichte der Stadt Nürnberg.

—, 1971b. Dürer als Mathematiker. In *Praxis der Mathematik* **13**, 85–91; 117–122.

—, 1971c. Dürer als Mathematiker. In *Humanismus und Technik* **15**, 1–16.

Hultsch F., Ed. 1965. *Pappi Alexandrini, Collectionis quae supersunt*, 3 vols. Reprint Amsterdam. A. M. Hakkert [Edition originale: Berlin 1876–1878].

Le Goff, J.-P. 1991. Aux confins de l'art et de la science: *Le De prospectiva pingendi* de Piero della Francesca. In *Destin de l'art. Desseins de la science. Actes du colloque A.D.E.R.H.E.M.* (*Université de Caen, 24–29 octobre 1986*), D. Bessot, Y. Hellegouarc'h, and J.-P. Le Goff, Eds., 185–254.

Mancini, G., Ed. 1916. L'opera "De corporibus regularibus" di Pietro Franceschi detto della Francesca, usurpata da fra Luca Pacioli. *Atti della Reale Accademia dei Lincei s.*5, **14**, 446–580.

Nicco Fasola, G., Ed. 1942. *Piero della Francesca: De prospectiva pingendi*, 2 vols. Firenze [= *Raccolta di fonti per la storia dell'arte* **5**].

Olschki, L. 1919. *Geschichte der neusprachlichen wissenschaftlichen Literatur* 1. Kraus Reprint Vaduz 1965 [Annexe sur Albrecht Dürer, pp. 414–459].

Pacioli, L. 1494. *Summa de arithmetica, geometria, proportioni e proportionalita.* Vinegia: P. de Paganinis.

—, 1509. *Divina proportione...*, Venetiis: P. de Paganinis.

Panofsky, E. 1915. *Dürers Kunsttheorie. Vornehmlich in ihrem Verhältnis zur Kunsttheorie der Italiener.* Berlin: G. Reimer.

—, 1987. *La vie et l'art d'Albrecht Dürer* (traduit de l'anglais par Dominique Le Bourg, Collection 35/37), Paris: Hazan. [Édition originale anglaise 1943. Princeton, NJ: Princeton University Press.]

Recht, R., Ed. 1989. *Les bâtisseurs des cathédrales gothiques* (Strasbourg, du 3 septembre au 26 novembre 1989), Catalogue d'exposition. Strasbourg: Editions Les Musées de la Ville de Strasbourg.

Rupprich, H., Ed. 1956–1969. *Albrecht Dürers schriftlicher Nachlaß*, 3 vols. Berlin: Deutscher Verein für Kunstwissenschaft.

—, 1971. Dürer und Pirckheimer. Geschichte einer Freundschaft. In *Albrecht Dürers Umwelt. Festschrift zum 50. Geburtstag Albrecht Dürers am 21. Mai 1971*, 78–100. Nürnberg: Selbstverlag des Vereins für Geschichte der Stadt Nürnberg.

Sakarovitch, J., 1989. *Théorisation d'une pratique, pratique d'une théorie. Des traités de coupe des pierres à la géométrie descriptive.* Thèse de l'Ecole d'Architecture de Paris La Villette (à paraître dans Science Network. Basel: Birkhäuser).

Schefer, J.-L., Ed. 1992. *Leone Battista Alberti, De la peinture, De Pictura* (1435),

Préface, traduction et notes par J.-L. Schefer, Introduction par Sylvie Deswarte-Rosa. Paris: Macula, Dédale.

Schmuttermayer, H. 1881. Fialenbüchlein. *Anzeiger für Kunde der deutschen Vorzeit. Organ des Germanischen Museums*, 67–78.

Shelby, L. R. 1972. The geometrical knowledge of mediaeval master masons. *Speculum* **47**, 395–421.

Staigmüller, H. 1891. Dürer als Mathematiker. *Programm des Königlichen Realgymnasiums in Stuttgart am Schlusse des Schuljahrs 1890 / 1891*. Stuttgart: K. Hofbuchdruckerei zu Guttenberg (Carl Grüninger).

Steck, M. 1948. *Dürers Gestaltlehre der Mathematik und der bildenden Künste*. Halle: Max Niemeyer.

Strauss, W. L. 1977. *The Painter's Manual. A Manual of Measurement of Lines, Areas, and Solids by means of Compass and Ruler assembled by Albrecht Dürer for the Use of all Lovers of Art with Appropriate Illustrations Arranged to be Printed in the Year MDXXV*, Translated and with a Commentary. New York: Abaris Books.

Taton, R. 1951. *L'œuvre scientifique de Monge*. Paris: Presses Universitaires de France.

Werner, J. 1522. *In hoc opere haec continentur: Libellus super vigintiduobus elementis conicis,*..., Norimbergae: F. Peypus, impensis L. Alantsee.

Wuttke, D. 1978. Aby M. Warburgs Methode als Anregung und Aufgabe. *Gratia. Schriften der Arbeitstelle für Renaissanceforschung am Seminar für Deutsche Philologie der Universität Göttingen*, Heft 2, 2. Auf. Selbstverlag der Arbeitsstelle für Renaissance forschung.

—, 1980. Dürer und Celtis: Von der Bedeutung des Jahres 1500 für den deutschen Humanismus: 'Jahrhundertfeier als symbolische Form'. *The Journal of Medieval and Renaissance Studies* **10**, 73–129.

Zamberti, B. Ed. 1505. *Euclidis megarensis philosophi platonici... elementorum libros XIII cum expositione Theonis*, Venetiis: Joh. Tacuinus.

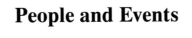

People and Events

Einige Nachträge zur Biographie von Karl Weierstraß

Kurt-R. Biermann

Lindenberger Weg 22, D-13125 Berlin, Germany

und

Gert Schubring

Institut für Didaktik der Mathematik,
Universität Bielefeld, D-33615 Bielefeld, Germany

Unfortunately, nobody has as yet written a standard biography of the outstanding mathematician Karl Weierstraß (1815–1897). This paper examines two aspects of his life. The first part deals with the "secrecy" with which Weierstraß enveloped his foster-son, Franz, since 1886. The paper shows that Weierstraß deliberately covered up tracks and falsified information. It refers to Weierstraß's relationship to Franz, the child's biography, and, in particular, the question of who his mother was. With regard to the considerable amount of money spent by Weierstraß on Franz, the second part of the paper concentrates on how Weierstraß obtained the highly endowed honorary prize for scientific achievement awarded by the Peter-Wilhelm-Müller Foundation in 1895. The paper describes the foundation, which has sunk into obscurity today, and reconstructs the negotiations related to the award in 1894/1895. Originally, H. v. Helmholtz coveted the prize for the physicist H. Hertz, but on the initiative of L. Koenigsberger, one of Weierstraß's disciples who had become a new adviser to the foundation, the prize was awarded to Weierstraß.

1. WEIERSTRAß' GEHEIMNIS

1965 hielt ich auf der Gedächtnisfeier aus Anlaß des 150. Geburtstages von Karl Weierstraß (1815–1897) in Münster (Westf.) einen Vortrag [Biermann 1966a]. Auf dem Postkolloquium setzte ich die in Berlin auf der vorangegangenen Gedenkfeier der Berliner Akademie der Wissenschaften für den großen Mathematiker begonnene Diskussion mit dem Initiator der

Münsteraner Veranstaltung Prof. Heinrich Behnke (1898–1979) über offene Fragen in der Biographie von Weierstraß fort. Bei der Vorbereitung des Berliner Vortrages [Biermann 1966b] hatte ich zwar eine ganze Anzahl von Unrichtigkeiten und Widersprüchen in der vorliegenden Literatur klären, auch manche Lücke im Lebensablauf von Weierstraß schließen können, aber einige Fragezeichen waren geblieben. Der Grund hierfür lag darin, daß zwar relativ viele Erinnerungen an Weierstraß überliefert worden sind, aber keine Biographie im eigentlichen Sinne geschrieben und in der Berliner Akademie erstaunlicherweise kein Nekrolog auf ihn gehalten worden ist. Da keine Zeitgenossen von Weierstraß mehr leben, so führte ich aus, sei es unumgänglich, seinen schriftlichen Nachlaß aufzufinden, wolle man die verbliebenen Rätsel lösen. Professor Behnke stimmte dem zu und bemerkte, daß Weierstraß sicherlich in seinem Letzten Willen bestimmt haben dürfte, was mit seiner wissenschaftlichen und brieflichen Hinterlassenschaft geschehen solle. Der Weg zum Nachlaß führe also vermutlich über das Testament. Testamentsvollstrecker sei der ehemalige Weierstraß-Schüler Ferdinand Rudio (1856–1929) in Zürich gewesen. Übrigens, fügte Behnke hinzu, habe Rudio ihm anvertraut (ohne jedoch Einzelheiten preiszugeben), Weierstraß habe "ein großes Geheimnis mit in das Grab genommen".

Mit Unterstützung der Herren Prof. Behnke, Prof. J. J. Burckhardt (Zürich), Dr. H. Balmer (Konolfingen) und anderer Mathematiker und Mathematikhistoriker versuchte ich ausgangs der 60er Jahre festzustellen, ob sich in Rudios Nachlaß das Testament von Weierstraß oder vielleicht Teile seines wissenschaftlichen Vermächtnisses befunden hätten bzw. ob das Testament bei einem Schweizer Gericht deponiert worden sei.

Es zeigte sich, daß damals in Zürich noch eine Tochter Rudios, Fräulein Alice Rudio, lebte. Die alte Dame gab ihren Besuchern widersprüchliche Auskünfte. Nachdem es zunächst den Anschein hatte, als verfüge sie tatsächlich über gewisse Dokumente, zeigte sich später, daß ihr die Erinnerung wohl einen Streich gespielt hatte. Anfangs hieß es, aus dem Nachlaß von Weierstraß seien zwei Pakete an Rudio gegangen, eines sei zur Vernichtung bestimmt gewesen, durch ein Versehen sei jedoch das falsche verbrannt worden. Es war die Rede von einer Kiste, deren Inhalt die Besitzerin nicht kenne. Dann aber wurde das alles ins Reich der Fabel verwiesen und mit Bestimmtheit versichert, es gäbe außer einigen Erinnerungsstücken nichts im Nachlaß von Rudio, was mit Weierstraß in Verbindung zu bringen sei. Ihre Schwestern, erklärte Fräulein Rudio, hätten aus dem Nachlaß ihres Vaters, von einigen Erinnerungsbriefen abgesehen, nichts bekommen. Herr Dr. E. Neuenschwander (Zürich) ermittelte, von mir darum gebeten, nach dem Ableben von Fräulein Rudio, daß in der Tat keine Papiere mit wissenschaftlichem Wert von oder über Weierstraß in ihrem Besitz gewesen waren. Über den Verbleib des Weierstraßschen Testaments war nichts in Erfahrung zu bringen. Inzwischen hatte Prof. I. Grattan-Guinness (Bengeo, Hertfordshire) die wesentlichsten Teile des Weierstraßschen Nachlasses im Mittag-Leffler-Institut in Djursholm, Schweden, aufgefunden [Grattan-Guinness 1971].

Damit aber war die Frage nach dem Testament zu einer sekundären geworden, und ich verwandte in den folgenden Jahren, ohnehin durch ganz andere Aufgaben in Anspruch genommen, keine Zeit mehr auf die Suche nach ihm. Es war ein reiner Zufall, der mich 1980 schließlich doch zu dem Dokument führte; am 2.12.1980 erhielt ich aus dem Archiv des damaligen Staatlichen Notariats in (Ost-)Berlin das Testament von Weierstraß vom 21.12.1896 zur Einsichtnahme [1]. Was ergab nun das Studium dieses so lange gesuchten Schriftstücks?

Zunächst eine Enttäuschung: Über seinen wissenschaftlichen Nachlaß hat Weierstraß keine besondere Verfügung getroffen. Das aber ist umso verwunderlicher, als bekanntlich seine letzten Lebensjahre von der großen Sorge verdüstert wurden, die begonnene Herausgabe seiner Werke werde nicht vollendet werden und damit das von ihm in der Mathematik Errungene zugrunde gehen, bevor die weiterschreitende Forschung Vollkommneres an seine Stelle gesetzt haben werde. Wir wissen, daß er alles daransetzte, um die Fortsetzung der Edition nach seinem Tode zu gewährleisten. Warum also fehlt jedes Wort über die Vorhaben in seinem Testament? Hatte er die für die Werkausgabe erforderlichen Aufzeichnungen bereits Johannes Knoblauch (1855–1915) übergeben, der bis zu seinem Tod die Hauptlast der Redigierung der (dann doch unvollendet gebliebenen) Edition [Biermann 1968] getragen hat? Weierstraß setzte als Generalerben seine unverheiratete Schwester Elise Weierstraß ein, substituierte ihr einen am 2.8.1882 geborenen Pflegesohn Franz Weierstraß [Biermann 1966b, 194] und diesem seinen leiblichen Bruder, den damals in Breslau lebenden unverheirateten Gymnasialprofessor Peter Weierstraß. Als Vormund seines Pflegesohns setzte Weierstraß den eben genannten Berliner Mathematiker Knoblauch ein. Rudio wird überhaupt nicht erwähnt. Ob er vielleicht der vorgesehene Vollstrecker eines am 21.11.1893 errichteten Testaments gewesen ist, das jedoch durch die genannte "Letztwillige Verfügung" vom 21.12.1896, also zwei Monate vor Weierstraß' Tod, umgestoßen worden ist, muß dahingestellt bleiben, weil es noch auf Weisung von Weierstraß aus der Verwahrung des Gerichts herausgenommen wurde und daher nicht erhalten geblieben ist. Oder war es eine Legende, die die vergebliche Suche in Zürich veranlaßt hatte? Immerhin könnte, wie wir sehen werden, eine Spur zu Rudio führen.

Eine Passage in dem Testament vom 21.12.1896 erregte Aufmerksamkeit, nämlich die, wonach der Pflegesohn Franz "in der Gemeinde Aussersihl im Kanton Zürich eingebürgert" worden sei und Weierstraß ihm am 15.11.1886 ein Vermögen geschenkt habe. Vom Üblichen wich auch die Bestimmung ab, daß dieser Franz die Verfügung über sein Vermögen nicht mit Erreichen der Volljährigkeit erhalten sollte, sondern die Verwaltung der Mittel bis zur Vollendung des 30. Lebensjahres unter der Pflegschaft des eingesetzten Vormunds J. Knoblauch zu bleiben habe. Es erhob sich die Frage, was es mit diesem Pflegesohn für eine Bewandtnis gehabt hat.

In meinem Vortrag auf der Weierstraß-Gedenktagung in Berlin hatte ich am 19.10.1965 aufgrund von Notizen im Archiv der (Ost-)Berliner Akademie der Wissenschaften erwähnt, daß Weierstraß Ende 1884 ein Kleinkind namens Franz, Nachkomme eines Onkels von Weierstraß, zu sich genommen hat, um dessen Erziehung zu leiten. Diese Aufzeichnungen aus dem Akademie-Archiv hat dann 1972 der französische Mathematikhistoriker Prof. Pierre Dugac in einem Preprint im vollen Wortlaut wiedergegeben [Dugac 1972] und 1973 veröffentlicht [Dugac 1973]. Ich zitiere hiernach:

> Der zum Katholizismus übergetretene Vater von Weierstraß "hatte einen Bruder, welcher Protestant blieb und in erster Ehe mit einer nicht sehr gesunden Frau verheiratet war, von der er eine große Zahl Kinder erhielt, die jedoch größtenteils an der Schwindsucht starben. Er selbst war 14 Jahre lang in einer Anstalt für Geisteskranke. Ein Sohn war verheiratet mit der Tochter eines Malers aus Soest. Von den Nachkommen lebt jetzt nur noch eine kleine Zahl. Einer derselben, dessen Vorname Oskar ist, war in erster Ehe, welche nicht sehr glücklich war, verheiratet. Die Ehe blieb 7 Jahre kinderlos. Dann wurde ein Sohn Franz am 2ten Oktober 1882 auf dem Gut Asbeck bei Neviges in der Nähe von Jülich geboren. Diesen kleinen Jungen nahm, als seine Mutter gestorben war, und der Vater zu einer neuen Ehe schritt, Professor Weierstraß zu sich, um seine Erziehung zu leiten." [Dugac 1972, 191; 1973, 167]

Demzufolge war Oskar Weierstraß ein Enkel des Bruders von Weierstraß's Vater; er war also ein Neffe, und sein Sohn Franz war ein Großneffe des Mathematikers.

Die eben wiedergegebenen, ohne Zweifel auf Angaben von Weierstraß selbst beruhenden biographischen Notizen finden sich im Nachlaß [2] des Weierstraß-Schülers und späteren Nachfolgers Hermann Amandus Schwarz (1834–1921), der sie im August 1888 in Wernigerode gesammelt hat, wo er und einige andere Mathematiker dem Erholung suchenden Weierstraß Gesellschaft leisteten. Schwarz sah wohl voraus, daß er in absehbarer Zeit den Nachruf auf Weierstraß in der Berliner Akademie werde halten müssen, denn dessen Erkrankungen (Bronchitiden und Venenentzündungen) nahmen an Dauer und Intensität zu. (Als Weierstraß 1897 gestorben war, wartete freilich die wissenschaftliche Welt vergeblich auf den Nachruf, und, wie Gösta Mittag-Leffler (1846–1927) 1904 prophetisch voraussagte [3], hat Schwarz den Nekrolog nie gehalten.)

Dugac hat am angegebenen Ort noch eine weitere Information über Franz Weierstraß veröffentlicht, auf die ich später wiederholt zurückkommen werde.

In seinen Briefen an seine Schülerin Sonja Kowalewsky (1850–1891) [Kočina-Polubarinova 1973] hat Weierstraß zuerst während seines langen Erholungsaufenthaltes 1885/86 in der Schweiz den Franz Weierstraß erwähnt. Am 14.12.1885 schrieb er ihr: "Seit einigen Tagen bin ich mit meinen Schwestern und dem kleinen Pflegesohn, von dem Dir Mittag-Leffler erzählt haben wird, hier." [Kočina-Polubarinova 1973, 131]

Da die Vermutung nahe lag, daß Weierstraß die Einbürgerung von "Fränzchen", wie ihn sein Pflegevater häufig nannte, während einer seiner Schweizreisen in die Wege geleitet hat, und ich zudem aus dem erwähnten Testament den Ortsnamen Aussersihl (bis 1892 selbständig, dann von Zürich eingemeindet) kannte, bat ich Herrn Dr. Neuenschwander erneut um Unterstützung. Er hat sich daraufhin mit Hilfe von Frau Bizzozero vom Stadtarchiv Zürich sehr mühevollen archivalischen Nachforschungen unterzogen, für die ich an dieser Stelle noch einmal aufrichtig danken möchte und die nach und nach folgendes an den Tag brachten.

Am 21.5.1886 hat ein Advokat Goll an den Gemeinderat Aussersihl das nachstehende Schreiben gerichtet:

"Namens und im Auftrag des Herrn Professor K. Weierstraß in Berlin erlaube ich mir hiermit, das Gesuch an Sie zu stellen, den unterm 2. August 1882 geborenen Knaben Franz, mit Namen Weierstraß, in das dortige Gemeindebürgerrecht aufnehmen zu wollen. Der gedachte Knabe gehört keinem Staatsverbande an und ist daher heimatlos. Um die Gemeinde gegen jede Gefahr der Verarmung u. s. f. zu decken, bin ich ermächtigt zu erklären, daß der Aufzunehmende mit einem Baarvermögen von fs. 50000 ausgerüstet werden wird. Dieses Vermögen wird der dortigen Schirmlade einverleibt und unter die Verwaltung eines obrigkeitlichen Vormundes kommen.

Außer der gesetzlichen Einkaufsgebühr bin ich ermächtigt, als freiwillige Gabe an Ihre Gemeinde fs. 2500 (sage zweitausend fünfhundert Franken) für den Fall der Einbürgerung auszubezahlen.

Diese Summe sowohl, wie das dem Knaben zuzuwendende Vermögen von fs. 50000 werden in den nächsten Tagen bei mir in baar deponirt und zu dem genannten Zwecke zu Ihrer Verfügung gehalten werden.

Zürich, 21. Mai 1886

Hochachtungsvoll
A. Goll, Fürspr [ech]" [4]

Wie mag Weierstraß an diesen Rechtsanwalt August Goll [5] geraten sein? Hier bietet sich natürlich die Annahme an, daß Rudio, seit 1881 in Zürich Mathematikdozent, als Mittelsmann fungiert hat und von daher in das "Geheimnis" eingeweiht gewesen ist. Auf jeden Fall weilte Weierstraß, wie oben erwähnt, bis etwa zu dem Tag, an dem Golls Brief abgefaßt wurde, in der Schweiz. In dem Gollschen Antrag fällt auf, daß nichts über den Geburtsort des Kindes und nichts über seine Eltern gesagt wird. Einigermaßen unverständlich ist die Behauptung Golls, Franz gehöre "keinem Staatsverband an" und sei "daher heimatlos", wenn man sich an die zitierten Notizen aus dem Nachlaß von H. A. Schwarz erinnert. Sollte jene Aufzeichnung etwas Unzutreffendes besagen? Ich bat den Herrn Dr.-Ing. Th. Gerardy (†) in Hannover um Nachforschungen nach einer Geburtsregistrierung des Franz W., da mir nun die Ortsangabe "Asbeck bei Neviges in der Nähe von Jülich" dubios erschien. Herr Dr. Gerardy stellte daraufhin dankenswerterweise eingehende Recherchen an. Ihr Resultat: Weder für das Einzelgehöft Asbeck nördlich von Neviges in der Gemeinde Velbert noch in einem der vier weiteren Wohnplätze, die um 1882 in Nordrhein-Westfalen den Namen Asbeck führten (für keines der fünf Asbecks kann die "Nähe von Jülich" in Anspruch genommen werden!), ist die Eintragung einer Geburt eines Knaben Franz unter dem 2.10.1882 erfolgt, auch der Name Weierstraß tritt nicht auf [6]. Die gerade durch die detaillierte Ortsangabe überzeugend wirkende Notiz im Nachlaß von Schwarz entbehrt demnach der Grundlage. Spätestens jetzt kam mir der Verdacht, daß absichtlich Spuren verwischt worden sind.

1.1. Aber Zurück zu den Züricher Quellen

Der Gemeinderat Aussersihl befaßte sich am 30.5.1886 mit dem Gollschen Antrag und beschloß einstimmig, den "heimatlosen Knaben Franz Weierstraß" unter den angebotenen Bedingungen als Bürger aufzunehmen [7]. In das Bürgerbuch von Aussersihl wurde auf S. 1135 Franz Weierstraß eingetragen und daß er am 30.5.1886 als Heimatloser gegen Einkauf aufgenommen sei. In die Spalte mit den Namen der Eltern wurde eingesetzt "Unbekannt"; die Angabe eines Geburtsortes fehlt [8]. Offensichtlich ging alles reibungslos und nach Wunsch vonstatten.

Auch die Klippe eines geforderten persönlichen Erscheinens des kleinen Franz vor den Häuptern der Gemeinde Aussersihl wurde glücklich umschifft. Zunächst wurde die Präsentation aufgeschoben, dann geriet sie offenbar in Vergessenheit.

Die Vormundschaft über Franz Weierstraß übernahm der Gemeinderatspräsident von Aussersihl, Heinrich Walcher. Seiner Inventur des Besitzstandes von Franz Weierstrass vom 4.4.1887 entnehmen wir, daß Weierstraß in Aussersihl 3 Obligationen auf die Schweizerische Eidgenossenschaft in Höhe von je 10000 Franken und 20 weitere derartige Obligationen von je 1000 Franken deponiert hat. In einer zusätzlichen Bemerkung heißt es:

"Laut Schenkungs-Urkunde des Pflegevaters Hrn. Professor Karl Weierstraß in Berlin, vormundschaftlich genehmigt am 8./9. Dezember 1886, werden die Zinsen dieses Capitals, solange der Bevogtete minderjährig ist, nach Abzug der daraus zu bestreitenden Steuern und Bräuche resp. Vogtgebühren u. drgl. an den Pflegevater des Beschenkten, Professor K. Weierstraß halbjährlich ausgefolgt, der hinwiederum die Verpflichtung auf sich nimmt, den Unterhalt, die Verpflegung und Erziehung des Knaben auf eigene Kosten zu übernehmen." [9]

Franz lebte weiter bei seinem Pflegevater, bis sich ab 1896 die Todesfälle im Hause Weierstraß häuften: Am 23.3.1896 starb Clara Weierstraß, die gemeinsam mit ihrer Schwester Elise den Haushalt ihres Bruders Karl geführt hatte. Ein Jahr darauf, am 19.2.1897 starb Weierstraß selbst. Wiederum ein Jahr später erkrankte Elise Weierstraß an Krebs und mußte sich am 1.3.1898 operieren lassen. Um den nunmehr 15-jährigen aus der Atmosphäre von Krankheit und Sterben herauszubringen, wurde Franz in Freienwalde in Pension gegeben, wo er das dortige Gymnasium besuchte. Am 3. Mai 1898 starb er plötzlich, 15 1/2 Jahre alt. Zwei Monate danach verschied am 9. Juli auch Elise Weierstraß.

P. Dugac hat einen bemerkenswerten Brief des schon genannten schwedischen Mathematikers Mittag-Leffler veröffentlicht [Dugac 1972, 189, 1973, 165]. Ebenfalls ein Schüler von Weierstraß und mit der ganzen Familie (wie auch mit Sonja Kowalewsky) befreundet, unterrichtete er am 21.7.1898 den französischen Mathematiker Charles Hermite (1822–1901) über die Tragödie im Hause Weierstraß. Auf der Rückreise von Gastein nach Stockholm hatte er Elise Weierstraß noch kurz vor ihrem Tode besuchen wollen, konnte aber nicht mehr empfangen werden. Er teilte aus Stockholm Hermite mit, was er in Berlin erfahren hatte: Franz habe sich in jugendlicher Hysterie mit einem Revolver erschossen, nachdem er seiner Tante Elise geschrieben habe, er könne ihr den Grund für seinen Freitod nicht mitteilen. Offiziell werde es heißen, Franz sei bei einem Unfall ums Leben gekommen.

Ich habe das nachgeprüft:

Im Stadtarchiv Zürich liegt eine Sterbe-Urkunde für Franz Weierstraß. Es fällt auf, daß der vorgedruckte Text des auszufüllenden Formulars durchkreuzt ist und die Beurkundung am Rand handschriftlich vorgenommen wurde. So wurde die Angabe der Todesursache umgangen.

> "Sterbe-Urkunde. Freienwalde, Oder am fünften Mai eintausendachthundertneunzig undacht.
> Auf Mitteilung der hiesigen Polizeiverwaltung ist heute eingetragen worden, daß der Gymnasiast Franz Weierstraß, 15 Jahre alt, evangelischer Religion, wohnhaft zu Freienwalde, Oder, geboren zu Aussersihl bei Zürich, Sohn des Landwirthes Oscar Weierstraß und dessen Ehefrau—Namen unbekannt—beide unbekannten Orts verstorben, zu Freienwalde, Oder im städtischen Krankenhaus am driten [sic!] Mai des Jahres tausendachthundertneunzigundacht Nachmittags um elf Uhr verstorben sei.
> Der Standesbeamte O. Hesse". [10]

Aussersihl wurde damit amtlich zum Geburtsort gemacht.

Der Direktor des Oderlandmuseums in Bad Freienwalde, Herr Dr. R. Schmook, unternahm es freundlicherweise auf meine durch seinen Amtsvorgänger Herrn H. Ohnesorge (†) in Altranft vermittelte Bitte, die Lokalpresse auf den Reflex des Todes von Franz Weierstraß durchzusehen. Das Ergebnis: Das Oberbarnimer Kreisblatt Nr. 104 vom 5.5.1898 machte zwar Mitteilung, daß sich ein Freienwalder Gymnasiast aus unbekannten Gründen erschossen habe, nannte jedoch keinen Namen. Fünf Tage später wurde unter "Kirchliche Nachrichten" am 10. Mai in der Nr. 108 des gleichen Blattes der Tod von Franz Weierstraß angezeigt, ohne daß eine Todesursache erwähnt wurde. Die Zeitungsleser in Freienwalde wußten natürlich ohnehin Bescheid, aber die Leser außerhalb werden kaum einen Zusammenhang zwischen den beiden Meldungen hergestellt haben. Mittag-Lefflers Behauptung, es liege Selbstmord vor, der vertuscht werde, war also begründet.

Sollte von keiner Berliner Behörde eine Frage nach der Herkunft von Franz Weierstraß gestellt worden sein? *Ein* Amt wenigstens hat nachgefragt:

Am 18.5.1898 legte das "Königliche Stempel- und Erbschaftssteuer-Amt", Abt. IV, zu Berlin dem Waisenamt der Stadt Zürich einige Fragen im Zusammenhang mit der Weierstraßschen Erbschaft vor. Unter anderem wolle man wissen, "ob der Franz Weierstraß mit seinem Pflegevater Professor Dr. Karl Weierstraß blutsverwandt ist" [11]. Hierzu antwortete der uns schon bekannte Heinrich Walcher am 3.6.1898, daß Franz "als Heimatloser eingebürgert" worden sei, "über seine Herkunft sind keine Angaben vorhanden" [12]. Er setzte noch hinzu, die Schenkung sei seinerzeit nicht in Berlin erfolgt, sondern "hier in Zürich bei Anlaß eines längeren Aufenthaltes des Hrn. Professor Weierstraß am Zürichsee." Dadurch erhält unsere Hypothese der Einbeziehung Rudios in den ganzen Vorgang eine zusätzliche Stütze.

Ein weiterer Brief Walchers vom 28.8.1898 enthält eine beachtenswerte Bemerkung. Inzwischen war, wie erwähnt, auch Elise Weierstraß verstorben und laut Testament von Karl Weierstraß der einzige noch Überlebende der Weierstraß-Geschwister, der bereits genannte Pensionär Peter Weierstraß in Breslau, als Erbe an die Stelle von Franz und Elise getreten. Walcher konstatierte nämlich, "daß der unrichtig ausgestellte Todtenschein des Franz, worin er als Sohn eines Landwirthes figuriert, noch Anlaß zu verschiedenen Umständen" geben werde [13]. Er dürfte also gewußt haben, wer der Vater von Franz gewesen ist. Indessen erkannte das Waisenamt Zürich die Berechtigung von Peter Weierstraß an, das Erbe von Franz anzutreten. In dem Protokoll der Verhandlung hierüber wird Franz nunmehr als "ein Findelkind" bezeichnet [14].

Eines dürfte aus all den sich widersprechenden Verlautbarungen klar geworden sein: Der Kauf des Schweizer Bürgerrechts für Franz Weierstraß ist nicht erfolgt, um einem aus Asbeck gebürtigen, "keinem Staatsverband" angehörenden Sohn eines Oskar Weierstraß, bzw. einem "Heimatlosen", einem "Findelkind," dessen Eltern unbekannt waren, Heimatrechte zu verschaffen, sondern um alles, was auf die wahre Herkunft des Kindes hindeuten könnte, für immer zu verbergen. Dazu war zunächst einmal die Vorlage einer Geburtsurkunde zu umgehen.

Wann hätte dies Dokument vorgelegt werden müssen? Bei der Einbürgerung in Aussersihl war, wie wir gesehen haben, diese Klippe elegant umschifft worden. Bei einer Adoption wäre die Geburtsurkunde gefordert worden. Also beließ man es bei dem Status eines "Pflegesohns", der als Verwandter schon von Geburt an den Namen des Pflegevaters trug. In Preußen wäre dann natürlich bei der Musterung zum Militärdienst die Geburtsurkunde verlangt worden. Das Vorweisen des Schweizer Einbürgerungsdokuments hätte jedoch alle weiteren Fragen überflüssig gemacht und von der Dienstpflicht befreit. Für die Inanspruchnahme des Erbes bei Erreichen der Volljährigkeit wäre ein weiteres Mal der

Geburtsschein gefordert worden. Auch an diese Hürde war gedacht worden: Sie wurde durch die testamentarische Verfügung genommen, daß Professor Knoblauch als Vormund die Verwaltung der Mittel behalten würde, bis Franz das 30. Lebensjahr vollendet haben würde. Und wie geschickt Knoblauch zu handeln wußte, zeigt sich an der geschilderten Ausstellung der Sterbeurkunde in Freienwalde, die wohl kein anderer als er arrangiert haben dürfte. Dabei gelang es zum ersten Male, eine *amtliche*, und dennoch falsche Beurkundung der Vaterschaft von Oskar Weierstraß und eines Geburtsortes Aussersihl zu erreichen. Niemand hatte freilich den tragischen Anlaß, der zugleich den Schlußpunkt setzte, voraussehen können.

Nach allem, was wir hier dargelegt haben, dürfte kaum ein Zweifel daran bleiben, daß das "Geheimnis" von Weierstraß, das Rudio gegenüber Behnke erwähnt hat, in der Herkunft des Franz Weierstraß bestanden hat. Wer waren nun aber tatsächlich die Eltern dieses zweiten Kaspar Hauser? Sollten wir diese Frage zu beantworten suchen, oder sollten wir nicht vielmehr die ganze Angelegenheit auf sich beruhen lassen und das systematisch mit Scharfsinn und Umsicht über die Eltern verbreitete Dunkel respektieren?

Ich würde die letztgenannte Alternative wählen, wenn nicht durch P. Dugac bereits ein Brief veröffentlicht worden wäre [Dugac 1972, 189, 1973, 165], in welchem Mittag-Leffler die Hälfte der Frage beantwortet hat. In seinem Schreiben vom 21.7.1898 an Hermite, aus dem wir schon zitiert haben [Dugac 1972, 189, 1973, 165], behauptete Mittag-Leffler nämlich:

> "Le petit Franz était le fils de Mme Borchardt je le sais, aussi bien que je sais qu'il n'était pas le fils de Weierstraß." [Dugac 1972, 189; Dugac 1973, 165]

Da, wie gezeigt wurde, die im gleichen Brief enthaltene Nachricht vom Selbstmord des Franz Weierstraß und die implizierte Feststellung, der Freitod werde verheimlicht, richtig waren (es findet sich in dem Brief allerdings auch eine unrichtige Angabe: Als Todestag von Franz wird der Tag angegeben, an dem Elise Weierstraß operiert wurde, also der 1.3.1898, während Franz tatsächlich am 3.5.1898 starb), Mittag-Leffler überdies mit der Familie Weierstraß eng befreundet war, ist es erforderlich, zu dieser Behauptung etwas zu sagen.

Mit "Madame Borchardt" ist die Witwe des bekannten Berliner Mathematikers Carl Wilhelm Borchardt (1817–1880) gemeint. Borchardt [Biermann 1988, 80–81; Scriba 1970] stammte aus einer reichen jüdischen Familie, wurde selbst anderthalb Monate nach der Geburt getauft. Finanziell ganz unabhängig, hatte er keine Professur an der Berliner Universität inne, hielt aber dort gelegentlich Vorlesungen in seiner Eigenschaft als Mitglied der Berliner Akademie der Wissenschaften. Die Bedeutung von Weierstraß hatte er schon erkannt, als dieser noch Lehrer in der Provinz war. Ihre Beziehungen hatten sich nach der Berufung Weierstraß' nach Berlin 1856 zu einer engen Freundschaft gestaltet. Die mathematischen Arbeiten Borchardts sind nicht zahlreich, aber Weierstraß schätzte sie und sorgte nach seinem Tode gegen den Widerstand Leopold Kroneckers (1823–1891) [Kočina-Polubarinova 1973, 132–133] für ihre Zusammenfassung in einer Werkausgabe. Das Hauptfeld der Wirksamkeit Borchardts war die Herausgabe des Journals für die reine und angewandte Mathematik 1856 bis 1880 (Bd. 57 bis 90), wobei er für ein hohes wissenschaftliches Niveau dieser angesehenen Fachzeitschrift sorgte. Verheiratet war er seit etwa 1863 mit Rosa Oppenheim, Tochter eines wohlhabenden Königsberger Bankiers. Frau Borchardt war 23 Jahre jünger als ihr Mann. Aus der Ehe gingen 6 Kinder hervor, die zwischen 1865 und 1873 geboren wurden [15].

Weierstraß wurde von Borchardt zum Vormund seiner Kinder bestimmt; er erklärte: "Bei den innigen Beziehungen, in denen ich seit 25 Jahren zu Borchardt und seiner Familie gestanden habe, kann ich mich der Pflicht nicht entziehen, Frau Borchardt bei der Verwaltung ihres bedeutenden Vermögens mit Rat und Tat beizustehen, da sie selbst in Geschäften ganz unerfahren ist" [Kočina-Polubarinova 1973, 84]. Wenn Rosa Borchardt die Mutter von Franz Weierstraß war, dann fragen wir uns natürlich, wie Weierstraß als betagter Mann dazu kam, ihr Kind als seinen Pflegesohn bei sich aufzunehmen, ihm seinen Familiennamen zu geben und mehr als 52.500 Franken zu opfern, um seine Zukunft zu sichern. Man könnte vermuten, Frau Borchardt habe ihm die Mittel zur Verfügung gestellt. Daß es nicht Weierstraß' Geld war, wird dadurch wahrscheinlich, daß er geklagt hat, seine Besoldung reiche nicht aus, um die ständig steigenden Ausgaben zu bestreiten [Kočina-Polubarinova 1973, 84].

Der Borchardtschen Herkunft des namhaften Betrages scheint der Umstand zu widersprechen, daß laut letztwilliger Weierstraß'scher Verfügung jene 50000 Franken im Falle des Todes von Franz im Besitz eines Weierstraß-Erben bleiben sollten. Wären sie indessen an die Familie Borchardt zurückgefallen, hätte das Fragen nach dem Grund gegeben, und gerade solche sollten eben zur Wahrung des Geheimnisses vermieden werden.

Wie konnte aber Frau Borchardt die Schwangerschaft vor ihren Berliner Bekannten und ihren Kindern, deren ältestes bereits 17 Jahre alt war, verbergen? Darauf gibt es Antworten in Briefen von Clara Weierstraß, deren Kenntnis ich Mitteilungen von Herrn Dr. Reinhard Bölling (Berlin) 1990 zu verdanken habe, der im Djursholmer Archiv [16] gearbeitet hat und von meinen Nachforschungen wußte. Danach litt Rosa Borchardt seit Anfang 1882 an einer rätselhaften Krankheit. Sie suchte Heilung in Italien. Eine deutsche Begleiterin wurde bald in die Heimat zurückgeschickt, und es wurde eine neue Gesellschafterin engagiert. Clara Weierstraß stellte in ihren Briefen an Sonja Kowalewsky der Frau Borchardt das denkbar schlechteste moralische Zeugnis aus, ohne freilich je die "Erkrankung" beim richtigen Namen zu nennen. Fest steht jedoch, daß sich Rosa Borchardt an dem Tage der Geburt von Franz (2.8.1882) in Italien befand. Es gibt also starke Indizien dafür, daß Rosa Borchardt tatsächlich die Mutter von Franz gewesen ist.

Es könnte die Annahme auftauchen, Weierstraß' Schülerin und Freundin Sonja Kowalewsky sei die Mutter des Franz gewesen. Indessen muß eine solche Hypothese spätestens nach der Lektüre der ihr von Weierstraß geschriebenen Briefe [17] verworfen werden: Die Erwähnungen des Franz sind nach der ersten, oben wiedergegebenen vom 14.12.1885 völlig beiläufig, wie etwa Grüße von ihm an Sonjas Töchterchen Fuffi (1878–1952), keineswegs solche, wie man sie einer um das Ergehen ihres Kindes besorgten Mutter macht.

Überdies geht aus den Erwähnungen des Franz in den Briefen von Clara Weierstraß an Sonja Kowalewsky, zuerst bereits am 29.10.1884 (!), ganz eindeutig hervor, daß Sonja bis dahin nicht von der Existenz des Jungen informiert worden war (Mitteilung von Herrn Dr. Bölling). Bemerkenswert ist an dem eben zitierten Brief auch, daß Clara sich darin als diejenige bezeichnet, von der die Idee stammte, das "elternlose Kind aus der Verwandtschaft" aufzunehmen, und zwar anfänglich gegen den Widerstand ihrer Geschwister, die dann aber zustimmten, weil es eine "Gewissenssache" gewesen sei, den Kleinen aufzuziehen. Es hat den Anschein, als habe zuerst Clara als Vormund des Knaben fungieren sollen. Weierstraß unterrichtete nämlich diese Schwester am 23.5.1886, er selbst habe zur Beschleunigung die Funktion eines zweiten Vormunds (neben dem Vertreter der Gemeinde Aussersihl) übernommen, weil einer Frau nach den Schweizer Bestimmungen nur dann diese Aufgabe übertragen werden könne, wenn sie von ihrem Mann dazu ausdrücklich "ernannt" worden sei. (Die Kenntnis dieses Briefes verdanke ich Herrn Reinhard Blum aus Essen, der ebenfalls in Djursholm gearbeitet hat und mir 1983 davon Mitteilung machte, und Herrn Dr. Bölling.) Seit wann sich Franz in der Obhut von Clara Weierstraß befand, läßt sich übrigens eindeutig nicht ermitteln. Für den Sommer 1884 ist dies wenigstens bereits belegt. Aber noch zwei Jahre später war Weierstraß besorgt, der Kleine könne etwas ausplaudern. Dies zeigt sein Brief vom 23.5.1886 an seine beiden Schwestern, die damals mit Franz in einer Schweizer Pension weilten, während er nach Forcierung der Einbürgerung mit Hilfe des erwähnten Advokaten Goll bereits nach Berlin zurückgekehrt war. (Auch diesen Brief verdanke ich den Herren Blum und Dr. Bölling). In diesem Schreiben ermahnt Weierstraß seine Schwestern nämlich, den Knaben "vor zu vertraulicher Annäherung an die Mitbewohner" zu hüten.

Es spricht also viel für die Wahrheit der Behauptung Mittag-Lefflers, die Mutter des Franz Weierstraß sei Rosa Borchardt gewesen, aber ein Beweis dafür kann nicht geliefert werden. Nun noch einige Worte zum zweiten Teil seiner Aussage, Weierstraß sei nicht der Vater des Kindes gewesen.

Weierstraß war 1882 67 Jahre alt und seit langem ein leidender Mann. Andererseits war sein geschildertes Vorgehen zur Verbergung der Herkunft des kleinen Franz so zielbewußt, umsichtig, in mehrfacher Hinsicht aufopfernd (von der pekuniären Seite der Sache abgesehen, denke man nur an die Unruhe, die ein Kleinkind in den kinderlosen Haushalt alter Leute bringt, oder daran, daß die Geschwister Weierstraß 1886 eine größere Wohnung in Berlin nehmen mußten, um Franz und ein Kindermädchen logieren zu können) und, dazu noch, nicht ohne Risiko, daß man nicht umhin kann, stärkstes persönliches Interesse an der Geheimhaltung, und nicht nur freundschaftliche Hilfsbereitschaft, zu unterstellen.

Ein interessanter Hinweis dazu findet sich in den Lebenserinnerungen von Ferdinand Lindemann (1852–1939), die 1971 von seiner Enkelin als Privatdruck vervielfältigt wurden. Lindemann, ein passionierter Bergsteiger, erwähnt dort auf Seite 82, innerhalb eines Berichts über Klettertouren in Schweizer Gletschern im August 1881: "Auf der Rückreise übernachteten wir in Samadan, und dort sah ich zum ersten Male Weierstraß mit der Frau Borchard, allerdings ohne ihn kennenzulernen."

Eine Frage bedarf wohl noch der Antwort: Warum, so fragt sich der Leser von heute, warum überhaupt der ganze Aufwand zur Geheimhaltung? Heute freilich nimmt niemand Anstoß an einem Kind eines ledigen Vaters, einer ledigen Frau bzw. einer Witwe; kein Elternteil hat in einer solchen Situation einen wirklich plausiblen Grund, seine Vater- oder Mutterschaft zu verleugnen. Ganz anders vor 100 Jahren, insbesondere aber in der Hauptstadt Preußens und des deutschen Kaiserreiches. Damals galt so etwas als ein "Skandal", und den mußten beide Eltern fürchten.

Weierstraß glaubte zudem, sich einer von Kronecker geführten Allianz gegenüber zu sehen, die sein Lebenswerk in Frage stellen wolle (siehe [Biermann 1966b, 212–214]). Umso dringender war in seinen Augen die Notwendigkeit, sich im Privatleben keine Blößen zu geben, auch wenn er oft gezeigt hat, daß er sich nicht zum Knecht der bürgerlichen Konventionen machen ließ.

Nicht ohne Grund dürfte Clara Weierstraß am 16.7.1882 von Sonja Kowalewsky gefordert haben, "bei andern Leuten ihren [d. h. Frau Borchardts] Namen nicht mit dem meines Bruders [Karl] zusammen" zu nennen, "auch nicht mit unserm [der Schwestern Weierstraß]".

Wir glauben Mittag-Leffler, wenn er die Vaterschaft von Weierstraß in Abrede stellt, aber auf ein Fragezeichen hinter dem zweiten Teil seiner Behauptung können wir noch nicht verzichten.

Damit schließen wir die Akten und lassen das so sorgsam gehütete Geheimnis letztenendes ein Geheimnis bleiben, auch wenn eine systematische Durchsicht der in Djursholm aufbewahrten Hinterlassenschaft sicherlich zur Lüftung des noch verbliebenen Restschleiers führen würde.

Nach Kenntnisnahme der hier zitierten Archivalien können wir nun besser verstehen, welche Sorgen die letzten Lebensjahre von Weierstraß verdüstert haben und warum er 1885 nahezu entschlossen war, Deutschland zu verlassen und in die Schweiz überzusiedeln [Kočina-Polubarinova 1973, 126].

2. WEIERSTRAß ALS PREISTRÄGER DER PETER-WILHELM-MÜLLER-STIFTUNG

Wie wir aus der Darstellung zu Weierstraß' Geheimnis ersehen, hat Weierstraß seinen Pflegesohn Franz mit dem bedeutenden Vermögen von 50.000 Schweizer Franken ausgestattet. Diese Summe entspricht etwa 40.000 Mark der damaligen Währung des Deutschen Reiches [Klimpert 1896, 107]. Immerhin ungefähr das Achtfache eines normalen Jahresgehalts eines Ordinarius in Preußen [18]! Häufig erhält man in Biographien von Wissenschaftlern gerade über die ökonomische Seite ihres Lebens nur dürftige oder keine Informationen. Daß z.B. Gauß—der seine Studien als armer Stipendiat begann—als reicher Mann starb, mit einem Kapitalvermögen von 150,000 Talern, das er u.a. in Börsenspekulationen erworben hatte, ist ja durchaus aufschlußreich für seine Persönlichkeit [19].

Es bildet daher eine Ergänzung der Biographie Weierstraß', über eine wissenschaftliche Ehrung zu berichten, die zugleich—im Gegensatz zum zumeist immateriellen Wert solcher Ehrungen—mit einer beträchtlichen finanziellen Zuwendung versehen war. Über diese Auszeichnung berichtet Leo Koenigsberger (1837–1921), Mathematik-Professor an der Universität Heidelberg und ein Weierstraß-Schüler der "ersten Stunde"—er gehörte zu dessen ersten Studenten an der Berliner Universität—in seiner Autobiographie, allerdings eher nebenbei, im Zusammenhang seiner eigenen Interessen für Physik und seiner Beziehungen zu dem berühmten Physiker und Physiologen Hermann von Helmholtz (1821–1894). Im Anschluß an einen Bericht über die Heidelberger Naturforscherversammlung von 1889 und die dabei zum Ausdruck gekommene Bewunderung für den Physiker Heinrich Hertz (1857–1.1.1894) und dessen Forschungsergebnisse zu elektromagnetischen Wellen erwähnt Koenigsberger einen Vorgang von 1894/95:

"Im Mai 1894 schrieb mir Helmholtz:
'Verehrter Freund!
Ich habe durch *Dr. Hans Müller*, einen der Curatoren der *Peter-Müller* Stiftung, erfahren, daß Sie eingewilligt haben, die Stellung als Preisrichter für die in diesem Jahre zu erteilende Vergebung eines mathematischen oder mathematisch-physikalischen Preises (von 15.000 Mark) zu übernehmen. Ich erlaube mir Ihnen als den mit dem Preise zu Krönenden den im Anfange dieses Jahres verstorbenen *Heinrich Hertz* vorzuschlagen. Was die Größe seiner Entdeckungen und deren wissenschaftliche Durchführung betrifft, so glaube ich ihn allen Zeitgenossen voranstellen zu müssen. Der Umstand, daß er gestorben ist, schließt, soweit ich aus den Statuten erkennen kann, die Ertheilung des Preises nicht aus, auch reichte sein Leben noch in dieses Jahr hinein. Ich habe darüber auch den vorgenannten *Dr. Hans Müller* befragt, der derselben Meinung war, und dem mein Vorschlag zu gefallen schien, er wollte nur keine definitive Antwort ohne Rücksprache mit den anderen Curatoren geben. Wenn Sie dem Vorschlage zustimmen, der mir auch eine Schuld unserer Nation zu tilgen scheint, insofern Hertz während seines Lebens von den deutschen Landsleuten viel weniger geehrt worden ist, als vom Auslande, viel weniger jedenfalls, als seinen Verdiensten entsprach, so können wir die Abstimmung kurz schriftlich abmachen. Haben Sie Zweifel oder wollen Sie einen andern Vorschlag machen, so bitte ich Sie, es mich wissen zu lassen. Wir müssen dann eine Zusammenkunft verabreden, wozu ich Bonn vorschlagen möchte, da *Lipschitz* [20] von uns dreien das Reisen wohl am schlechtesten verträgt.
Ich bleibe noch bis 2. August hier, gehe dann nach Gastein, Mitte September nach Wien zur Naturforscherversammlung. Unser Endurteil wünscht man nur vor Ende des Jahres zu haben.
Darf ich Sie bitten, mich zu benachrichtigen, welche Zeit Ihnen am besten passen würde ... .'
Lipschitz und ich stimmten dem Vorschlage von *Helmholtz* zu—unsere Antworten konnte er sich nur noch vorlesen lassen, am 8. September starb

er. Das Kuratorium mußte auf Grund der Statuten am Anfange des nächsten Jahres den Vorschlag ablehnen, *Lipschitz* trat aus Gesundheitsrücksichten aus der Kommission. Auf Wunsch des Kuratoriums schlug ich zwei neue Kommissionsmitglieder vor, *Warburg* und *Planck* [21], und *Weierstraß* erhielt auf meinen Vorschlag den Preis" [Koenigsberger 1919, 185–186].

Eine identische Darstellung hat Koenigsberger in seiner dreibändigen Helmholtz-Biographie von 1902/1903 gegeben, allerdings unter der Überschrift "Hertz von Helmholtz für den Preis der Peter-Müller-Stiftung vorgeschlagen"—mit dem einzigen Unterschied, daß er als Datum von Helmholtz' Brief nicht "Mai 1894," sondern den 11. Juli 1894 nennt. Das ist auch das zutreffende Datum, da in der Biographie Koenigsberger hier den Kontext von Helmholtz' letzter Lebensphase schildert: Der 11. Juli war der letzte Tag, an dem "er noch völlig geistig frisch" war. Am 12. Juli erlitt Helmholtz die Gehirnblutung, die zu seinem Tode acht Wochen später führte [22]. In seiner später verfaßten Autobiographie hat Koenigsberger offenbar einen früheren Brief der Stiftung mit dem Brief von Helmholtz an ihn verwechselt.

Koenigsberger konnte offenbar voraussetzen, daß die Peter-Müller-Stiftung ebenso bekannt war wie die Bedeutung ihrer Preise, so daß er keinerlei nähere Angaben über die Stiftung für nötig hielt. Bei den Recherchen zu Koenigsbergers Bericht stellte sich aber heraus, daß diese Stiftung heute völlig unbekannt ist. Sie ist in keinem der gängigen Verzeichnisse aufgeführt, und auch der historische Experte des Stifterverbandes für die Deutsche Wissenschaft, Herr Dr. Klaus Neuhoff, konnte keinen Hinweis geben. In keinem biographishen Nachschlagewerk war ein Peter Müller zu finden, der als Stifter eines so beträchtlichen *wissenschaftlichen* Preises in Frage gekommen wäre. Erschwerend wirkte nicht nur die Häufigkeit des Namens Müller, sondern auch die sehr große Anzahl von Stiftungen, die traditionell mit jeder Universität verbunden sind. Da der Verbleib des Nachlasses von Koenigsberger nicht bekannt war und der wissenschaftliche Nachlaß von Lipschitz (der sich in der Bibliothek des Mathematischen Instituts der Universität Bonn befindet) keine Hinweise zur Stiftung und zur Tätigkeit von Lipschitz als Preisrichter enthält, richtete sich meine Aufmerksamkeit auf den Helmholtz-Nachlaß im Archiv der Akademie der Wissenschaften in (Ost−) Berlin.

Hier fanden wir nun in der Tat die entscheidenden konkreten Hinweise und den auslösenden Briefwechsel: Der Name der Stiftung ergab sich als "Peter-Wilhelm-Müller-Stiftung" und als ihr Sitz Frankfurt am Main. Im Namen des Stiftungsrates wandten sich am 29.04.1894 dessen Vorsitzender L. August Müller und der Schriftführer Hans Müller an Helmholtz mit der

Bitte, das Preisgericht zur Verleihung des "Ehrenpreises" und der goldenen Medaille der Peter-Wilhelm-Müller-Stiftung zu bilden. Die zu prämierenden Leistungen sollten aus dem Bereich der "reinen" oder der "angewandten Mathematik" stammen und die Entscheidung Helmholtz' "freiem Ermessen ... überlasssen" sein. Helmholtz nahm das Angebot an und schlug in seinem Antwortbrief vom 12. Mai 1894 Lipschitz und Koenigsberger als weitere Jury-Mitglieder vor [23].

Da ich damit den Sitz der Stiftung kannte, wandte ich mich an das Frankfurter Stadtarchiv und erfuhr, daß dort nicht nur die Stiftung bekannt war, sondern daß dort auch drei Aktenbände über sie vorhanden sind. Als aufschlußreichstes Dokument erwies sich hier das Statut der Stiftung vom 10. Mai 1882, das gedruckt in einem Umfang von zwölf Seiten vorlag.

Die Stiftung, am 11. September 1882 vom preußischen König als Landesherrn genehmigt, und zwar mit dem enormen Stiftungsvermögen von über 1,5 Millionen Mark, war durch testamentarische Verfügung von Peter Wilhelm Müller eingesetzt worden. Sie führte als vollen Namen: Peter-Wilhelm-Müller-Stiftung für Wohlthätigkeit und Förderung von Kunst, Wissenschaft und Gewerbe. Diese gemeinnützigen Zwecke sollten in zwei Sparten gefördert werden: Erstens durch Unterstützung unbemittelter Familien, und zweitens für "künstlerische, wissenschaftliche und gewerbliche Bestrebungen": "durch Unterstützung junger, talentvoller Leute, welche mittellos sind, andererseits durch Ehrenbelohnung höchster Leistungen" (§1).

Die Mittel, die ausdrücklich keiner Diskriminierung von Religionsgemeinschaften unterliegen sollten, sollten im Verhältnis 1:2 für wohltätige bzw. kulturelle Zwecke aufgewendet werden. Für den ersten Zweck waren Einwohner von vier Orten bedenkbar, zu denen für den Stifter persönliche oder berufliche Verbindungen bestanden hatten: Mülheim/Rhein, Eupen, Bonn und Bodendorf an der Ahr (§6 A). Eine neuartige Stiftungsform sah insbesondere der zweite Zweck vor, mit dem den Kultus- und Handelsministerien von Preußen, Baden, Hessen und dem Elsaß umfangreiche Mittel zur Studien- und Wissenschaftsförderung zur eigenen Verwaltung übergeben wurden:

> "Für Schüler von Universitäten, polytechnischen, Kunst- und Kunstgewerbeschulen, Conservatorien, ferner für Privatdocenten, welche zu wissenschaftlichen Arbeiten einer Unterstützung bedürfen, wird den betreffenden Ressortministerien der Staaten Preussen, Baden, Hessen und der Reichslande nach Massgabe der Beschlüsse des Stiftungsrathes periodisch auf eine Reihe von Jahren ein Betrag bis zu zwei Drittheilen der jährlichen Stiftungserträgnisse ... zur Verfügung gestellt. Die Ministerien werden nach

Anhörung der zuständigen Anstaltsbehörden über die Vertheilung beschliessen. Der Beschluss wird der Stiftung kund gegeben. Die Auszahlung erfolgt an die betreffenden Ministerien nach Einlauf der Kundgebung." (§6, B I.)

Auch nicht-staatliche, also kommunale oder private Einrichtungen sollten Stiftungsmittel erhalten können.

Für den Weierstraß zuerkannten Preis und die Medaille ist eine ausführliche Regelung über den Ehrenpreis aussagereich:

"Es ist ein Reservefond von jährlich bis zu neun tausend Reichsmark zu bilden, aus welchem alle drei Jahre ein Ehrenpreis von neun tausend Reichsmark und eine goldene Medaille für höchste Leistungen auf einem Gebiete der Kunst und Wissenschaft innerhalb der letztverflossenen fünfzehn Jahre ertheilt werden soll.

Der Rest kann von dem Stiftungsrathe im Sinne dieses Absatzes B. frei vergeben werden.

Zu berücksichtigen sind bei der Vertheilung des Ehrenpreises in nachstehender Reihenfolge:

1. Bildende Kunst (Malerei und Plastik), 2. Dichtkunst und Musik, 3. Philosophie und historisch-philologische Wissenschaften, 4. Mathematik und Naturwissenschaften. Die Ertheilung des Preises wird Seitens der Stiftung in die Hände von drei hervorragenden Vertretern des betreffenden Faches gelegt, denen Reisespesen und Auslagen zu vergüten sind.—Theilhaftig der Auszeichnung können werden Angehörige des Deutschen Reichs, Deutsch-Oesterreichs und der deutschen Schweiz." (§6, B II.) [24]

Der zweite Teil der Stiftungsgelder stellte eine innovative Form der Bildungs- und Wissenschaftsförderung dar, die mit ihrer Verbindung zwischen privatem Mäzenatentum und staatlicher Kultusverwaltung ein frühes Modell für Stifterverband und Deutsche Forschungs-Gemeinschaft bildet. Es wäre sehr interessant, in den Archiven der betreffenden Ministerien Genaueres über die Förderungspraxis zu ermitteln. Wie ich im Zentralen Staatsarchiv Merseburg (jetzt Teil des Geheimen Staatsarchivs Preußischer Kulturbesitz) feststellen konnte, ist dort bei den Akten des Preußischen Kultusministeriums ein zweibändiger Bestand zu dieser Stiftung erhalten [25].

Natürlich drängte sich mir die Frage auf, wer der Stifter dieser so neuartigen und in ihrem Umfang wohl die neuen Positionen des Bürgertums im Kaiserreich kennzeichnenden Stiftung war. Leider sind die erreichbaren Daten wenig umfangreich. Peter Wilhelm Müller wurde am 5. Oktober 1788 in Mülheim am Rhein geboren. Über Ausbildung und die ersten beruflichen Tätigkeiten ist nichts bekannt. Von 1831 bis 1832 war er Hospitalmeister in Mülheim [26]. Er gründete zusammen mit seinem

Freund und Associé Laurent Hammel in Antwerpen die Firma "Hammel & Cie." Beide wurden stets als "Koopman/négociant", d. h. Kaufmann, bezeichnet, aber trotz aller Bemühungen des Stadtarchivs Antwerpen ließ sich nicht ermitteln, um welche Handelsware es in der Firma ging. Die Tatsache, daß der Stiftungsrat später von der Bethmannschen Bank geführt wurde (s. unten), läßt daran denken, daß es ein Bankgeschäft war—aber weder konnte das Stadtarchiv Antwerpen einen Hinweis dafür finden, noch konnte Johann Philipp Freiherr von Bethmann, Sohn des letzten Stiftungsratsvorsitzenden und bis 1983 geschäftsführender Gesellschafter der Bethmannschen Bank, diese Vermutung bestätigen [27]. Hammel starb unverheiratet am 23. Juni 1849; Müller führte danach die Firma allein weiter [28]. Vermutlich um 1860 hat Müller sich aus der Firma zurückgezogen: Er, der ledig und kinderlos geblieben war, zog jetzt nach Frankfurt/M. und lebte dort als "Rentner," d. h. von den Zinsen seiner Einkünfte. Sein Vermögen vermachte er testamentarisch der Stiftung, und als er am 20.1.1881 in Frankfurt starb, konnten deren Bestimmungen in Kraft gesetzt werden.

Trotz ihres enormen Vermögens erlitt die Stiftung in der Folge der Inflation von 1923 das gleiche Schicksal wie so viele Stiftungen in Deutschland und wie der Besitz des Bürgertums insgesamt: Das Vermögen wurde radikal entwertet. Nach der Inflation waren nur noch ca. 10.000 Reichsmark Vermögen erhalten geblieben. Das Depot-Verzeichnis der Wertpapiere vom 23.1.1933 weist den Bestand mit 13.605 RM aus. Die Wertpapiere waren Auslosungsschuldverschreibungen, die keine regelmäßigen Zinsen erbrachten. "Die Tätigkeit der Stiftung hat deswegen in den letzten Jahren zwangsläufig ruhen müssen," heißt es daher in den städtischen Unterlagen. Die Stadt Frankfurt begann schließlich Mitte der 1930er Jahre, einen sanften Druck auszuüben: Die Peter-Wilhelm-Müller-Stiftung solle mit der 1912 gegründeten Heussenstamm-Stiftung vereinigt werden, die ebenfalls wohltätige Zwecke verfolge und gewerbliche, wissenschaftliche und künstlerische Zwecke fördere, und eine unselbständige Stiftung der Stadt Frankfurt ist [29]. 1938 erklärte sich der Stiftungsrat schließlich mit dieser Auflösung einverstanden: Der Bankier Moritz Freiherr von Bethmann als Stiftungsratsvorsitzender und Dr. Hans Müller wählten als weiteres Mitglied Hans Loeffler, Justitiar des Bankhauses Bethmann. In einer weiteren Sitzung des Stiftungsrates vom 29.11.1938 wurde dann die Übertragung des Vermögens an die Heussenstamm-Stiftung beschlossen [30]. Die formelle Übertragung erfolgte 1940, das Vermögen wurde bei der Frankfurter Bank angelegt [31].

Die Heussenstammsche Stiftung ist noch heute aktiv; sie verfügt aber über keine Unterlagen zur Peter-Wilhelm-Müller-Stiftung [32]. Da ihr auch nur das Vermögen der älteren Stiftung übertragen worden war, ist daher anzunehmen, daß die Sachakten über die Förderungsmaßnahmen beim letzten Stiftungsratsvorsitzenden verblieben sind—und das heißt konkret bei Hans Löffler, der die Geschäfte führte. Leider sind aber alle von Löffler verwahrten Akten im Krieg verbrannt, und die Bethmannsche Bank besitzt keine Unterlagen mehr über die Peter-Wilhelm-Müller-Stiftung [33].

Ohne die Stiftungsakten läßt sich daher heute nichts Genaueres mehr ermitteln über die Beratungen des Stiftungsrates zur Verleihung des Ehrenpreises und der Medaille an Karl Weierstraß. Obwohl Koenigsberger den Übergang vom ursprünglich vorgesehenen Preisträger Heinrich Hertz auf Weierstraß als ganz problemlos schildert, dürfte darin doch einiger Konfliktstoff gelegen haben. Wie aus dem eingangs wiedergegebenen Brief von Helmholtz an Koenigsberger hervorgeht, hatte sich dieser bereits mit dem Stiftungsrat wegen der Wählbarkeit von Hertz abgesprochen. Lipschitz und Koenigsberger hatte er offenbar als Juroren vorgeschlagen, die—obwohl Mathematiker—dem Vorschlag des Physikers Hertz nicht widersprechen würden: der eine aus Freundschaft zu Helmholtz, der andere aus Bewunderung für Hertz. Daß *nach* dem Tod von Helmholtz plötzlich statutenmäßige Bedenken gegen Hertz sprachen, kann man daher auch einer Intervention von Koenigsberger zuschreiben, der nun den Weg frei sah für Weierstraß als Preisträger. Auch der Rücktritt von Lipschitz erscheint als Hinweis auf interne Konflikte, denn die praktisch nicht vorhandene Arbeit in der Jury kollidierte schwerlich mit "Gesundheitsrücksichten."

Der zugrunde liegende Konflikt betraf offenbar das Verhältnis zwischen Mathematik und ihren Anwendungen. Die zu ehrenden Leistungen sollten in der reinen oder in der angewandten Mathematik liegen. Helmholtz hatte diese Bestimmung bereits einseitig interpretiert, indem er mit Hertz einen Physiker, der vor allem durch seine experimentellen Forschungen berühmt war, als Preisträger vorschlug. Koenigsberger verfolgte ein völlig entgegengesetztes Ziel: Mit Weierstraß erhielt erstmals der bekannteste Exponent der mathematischen Strenge, der den neuen Stil reiner Mathematik gewissermaßen verkörperte, eine spektakuläre Auszeichnung. Das war eine umso bedeutsamere Demonstration, als gerade in der Mitte der 1890er Jahre die sog. antimathematische Bewegung der Ingenieure ihren Höhepunkt erreichte—von ihr wurde Wert und Nutzen der reinen Mathematik massiv in Frage gestellt (vgl. [Hensel 1989]). Daß Koenigsberger zwei führende Physiker zur Unterstützung seiner Demonstration für die reine Mathematik hatte gewinnen können, ist daher sehr bemerkenswert [34].

Allerdings konnte ich schließlich den größeren Teil der Korrespondenz zwischen den beteiligten Wissenschaftlern auffinden, so daß die Arbeit der Jury im wesentlichen rekonstruierbar ist. Ausgangspunkt war zunächst die Suche nach dem Nachlaß von Max Planck. Vom Archiv der Max-Planck-Gesellschaft, das wegen der Vernichtung von Plancks Unterlagen im 2. Weltkrieg eine umfangreiche Sammlung von Originalbriefen von Planck und seiner Korrespondenten ebenso wie von Kopien aus anderen Einrichtungen aufgebaut hat, erhielt ich Kopien von Briefen Plancks und Warburgs an Koenigsberger (siehe unten). Aus den Kopien ging hervor, daß die Originale sich in der Sammlung Darmstädter in der Staatsbibliothek Preußischer Kulturbesitz in Berlin befinden. Das legte mir die Vermutung nahe, daß sich in der Sammlung Darmstädter noch weitere Korrespondenz Koenigsbergers befindet. Eine Anfrage ergab, daß in der Tat mehrere hundert Briefe an Koenigsberger vorhanden sind und mithin ein beträchtlicher Teil seines wissenschaftlichen Nachlasses erhalten ist. Allerdings handelt es sich nicht um den vollständigen Nachlaß—z.B. sind keinerlei Briefe von Weierstraß vorhanden, die zweifellos zum Nachlaß gehörten. Die Autographensammlung Darmstädters, die bekanntlich selbst elementarste Regeln des Umgangs mit Archivgut mißachtet hat, hat leider auch hier den Nachlaß-Zusammenhang zerrissen und die Briefe nach den einzelnen Absendern der riesigen Sammlung zugeordnet, so daß es heute nicht einfach ist, alle an Koenigsberger gerichteten Briefe in der Sammlung zu identifizieren. Es sind dort zwar keine Briefe von Weierstraß und auch keine Briefe der P. W. Müller-Stiftung erhalten, jedoch drei einschlägige Briefe von Lipschitz, die—zusammen mit den Briefen Plancks und Warburgs—eine Rekonstruktion des Ablaufs ermöglichten und einen Teil des Berichts von Koenigsberger als nicht zutreffend erwiesen:

Nach Helmholtz' Antwort vom 12.5.1894 an die Stiftung, in der er Koenigsberger und Lipschitz als die weiteren Jury-Mitglieder vorgeschlagen hatte, hatte die Stiftung umgehend bei den beiden angefragt, ob sie die Nomination annehmen. Das ergibt sich aus einem Brief von Lipschitz an Helmholtz vom 25. Mai, in dem er ihn über seine Annahme informierte und zugleich anfragte, er möge ihn "über die Person des dritten Mitgliedes unterrichten, und mir die Schritte, welche Sie für zweckmäßig halten, angeben wollen, in dieser Angelegenheit zu ergreifen" [35].

Helmholtz reagierte hierauf zunächst nicht; offenbar erfolgte jetzt die Abklärung mit der Stiftung wegen der Wählbarkeit von Hertz. Erst am 11. Juli wandte er sich an Koenigsberger mit dem eingangs zitierten Brief und an Lipschitz in analoger Weise. Beide stimmten dem Vorschlag Hertz sofort zu; Koenigsberger erwiderte am 12. Juli,

"dass ich mit Freuden Ihren Vorschlag begrüße, den Namen unseres unvergesslichen Hertz mit dem Preise zu krönen, der—von einem abgesehen [36]—keinem ausgezeichneteren mathematischen Physiker und keiner edleren Gelehrtennatur zugesprochen werden könnte" [37].

Und Lipschitz antwortete am 13. Juli:

"Ihr Gedanke, den Preis und die Medaille der Peter Müller Stiftung an Heinrich Hertz zu verleihen, ist gewiß ein sehr schöner und glücklicher, dem ich herzlich gern meine Zustimmung gebe" [38].

Entgegen Koenigsbergers späterem Bericht konnten beide Antworten Helmholtz nicht mehr vorgelesen werden; er konnte auch keine Schritte mehr für die Preisverleihung unternehmen. Auf beiden Briefen hatte sein Assistent Wachsmuth vermerkt: "Antwort, dass Brief vorläufig nicht gelesen." Auch Lipschitz bestätigte in einem Brief an Koenigsberger vom 13./14. September 1894, daß Helmholtz seine "Antwort nicht mehr zu lesen bekommen" habe. Da Lipschitz von der Stiftung eilig befragt worden war, vor der Stiftungsratssitzung am 22. September das gemeinsame Beratungsergebnis zu übermitteln, bat er Koenigsberger in diesem Brief, ihm die Zustimmung zum Vorschlag Hertz offiziell zu bestätigen. Da nach dem Tode von Helmholtz wohl nicht mehr eine hinreichende Autorität hinter dem Vorschlag für Hertz als vorwiegend experimentellem Physiker stand, lehnte der Stiftungsrat in seiner Sitzung diesen Vorschlag ab. Sichtlich verärgert informierte Lipschitz am 1. Oktober Koenigsberger,

"dass unser mit dem verewigten Helmholtz vereinbarter Vorschlag innerhalb des Stiftungsrathes keine Annahme gefunden hat, wie ich doch geglaubt hatte erwarten zu dürfen."

Und er informierte ihn, daß er—offensichtlich aus dieser Verärgerung heraus—dem Stiftungsrat die Niederlegung seiner Stelle im Preisgericht mitgeteilt habe [39]. Die Ablehnung seitens der Stiftung erfolgte also nicht erst Anfang 1895. Nach der Stiftungsratssitzung vom September ruhte die Angelegenheit zunächst. Anfang 1895 unternahm dann die Stiftung einen neuen Anlauf und beauftragte offenbar Koenigsberger als letztes verbliebenes Jury-Mitglied, als neuer Leiter die Jury zu vervollständigen und einen neuen Vorschlag für Preis und Medaille zu unterbreiten. Das läßt sich aus einem Brief von Emil Warburg vom 21.1.1895 folgern, in der Warburg für Koenigsbergers Aufforderung vom 19. Januar dankt, "in die fragliche Kommission einzutreten", und sich dazu "gern bereit" erklärt.

Die entsprechende Annahmeerklärung Plancks ist nicht erhalten, aber die Reaktionen sowohl von Planck wie von Warburg auf die—nach somit erfolgter Konstituierung der Jury—Anfang Februar beiden Physikern von Koenigsberger zugesandte Vorschlagsliste. Aus den Antwortbriefen ergibt sich, daß Koenigsberger eine Liste mehrerer möglicher Preisträger erstellt und an erste Stelle Weierstraß, an zweite Stelle Ludwig Boltzmann (1844–1906), einen theoretischen Physiker mit intensiver Beziehung zur Mathematik, gesetzt hatte. Zugleich hatte Koenigsberger den Vorschlag Weierstraß mit einer besonderen Dringlichkeit versehen: Offenbar hatte er ein baldiges Ableben von Weierstraß aufgrund seiner Krankheit als möglich hingestellt und zur Vermeidung, daß wie im Fall von Hertz, der Preis nicht mehr zuerteilbar sei, zur Eile gedrängt. Warburg antwortete am 7.2.1895: "Mit Ihren Vorschlägen 1. Weierstraß, 2. Boltzmann erkläre ich mich völlig einverstanden." Planck war in seiner Antwort, ebenfalls vom 7. Februar, auf die Vorschlagsliste vom 4. Februar ausführlicher. Er erklärte sich nicht nur "mit dem Vorschlag von Weierstraß unbedingt einverstanden," sondern fügte noch hinzu,

(ich) "hoffe sehr, daß es gelingen wird, den Preis noch vor Eintritt der von Ihnen erwähnten, nach menschlichem Ermessen nicht mehr sehr lange ausbleibenden Katastrophe zur Verleihung zu bringen."

Auf Boltzmann wollte Planck sich allerdings noch nicht festlegen:

"Sollte jedoch wider Hoffen und Erwarten das Unglück zu früh eintreten, dann müßte ich mir allerdings noch eine bestimmte Erklärung über meinen Vorschlag vorbehalten. Soweit ich augenblicklich beurteilen kann, neige ich mich mehr dazu, Boltzmann zu nennen."

Dieser Fall ist aber nicht eingetreten. Wie aus dem zweiten Brief Warburgs hervorgeht, hat Koenigsberger nur einen Einer-Vorschlag an den Stiftungsrat geschickt. Er hat offensichtlich auch allein, ohne Mitberatung der beiden Jury-Mitglieder, die Laudatio bzw. Begründung des Vorschlags Weierstraß für die Stiftung verfaßt [40].

ANMERKUNGEN

Teil 1 wurde von K.-R. Biermann und Teil 2 von G. Schubring verfaßt.

[1] Staatliches Notariat (Ost-)Berlin: 95 V 4077.96.

[2] Zentrales Archiv der Akademie der Wissenschaften, (Ost-)Berlin: Abt. Nachlässe, Bestand H. A. Schwarz, Weierstrassiana.

[3] G. Mittag-Leffler an H. Schulz, 5.7.1904. Institut Mittag-Leffler, Djursholm, Schweden: Nachlaß Weierstraß Nr. 3666.

[4] Stadtarchiv Zürich: Akten Gemeinderat Aussersihl.

[5] Vgl. z.B. Verzeichnis der Stadtbürger von Zürich auf das Jahr 1855, S. 63; auf das Jahr 1864, S. 76; auf das Jahr 1892, S. 171.

[6] Briefliche Mitteilungen von Dr. Gerardy, Hannover, aufgrund amtlicher Auskünfte.

[7] Stadtarchiv Zürich: Protokolle der Bürgergemeinde Aussersihl 1867–1886.

[8] Stadtarchiv Zürich: Abt. VIII E, Bürgerbuch Aussersihl 1877–1888.

[9] Stadtarchiv Zürich: Akten Gemeinderat Aussersihl.

[10] Stadtarchiv Zürich: Auswärtige Todesakten 1898, Nr. 104.

[11] Stadtarchiv Zürich: Akten Vormundschaft Nr. 29.

[12] H. Walcher an das Waisenamt Zürich, 3.6.1898. Stadtarchiv Zürich: Akten Vormundschaft Nr. 29.

[13] H. Walcher an Bachofen, 28.8.1898. Stadtarchiv Zürich: Akten Vormundschaft Nr. 29.

[14] Protokoll vom 15.7.1898, gen. 23.9.1898, Ziff. 2. Stadtarchiv Zürich: Akten Vormundschaft Nr. 29.

[15] Durch Herrn Prof. E. Amburger, Heuchelheim, freundlicherweise vermittelte Auskünfte von Herrn F. W. Euler über die Familie Borchardt. Siehe [Kočina-Polubarinova 1973, 84].

[16] Im Archiv in Djursholm befindet sich auch eine Photographie von Franz, wie uns Herr Dr. Bölling mitgeteilt hat.

[17] Siehe [Kočina-Polubarinova 1973]. Vgl. jetzt auch die umfangreich kommentierte Edition [Bölling 1993].

[18] Vgl. etwa die Debatten in Preußen um 1896/97 über die Normierung von Professorengehältern und die Stellungnahme von Friedrich Paulsen in *Preußische Jahrbücher*, 87 (1857), 136–144.

[19] In der neueren Gauß-Biographie [Bühler 1981] findet sich diese Information nicht, aber in der älteren Arbeit von Dunnington [1955, 237].

[20] Rudolf Lipschitz (1832–1903), Schüler von P. G. Lejeune Dirichlet und Mathematik-Professor in Bonn seit 1864, enger Freund von Helmholtz (vgl. [Scharlau 1986, XI]).

[21] Emil Warburg (1846–1931) war ein führender Experimentalphysiker. 1895 erhielt er an der Universität Berlin die Professur für Experimentalphysik, die in Deutschland als die beste Position dieses Faches betrachtet wurde. Max Planck (1858–1947), Nobelpreisträger in Physik 1918, war schon damals als bedeutender theoretischer Physiker anerkannt; er war Direktor des Instituts

für Theoretische Physik der Berliner Universität und seit 1894 Mitglied der Berliner Akademie. Über den Nachlaß von Warburg konnte ich nichts ermitteln; der Nachlaß von Planck ist verbrannt, das Archiv zur Geschichte der Max-Planck-Gesellschaft (Berlin) hat aber durch umfangreiche Sammeltätigkeit einen beträchtlichen Teil seiner Korrespondenz erwerben können.

[22] [Koenigsberger 1902–1903, III, 120–123]. Das Datum des 11.7.1894 wird zusätzlich bestätigt durch die Antwortbriefe von Koenigsberger und Lipschitz vom 12. bzw. 13.7.1894 (vgl. dazu unten).

[23] Zentrales Archiv der Akademie der Wissenschaften (Ost-)Berlin: Abt. Nachlässe, Bestand H. von Helmholtz, Nr. 322. Der Antwortbrief vom 12.5. ist Helmholtz' Konzept. Der Anfrage der Stiftung vom 29.4. war wohl eine weitere vorausgegangen, die Helmholtz aber wegen einschränkender Bedingungen oder nicht genügender Vollmachten nicht sofort angenommen hatte.

[24] Zitiert nach: Statut der Peter-Wilhelm-Müller-Stiftung in Frankfurt a. M., Frankfurt a. M. 1882. In: Stadtarchiv Frankfurt/M., Bestand Magistrats- und Stiftungsakten: Nr. 477, Peter-Wilhelm-Müller-Stiftung 1893–1930.

[25] Rep. 76 V a, Sekt. 1, Abt. XI, Nr. 58: Die Peter Wilhelm Müller-Stiftung zu Frankfurt a. M. für Privatdocenten etc., Bd. 1 (1883–1896) und Bd. 2 (1896–1934).

[26] Auskunft des Historischen Archivs der Stadt Köln vom 28.8.1991 (Mülheim ist inzwischen ein Stadtteil von Köln). Seine Eltern waren Peter Müller und Anna Gertrudis Kochs. L. August Müller, der spätere Stiftungsratsvorsitzende und gleichfalls Kaufmann in Antwerpen, war ein Sohn seines Bruders Johann Georg. Peter Wilhelm Müller war ein Onkel des "rheinischen Poeten" Wolfgang Müller aus Königswinter, wie eine biographische Anfrage beim Frankfurter Stadtarchiv im Februar 1940 darlegte (ebda., Stiftungsabteilung, Sign. 325).

[27] Brief vom 26.10.1990. Er konnte aber den Hinweis geben, daß es eine verwandtschaftliche Beziehung zum Bankgeschäft gibt: Peter Wilhelm Müller ist nämlich offenbar ein Nachkomme von Johannes Müller gewesen, der zwischen 1793 und 1807 Teilhaber der Bethmannschen Bank war.

[28] Auskunft des Stadsarchief Stad Antwerpen, 19.4.1991. 1852 verlegte Müller die bisherige Wohnung (zugleich Firmensitz) Beddenstraat 53 nach Jodenstraat 14.

[29] Stadtarchiv Frankfurt/M., Stiftungsabteilung, Signatur 325: Peter-Wilhelm-Müller-Stiftung.

[30] Stadtarchiv Frankfurt/M., Stadtkanzlei, 8420/32, Nr. 2: Heussenstammsche Stiftung 1937–1950.

[31] Wie Anm. 30.

[32] Mitteilung der Geschäftsführerin der Heussenstamm-Stiftung in Frankfurt/M., Frau Chr. M. Mumm von Schwarzenstein, 13.2.1990.

[33] Wie Anm. 27.

[34] Planck hat 1877 bis 1878 mit offenbar großem Gewinn in Berlin Vorlesungen von Weierstraß besucht (vgl. [Dictionary of Scientific Biography 1975, 8]). Auch Warburg hat wohl bei Weierstraß Mathematik studiert. Bemerkenswert ist Warburgs Kritik an der einseitig experimentell-induktiven Methode der älteren Physiker-Generationen, auch von Helmholtz, und sein Plädoyer für die Einbeziehung von *Hypothesen* zum Aufbau von Theorien (vgl. [Schubring 1991, 319–320]).

   In den Statuten ist zwar die regelmäßige Höhe des Ehrenpreises mit 9.000 Mark angegeben, da aber offenbar Kumulierungen möglich waren und der Preis für 1894 nicht vergeben worden ist, gibt es keinen Grund, an der von Koenigsberger berichteten Höhe von 15.000 Mark für Weierstraß zu zweifeln.

[35] Wie Anm. 23, Nr. 281.

[36] Damit meinte er Helmholtz selbst.

[37] Wie Anm. 23, Nr. 239.

[38] Wie Anm. 35.

[39] Beide Briefe in: Sammlung Darmstädter (Handschriftenabteilung der Staatsbibliothek Preußischer Kulturbesitz Berlin). H 1877(4): R. Lipschitz. Erst in einem Brief vom 13.12.1902 hat Lipschitz—nach Erhalt des ersten Bandes von Koenigsbergers Helmholtz-Biographie und der begleitenden Anfrage, in Vorbereitung des dritten Bandes, nach den Gründen seines Austritts 1894 aus der Jury erklärt, "mein Gesundheitszustand (war) damals kein befriedigender" (ibid., H 1877(9)).

[40] Briefe von Emil Warburg vom 21.1. und 7.2.1895 und von Max Planck vom 7.2.1895 an Leo Koenigsberger, in: Archiv zur Geschichte der Max-Planck-Geseslichaft, Berlin, V. Abt. Rep. 13, Planck, Nr. 232–234. (Es handelt sich um Kopien aus der Sammlung Darmstädter in der Handschriftenabteilung der Staatsbibliothek Preußischer Kulturbesitz.)

LITERATURVERZEICHNIS

Biermann, K.-R. 1966a. Die Berufung von Weierstraß nach Berlin. In *Festschrift zur Gedächtnisfeier für Karl Weierstraß 1815–1965*, H. Behnke and K. Kopfermann, Eds., pp. 41–52. Köln und Opladen: Westdeutscher Verlag.

—, 1966b. Karl Weierstraß. Ausgewählte Aspekte seiner Biographie. *Journal für die reine und angewandte Mathematik* **223**, 191–220.

—, 1968. O nezaveršennom izdanii trudov K. Weierstraßa. *Actes du Congrès International d'Histoire des Sciences*, **11** (3), 235–239.

—, 1988. *Die Mathematik und ihre Dozenten an der Berliner Universität 1810–1933*, 2nd ed. Berlin: Akademie-Verlag.

Bölling, R., Ed. 1993. *Briefwechsel zwischen Karl Weierstraß und Sofja Kowalewskaja.* Berlin: Akademie-Verlag.

Bühler, K. W. 1981. *Gauss—A Biographical Study.* Berlin and New York: Springer-Verlag.

Dictionary of Scientific Biography 1975. Vol. 11, p. 8. New York: Scribners.

Dugac, P. 1972. *Eléments d'analyse de Karl Weierstraß*, Maschinenschriftl. vervielfältigt. Paris: Université de Paris VI.

Dugac, P. 1973. Eléments d'analyse de Karl Weierstraß. *Archive for History of Exact Sciences* **10**, 41–176.

Dunnington, G. W. 1955. *C. F. Gauss, Titan of Science.* New York: Hafner.

Grattan-Guinness, I. 1971. Materials for the history of mathematics in the Institute Mittag-Leffler. *Isis* **62**, 363–374.

Hensel, S. 1989. Die Auseinandersetzung um die mathematische Ausbildung der Ingenieure an den Technischen Hochschulen in Deutschland Ende des 19. Jahrhunderts. In *Mathematik und Technik im 19. Jahrhundert in Deutschland*, S. Hensel *et al.*, Eds., pp. 1–111. Göttingen: Vandenhoeck u. Ruprecht.

Klimpert, R. 1896. *Lexikon der Münzen, Maße und Gewichte*, 2nd ed. Berlin: Regenhardt.

Kočina-Polubarinova, P. Ja., Ed. 1973. *Pišma Karla Wejerstrassa k Sof'e Kovalevskoj 1871–1891.* Moskva: Nauka.

Koenigsberger, L. 1902–1903. *Hermann von Helmholtz*, 3 vols. Braunschweig: Vieweg.

—, 1919. *Mein Leben.* Heidelberg: Winter.

Scharlau, W., Ed. 1986. *Rudolf Lipschitz, Briefwechsel mit Cantor, Dedekind, Helmholtz, Kronecker, Weierstraß.* Braunschweig: Vieweg.

Schubring, G. 1991. Spezialschulmodell versus Universitätsmodell: Die Institutionalisierung von Forschung. In *'Einsamkeit und Freiheit' neu besichtigt. Universitätsreformen und Disziplinenbildung in Preußen*, G. Schubring, Ed., pp. 276–326. Stuttgart: Steiner Verlag.

Scriba, C. J. 1970. Carl Wilhelm Borchardt. In *Dictionary of Scientific Biography*, vol. 2, pp. 298–299. New York: Scribner's.

Mathematics at the University of Toronto: Abraham Robinson in Canada (1951–1957)

Joseph W. Dauben

*Department of History, Herbert H. Lehman College, and The Graduate Center,
City University of New York, 33 West 42nd Street, New York, New York 10036*

> *Mathematics may be likened to a Promethean labor, full of life,
> energy and great wonder, yet containing the seed of an over-
> whelming self-doubt.... This is our fate, to live with doubts, to
> pursue a subject whose absoluteness we are not certain of, in
> short to realize that the only "true" science is itself of the same
> mortal, perhaps empirical, nature as all other human under-
> takings.*
>
> —*Paul J. Cohen* [1971, 15]

Christoph Scriba was affiliated with the University of Toronto from 1959 to
1962, in the Department of Mathematics. This followed by only a few years
Abraham Robinson's appointment as the successor to Leopold Infeld in the
Department, to teach applied mathematics. Although what follows describes
primarily the political circumstances under which Infeld left Canada, and the
character of the Department of Mathematics at the University of Toronto in
the mid-1950s, it will provide a sense of what it was like to be in Canada, and
at the University, just prior to Christoph Scriba's appointment there in 1959.

1. MATHEMATICS AT TORONTO: THE BACKGROUND

When Samuel Beatty (1881–1970), Head of the Department of Mathematics, retired from the University of Toronto in 1952, he had been a member
of the Department for 45 years and its Chairman for the last 18 of these
[1]. By then he was a piece of Canadian history itself—the first Canadian
Ph.D. in Mathematics (in 1915), having done his graduate work with J. C.
Fields of Fields Medal fame (Beatty was his only student). Although
Beatty's research was devoted primarily to the theory of algebraic functions, largely to aspects of the Lagrange Interpolation Formula, he also

93

taught courses on calculus and complex variables. When Beatty was made Chairman of the Department in 1934, his first priority was to build up theoretical mathematics.

Seeking to provide a "broader and deeper approach," Beatty hired Richard Brauer in the spring of 1935 to cover algebra and a year later H. S. MacDonald Coxeter to teach geometry. These appointments, as Gilbert Robinson has said, were "Beatty's greatest contribution to mathematics in Canada" [G. Robinson 1971, 489].

Meanwhile, not to be outdone, John L. Synge was just as interested in developing his own Department of Applied Mathematics—at the time a separate department with its own brass plaque on the building at 47 St. George Street. It was Synge who achieved a coup of sorts by bringing the refugee Polish mathematical physicist and friend of Albert Einstein, Leopold Infeld, to Toronto in 1938 [2]. As one of his biographers later wrote, Infeld's life was a "microcosm" of the century [G. Robinson 1968, pp. 123–125] [3]. It was the gap left in part by Infeld's resignation in 1950 that led to Robinson's appointment.

2. INFELD AT TORONTO: THE POLITICS OF A MARXIST MATHEMATICIAN

Born in Cracow, Poland, in 1898, Leopold Infeld studied physics at the famous Jagiellonian University. It was during a year at the University of Berlin in 1920–1921 that he met Albert Einstein and wrote his first paper on "Light Waves in the Theory of Relativity," for which he received his Ph.D.—the first to be given for theoretical physics in Poland—from the University of Cracow.

Following hard times in the 1930s and the tragic death of his first wife Halina, Infeld left Europe to work with Max Born in Cambridge on a Rockefeller Foundation Fellowship. As anti-Semitism continued its alarming advance in Germany, Einstein (already safely in the United States at the Institute for Advanced Study) suggested that Infeld consider coming to Princeton, which he did in 1936. Infeld, once settled in Princeton, collaborated with Einstein on their popular account, *Evolution of Physics*. It was the success of this work that attracted Synge's attention, and led him to invite Infeld to Toronto in 1937.

The Department of Applied Mathematics (according to Infeld's unsparing eyes) was housed in a "shabby, small, two-story building" at 47 St.

George Street:

> On the ground floor was a small lecture room. One flight up was the office
> of the head of the department, which was also used for conferences. Above
> this, under the roof, was my room. In addition, the building contained three
> offices of the other professors and small cubby holes for the assistants.
> [Infeld 1978, 17]

Beatty, to Infeld, seemed "more an administrator than a mathematician,
in general very decent but capable of doing something unfair more
through stupidity than bad character. Dean Beatty had little understanding
for the scientific work done by those under him" [Infeld 1978, 24].

Infeld was anxious "to build a strong centre of theoretical physics in
Toronto.... . But no one in Canada cared about it." Despite an "urgent
request that our group be enlarged," Infeld was clearly put off by the
answer he got from "this very rich university: 'No, we can't. We haven't the
money'" [Infeld 1978, 24]. Indeed, as Lewis Pyenson has observed, applied
mathematics "withered" because the University of Toronto "expressed
little interest in Infeld's desire to create a strong centre of theoretical
physics" [Pyenson 1978, 3].

Infeld was especially concerned about the Canadian "brain drain" to the
United States:

> If the most brilliant people continued to leave Canada for the United States,
> then, eventually, according to biological laws, the intellectual level in Canada
> would go down and I added, in jest, "if for thousands of years this trend does
> not change, then Canada will become a nation of morons." That very day the
> local paper carried the headline: "Infeld Predicts Canada to Be Country of
> Morons." [Infeld 1978, 24–25]

What especially bothered Infeld, however, about Canada and the United
States, was the lack of respect he felt in both countries for scientific work
—at least until the time of Sputnik in 1957.

After the war, the Polish ambassador to Canada began making increasingly compelling overtures to Infeld about his returning to Poland. Infeld
did so on two occasions—for lecture tours and consulting, first in the
summer of 1949 and again in 1950.

Infeld also took to lecturing widely in Canada and the United States
about the atomic bomb. He was well known for saying that there was really
no secret about it—"The real secret, I explained, was that the thing was
possible, and that circumstance ceased to be a secret with the explosion on
Hiroshima." He predicted the Soviet Union would be able to build an

atomic bomb in three or four years, an unfortunate (but as it turned out all too accurate) prediction, because some critics immediately took this to be a sign that Infeld must have secret information—and presumably secret contacts as well—with the Russians:

> If the Soviet Union could really make an atomic bomb in the time I predicted, then how did I know it? Didn't that mean that I had some kind of dangerous contact with them? [Infeld 1978, p. 26]

At the same time, and complicating Infeld's position considerably, he was a strong supporter of the new Workers' State in Poland. He was also vocal in denouncing what he called "American nuclear blackmail." Because of the two visits he had made to the new communist Poland late in 1940, he was denounced by local conservatives as a "potential traitor to the Canadian people," claims that were fueled in part by McCarthyism in the United States and the Gouzenko affair [4] in Canada [Peynson 1978, 3]. As a result, opposition in Canada began to swell against him. When Infeld asked permission to return to Poland in 1950, he was accused of wanting to give away atomic secrets to a country "behind the iron curtain":

> As a result, the Royal Canadian Mounted Police dogged my footsteps and those of my family, we were annoyed by vicious telephone calls, and finally the University of Toronto suddenly cancelled my leave of absence to visit Europe. [Infeld 1978, 26]

The truth, however, was that after returning to Toronto from his trip to Poland in 1949, Infeld's attitude towards Canada had begun to change:

> I myself recall how the wealth of Canada upset me—the rich but tasteless buildings, the people who had no idea what was happening in Europe, who did not appreciate scientific work—all this upset me and depressed me more and more. [Infeld 1978, 40] [5]

Due to an article in *The Ensign*, a Catholic weekly published in Montreal, questions about Infeld's knowledge of atomic secrets and his apparent Communist sympathies soon provoked "waves of persecution that quickly rose around me" [Infeld 1978, 46] [6]. His plans to go back to Poland for a year of teaching were suddenly in jeopardy. As he became increasingly the focus of political debate, the University of Toronto did its best to keep him in Canada. On this score Infeld had particularly unpleasant memories of the University's new president, Sidney Smith. A former law school lecturer, Smith had served as President of the University of Manitoba in Winnipeg before advancing to head the University of Toronto.

As Infeld saw him:

> He was plump, rather good-looking, a Rotary club type...a pillar of the Conservative Party...I had many talks with him and Dean Beatty. In one of these, the President told me that he would offer me twice my salary not to go to Poland. He imagined that for a few thousand dollars he could buy me. Finally he told me that, though it would be a loss if I were to go on leave, or even stay away for good, the university would not collapse. [Infeld 1978, 53] [7]

Unable or unwilling to bear the pressure on himself and his family, Infeld decided to leave Canada as soon as the academic year (1950) was over. In May he went to England and then to Poland by ship on the *Batory*. Two months later his wife and children joined him, sailing on the same ship from New York and arriving in Poland on the country's national holiday, July 22, 1950.

Sometime during the summer of 1950, Beatty wrote to Infeld warning him to return for the coming academic year or lose his job. Infeld's reply, after describing his own version of the University's actions, was blunt: "Please, therefore, consider this letter to be my resignation" [8].

Thus Infeld remained permanently in Poland, but he prospered as a member of the country's intelligentsia. Not only was he made a member of the *Praesidium* of the new Polish Academy, where he headed the Institute of Theoretical Physics, but he also served as Director of the Physics Institute at the University [9].

Although difficulties were to be expected in trying to live in Poland, especially as he tried to administer the Institute, Infeld was also insulated, perhaps protected, from the harsh realities of the Communist régime. Vehemently despising the political persecution he had experienced in Canada, he later confessed that had he known the extent of political persecution in Poland at virtually the same time, he would never have left North America. Instead, Infeld said he had no knowledge of "what was really happening during the Stalin period, which I learned only after the Twentieth Congress." Otherwise, he insisted, he "would not have gone to Poland" [Infeld 1978, 54] [10].

Infeld's departure from the University of Toronto was compounded by the loss of another of its applied mathematicians, A. F. C. Stevenson, at virtually the same time [Duff 1969, 103–107]. Beatty had already appointed John Coleman in 1949 (a former student of Synge's) and soon added J. A. Steketee to the Department as well. The staff in Applied Mathematics was not back to full strength, however, until Abraham Robinson was appointed in 1951, along with J. A. Jacobs the following year (just as Beatty retired

both as Chairman and from the Department of Mathematics in 1952
[G. Robinson 1979, 57]) [11].

3. THE ROBINSONS ARRIVE IN TORONTO: 1951

J. A. Steketee remembers picking up Abraham Robinson and his wife,
Renée, and driving them across town with their luggage in his De Soto
when they got to Toronto in September of 1951. For anyone arriving from
London, metropolitan Toronto struck most visitors soon after the war as
"an urban disaster":

> ...congeries of dull, small towns, dangling from a dark entanglement of
> overhead wires. And in those days, the official social life in the university was
> stiff with protocol and inhibitions, a trial to [anyone] who had known the
> gaiety that was always erupting amid the grimness of the wartime British
> captial. By the fifties, the Toronto environment had begun to change: a
> genuine city was beginning to emerge with some sense of style and a growing
> pride and confidence in itself. [Bissell 1974, 36]

It was this "new" Toronto that the Robinsons found most congenial.
They did not encounter the political problems that had driven Infeld away,
but instead were welcomed by their colleagues and soon discovered the
physical beauty of Canada in the course of many vacations, including trips
they took to the United States and Mexico in the early 1950s.

In fact, it was an exciting time to be in Canada. Toronto had grown
dramatically from its early days as a French trading post and somewhat
later as a fort to deter invasion by the Americans. The city eventually
replaced Niagara-on-the-Lake as the seat of Canadian government
(Toronto being, presumably, a safer distance from the U.S. border). By the
early 1950s, it was the fastest growing city in North America, with a
population well in excess of one million [12].

The Robinsons soon found a new home on Winnett Avenue [Robinson
I–1990]. Although the house was small, it had its own garden, "and when
you opened the door it was like entering a new world, cozy, small,
attractive. Sometimes, theater groups would meet there, actors and pro-
ducers together" [Robinson I–1983]. It was not long before Renée
Robinson's experience as an actress brought her to the attention of the
Canadian Broadcasting Corporation, where she was able to make good use
of her dramatic talents:

> Soon after having established ourselves [in Canada], as I had an actor's
> equity card from Vienna and then one from London, I tried to get an

audition at the Toronto Radio Station. I was lucky to have one of the most established Canadian actors present, an elderly man who became, more or less, my patron. Soon they asked me to take parts where my accent was either required or did not matter, and I became one of the actors circle [Robinson I–1990].

When the Austrian actor Josef Fürst came to Canada, he and Renée soon became good friends, and they were frequently cast together. In one film they played an Estonian couple, and for two years they were the "von Hohenfelds" in a daily radio series, "The Craigs," in which she played the part of an Austrian noblewoman who had supposedly moved to rural Canada [Ross I–1990, Robinson I–1990]. When the Robinsons left Canada in 1957, she was dramatically written out of the script (although she cannot recall whether she met an untimely soap-opera death or not).

In addition to radio, Renée made several television appearances, as well as a motion picture in which she played a psychiatrist for a production by the Film Board of Canada, capitalizing again on her Viennese heritage [Robinson I–1980]. She recorded several readings, one a part she had always dreamed of playing—"Medea." Another was a comedy, for which she had a special flair. She also made a 78 rpm record with Fürst of excerpts from Shakespeare's "Othello," Renée playing the part of Emilia, another role she had always hoped to play. She was disappointed not to have played the part opposite her friend Josef Fürst when he was cast as Iago for a CBC television production directed by David Green. (Instead, Green cast his wife in the part. This time, Renée knew it was not her accent that was the problem—after all, "If Iago has an accent [Fürst], then why not Emilia?" [Robinson I–1980].

At home on Winnett Avenue, Renée would often combine her friends from show business with the mathematicians from the University, "and they enjoyed each other very much." It was just this sort of openness and genial social grace that impressed Robinson's colleague and former graduate student R. A. Ross, who was a guest on several occasions at their home. "Pleasant but small," is how he described it, adding that the Robinsons' generosity in entertaining faculty and graduate students was unusual at the time [Ross I–1990] [13].

It was also a place where Robinson liked to work even after he had come home from the University. And he worked hard, systematically. Often when Renée's sister would stop by to see them in the evenings, Abby would be sitting on the sofa in the living room with his writing pad, working on his mathematics [Steketee R–1976].

4. COLLEAGUES IN APPLIED MATHEMATICS

The group of applied mathematicians worked together in what had been a private residence on St. George Street. Robinson was given one of the better rooms on the first floor, but it was nevertheless rather sparse:

> The furniture was restricted to some shelves, a filing cabinet, some chairs and a desk in the middle of the room. The desk was usually piled with a disorderly variety of papers leaving a few square feet for his actual writing. He was not the type who carried big businessmen's briefcases around, full with clobber. The work he took home was usually contained in a thin briefcase, a writing pad and some papers, and occasionally a book. [Steketee R–1976]

Robinson was a pipe smoker, and there was usually a tin of Players Navy Cut tobacco on hand, and occasionally cigars. His graduate student R. A. Ross remembers that Robinson was always available, but also very disciplined, allocating his time very carefully.

The few times Ross recalls seeing Robinson annoyed was when the phone rang at 47 St. George Street. There was only a single telephone for the entire building, and it had been installed in a clothes closet just outside Robinson's office. "He seemed to think he was the official telephone answerer—and would have to shout out names from one floor to another whenever the phone rang" [Ross I–1990] [14].

Just downstairs from Robinson, George Duff was assigned an office in the back room on the bottom floor. Duff remembers first meeting Robinson in the summer of 1952. In those days he saw a good deal of Robinson because Duff "sat on his doorstep a lot." Duff liked to discuss his work on partial differential equations, and Robinson was always willing to listen and be helpful.

Duff was a specialist on the subject of differential equations, about which Robinson—to Duff's surprise—also knew a great deal. Duff took it as a reflection, again, of Robinson's wide-ranging abilities that, when he was working on a book on partial differential equations, Robinson read over the manuscript and offered a number of suggestions about how to improve the book, making it clearer in various ways for prospective readers [15]. In fact, it was due to his versatility that Robinson soon knew everyone well. H. S. M. Coxeter was likewise impressed by Robinson's diverse interests, especially his wide-ranging knowledge of geometry, Coxeter's own field [16].

5. EARLY YEARS AT TORONTO: TEACHING AND LECTURING

In 1952 Robinson was asked to offer a short course of lectures to the National Research Council on "Supersonic Wing Theory" [Steketee R-1976]. Clearly the quality of his work and his importance in Canada for aerodynamic research were already being recognized. The following year, in 1953, he was also invited to give a lecture on "Wave Propagation near the Surface of an Elastic Medium" at New York University, as well as a paper on "Flow around Compound Lifting Units" at a Symposium on High Speed Aerodynamics at the National Aeronautical Establishment in Ottawa [Robinson 1953b].

Meanwhile, his teaching at Toronto was limited primarily to traditional subjects of the basic sort he had already taught as an instructor at the Royal College of Aeronautics in Cranfield (England): differential equations, fluid mechanics, and aerodynamics. He usually taught a course on partial differential equations in mathematical physics and on fluid dynamics in the fourth-year applied mathematics program. And he often taught a graduate course on supersonic wing theory, to which he had already contributed a great deal of research on his own while at the Royal Aircraft Establishment at Farnborough during the War, and later, at Cranfield afterwards. Robinson also taught some of the "service courses," including calculus, analytic geometry, and differential equations for engineering students.

6. ROBINSON'S FIRST PH.D. IN APPLIED MATHEMATICS: 1953

Robinson's first graduate student at Toronto was J. F. Hart, who had begun his doctoral work under Arthur Stevenson [Ross I-1990]. His chosen subject was the "Theory of Spin Effects in the Quantum Mechanics of a Many-Electron Atom with Sepcial Reference to Pb^{III}" [Hart 1953] [17]. Mathematically, Hart's thesis was of interest because it calculated for the first time the matrix elements for interconfiguration perturbations within the atom involving mutual spin interactions. This in turn showed that the orbital and spin coefficients were large enough to suggest that spin effects were not negligible, but comparable to electrostatic effects [Hart 1953, pp. 1–2] [18].

Hart's conclusions must have been appealing to Robinson, for they demonstrated how presumably negligible factors in a physical problem could actually take on considerable importance in reality. This was just the sort of conclusion that Robinson's work in fluid dynamics had impressed upon him time and again [Robinson 1945, 1947, 1948a, 1948b, 1951a].

7. ROBINSON AND AERODYNAMICS AT THE UNIVERSITY OF TORONTO

The Cold War ensured that Western governments, including Canada, would maintain substantial interest in aeronautics, and Robinson, along with his graduate students, were among those to benefit from increased support and facilities. By the early 1950s there was near universal agreement that gas turbine development was essential, and thus both gas dynamics and high-speed aerodynamics were given particular emphasis. The Korean War also served in a direct way to increase military funding, even in Canada, for both applied aeronautical development and theoretical research. Continuing economic expansion after the war also did its share, and the explosive growth in commercial aviation all over the world was another factor in maintaining interest in aerodynamics.

Consequently, the University of Toronto was anxious to expand its courses for both undergraduate and graduate students in aeronautics. The undergraduate courses were given as the "aeronautics option" in engineering physics, whereas graduate studies and research were collectively overseen by a new Institute of Aerophysics (later to become the Institute for Aerospace Studies) with G. N. Patterson as its Director. The Institute was also supported by the Canadian government with grants from the Defense Research Board (additional funding came from U.S. agencies). Students were enrolled for both the M.Sc. and Ph.D. degrees. Initially, emphasis was given to the physics of gases, applied aerodynamics and ballistics, with a special interest in supersonic flight, an area in which Robinson had developed a strong reputation of his own [19].

The first of Robinson's aerodynamic papers to appear after his move to Toronto was devoted to an analysis of nonuniform supersonic flow [Robinson 1953a]. This paper calculated the perturbations produced by a two-dimensional thin airfoil in a nonuniform (i.e. accelerating or decelerating) supersonic flow for the two-dimensional case based on linearized theory. In order to consider the problem of acceleration in isolation, Robinson assumed that the airfoil was placed in a plane of symmetry to the main flow so that the velocity component of the main flow (in a direction normal to the plane) was uniformly zero. Using linearized theory, Robinson then derived formulas for the perturbation potential which involved elliptic integrals (to which approximations could be made for nonuniform streams). These in turn gave correction terms for the case of uniform stream and also provided corrections for both lift and drag.

Robinson's interest in such problems went back to a number of earlier papers in which he also dealt with airfoils in nonuniform incompressible flow. The physical significance of this class of problems, Robinson noted, warranted its investigation. He added that one referee had even called his attention to the case of flow around jet vanes, where nonuniform supersonic flow was clearly of practical interest [Robinson 1953a, 183].

On the subject of supersonic airfoil design in general, Robinson once told Ross that he had been one of the pioneers of the subject at Farnborough. Although he had been the first there to work out the details of supersonic flow for delta wings during the war, all of the research was classified, and therefore he never got credit for any of his efforts because none of his results was published at the time [Ross I–1990].

8. THREE PH.D.'S IN APPLIED MATHEMATICS: A BUMPER CROP, 1955

J. A. Steketee had come to Toronto in 1950 from Delft, his interest in aerodynamics having already been stimulated there by J. M. Burgers. Robinson suggested the topic of his thesis on boundary layer transitions and provided "lively interest and encouragement." Steketee found that, due to the usual separate treatment of laminar and turbulent flows, he was unable to formulate the problem as clearly, mathematically, as he would have liked. Instead, he examined several models reflecting various aspects of concrete situations. In particular, he managed to account for "the importance of the velocity gradient" and showed that "some solutions of the perturbation equations can be made to agree with measurements of turbulent fluctuations in a shear flow" [Steketee 1955].

Once again, Robinson had set one of his students upon a problem in which usually neglected features of aerodynamic modeling could not be ignored. As Steketee wrote in the introduction to his thesis, although in many studies of fluid dynamics viscosity might be neglected, "in the neighborhood of walls and of obstacles placed in the flow, where the velocity gradients are generally large, one has to take viscosity into account." It was the problem of transitions at these boundary layers that provided the substance of Steketee's dissertation.

L. R. Fowell's dissertation was devoted to "An Exact Theory of Supersonic Flow around a Delta Wing," [20] and was concerned primarily with finding a solution for the equations of inviscid supersonic flow in the particular case of a delta wing. In analyzing flow over the expansion surface, he considered two cases for delta wings with supersonic leading edges and found that below a critical angle of attack a continuous solution might exist, while above this angle the solution was discontinuous. Fowell provided an exact solution for the case of flow over the expansion surface, but in the discontinuous case he only proposed an approximate method.

L. L. Campbell recalls that originally he was "assigned" to Robinson for supervision [21]. Robinson had worked on certain aspects of hyperbolic differential equations during the war, and his report on "Shock Transmission in Beams" dealt with some particular equations governing shock transmission in wings of aircraft [Robinson 1945]. He was especially interested in the initial value problem for hyperbolic partial differential equations in two variables and of general order, and proposed that Campbell try to extend his methods to initial boundary value problems for the same equations. Campbell succeeded in doing so (with appropriate guidance, he adds, from Robinson) and thus produced his thesis [22].

As Campbell worked on his dissertation, he met regularly with Robinson, who provided "much helpful discussion during the course of the investigation" [Campbell 1955, "Acknowledgments"]. The principal objective of his thesis was to develop the mixed initial and boundary value problem for a linear hyperbolic partial differential equation of arbitrary order in two independent variables. Campbell was also able to give some results on the mixed problem for a semilinear hyperbolic system of first order equations [23].

Campbell succeeded in solving the linear, hyperbolic equation of order n (assuming that the unknown function and its first $n - 1$ normal derivatives took specified values on a segment of the positive y-axis) by extending a method Robinson had already used to solve the initial value problem for an equation of general order [Robinson 1950a] [24]. The method basically reduced to solving mixed problems for a succession of semilinear systems of first-order equations. Campbell also solved a mixed problem for the linear hyperbolic equation of order n by using an extension of the Riemann method.

9. WING THEORY

Robinson's major piece of work in applied mathematics at Toronto (and the most comprehensive work he ever did in aerodynamics) was his book on *Wing Theory*. This drew upon courses he had taught at Cranfield—and new material he was teaching at Toronto. What Robinson intended to provide was a comprehensive survey of the subject of airfoils in general, a book designed to serve as a textbook for advanced courses on the subject [25]. As one of Robinson's colleagues at Cranfield put it, "the book took some three years to prepare and it developed into an impressive work of comprehensive scholarship and authority" [Young 1976, 311] [26].

It was E. F. Relf, first Principal of the College of Aeronautics (Cranfield), who invited Robinson to write the volume for the Cambridge Aeronautics Series [27]. Robinson agreed and began to work on *Wing Theory* during his last few years at Cranfield. Only later did he decide to ask his former student from Cranfield, John Laurmann, to join him in the venture as coauthor:

> It so happened [Robinson] and I moved to Canada at about the same time, I to the National Research Council in Ottawa, he to the University of Toronto. It was there that he suggested that I help him with its authorship, and I was of course delighted to accept. Actually, the book was finished a little later when I was at the University of California in Berkeley, finishing my graduate work, and we had some rather complicated mailings back and forth in finalizing the book and in proof reading. I was the one doing the routine stuff in getting properly typed manuscript to the University Press [28].

Wing Theory covers both subsonic and supersonic airfoil design, considered under conditions of both steady and unsteady flow and all presented with "that heightened sense of structure and unity that was a characteristic of Abby's work" [Young 1976, 311] [29]. As the authors explained in their introduction, the basic object of wing theory was

> ...the investigation and calculation of the aerodynamic forces which act on a wing, or on a system of wings, following a prescribed motion in a fluid medium, usually air. The theory is based on the assumption that the medium is continuous, and it can be shown that this assumption does not lead to any appreciable errors except at very high speeds or at very low pressures. [Laurmann and Robinson 1956, 1]

In writing *Wing Theory* Robinson was often torn between conflicting demands: mathematical rigor (involving at times the introduction of subtle "techniques which might be a heavy burden for those who are interested in the subject chiefly for practical reasons") versus presentation of the most efficient methods (as he admitted, for the solution of a particular problem, early and less efficient methods might provide better insight into the topic).

M. J. Lighthill, one of the world's authorities on the subject, regarded *Wing Theory* as "an admirable compendium of the mathematical theories of the aerodynamics of aerofoils and wings. Almost all the important results are referred to, even though there can be only a brief reference to literature in connection with the more difficult topics" [Lighthill 1957, 529]. But he found the chapter on compressible flow (Chapter 4, also the longest) troublesome due to "long stretches of the difficult Hadamard theory for three-dimensional flows being put before the simple two-dimensional theory." Even so, he was impressed by the "overall well-balanced nature" of the account.

Lighthill was more critical of the extent to which the book depended on linearized theory, which relied upon certain restrictive assumptions. Although the assumptions serve to simplify the mathematical analysis, they are not very satisfactory for actual physical applications (and this was especially true for the case of supersonic aerodynamics in compressible flow theory) [30]. Although Robinson and Laurmann did try to cover parts of nonlinearized theory, Lighthill was also disappointed that the subject of two-dimensional supersonic aerofoil theory was not given more attention. But after reading the book, his judgment remained high, and he regarded *Wing Theory* as "an invaluable introduction to wing aerodynamics for mathematically-minded students, as well as a solid stand-by for purposes of reference for all workers in this and allied fields" [Lighthill 1957, 529].

Thus Lighthill neatly sketched the work's major strengths and weaknesses: heavy on mathematics, short on data. This may come as something of a surprise, because Robinson's aerodynamic experience during the war was of a very concrete nature, and at Cranfield he was always interested in wind tunnel studies and the role models played in providing data against which to check his own theoretical work.

Indeed, as Robinson was well aware, models and wind tunnel experiments only approximated physical conditions in hopes of explaining actual free-flight conditions. There were enough complications in reconciling the two, but at high speeds various nondimensional parameters came into the picture. Two of the fundamentals of aerodynamics—the Navier–Stokes equations and the equation of continuity—provided the fundamental system of differential equations for the motion of an incompressible viscous fluid. But many complications arose in the exact theoretical description of this flow around a given particular body.

As Robinson's own research had shown, for fluids of relatively small viscosity, like air, approximate theoretical methods were generally sufficient. When checked against experimental data, they usually permitted (with any necessary adjustments) a quantitative explanation and prediction of the behavior of phenomena occurring in a viscous fluid, which was essential for calculating the important aerodynamical problem of the drag of an airfoil.

The Robinson–Laurmann book on *Wing Theory*, however, may have reflected Robinson's deeper and surpassing interest in theory itself, especially the power he believed mathematics had to elucidate the fundamental features of physical nature. In fact, the aerodynamics of wing theory was unusual in the extent to which purely theoretical considerations led to applicable results:

> This branch of the physics of fluids had a remarkable short-term history as one of the few in which theory alone gave immediately engineering applicable results without the need for much empirical amendment. The book on wing theory that I wrote as junior author with Abby (plus I suppose books by others) in a sense represents the close of a chapter in the fulfillment of the theoreticians contribution in this domain—a very neat package. In fact it was this completion of a field of theoretical fluid dynamics that later led me to new fluid dynamic challenges in geophysical application, and a new world of fundamental difficulties in trying to understand atmospheric and oceanic flows [31].

Thus it was perhaps inevitable that mathematics should dominate the development of *Wing Theory*. Robinson always looked for solutions that were mathematically elegant and simple if possible yet suggestive or profound. As his former colleague, Alec Young, put it:

> Abraham Robinson was an exceptionally good applied mathematician, and more than that, he had an outstanding ability for understanding in physics and the problems that he dealt with, so that he could apply mathematics to it in a way that was extraordinarily stimulating as well as elegant [Young R-1975].

This was a feature of Robinson's work that was typical even at Farnborough, where mathematical sophistication was a notable characteristic of his approach to even the most practical problems he was given. This was all clearly evident in his work with Laurmann on *Wing Theory*. As one of Robinson's major contributions to aerodynamics, and certainly his most comprehensive account of the elements of wing theory, it remained characteristically "Robinsonian" in its deft handling of the mathematics above all else.

10. ROBINSON'S LAST GRADUATE STUDENT
IN APPLIED MATHEMATICS

Robinson's last graduate student at Toronto, R. A. Ross, did not actually finish his dissertation until after Robinson had left Toronto. Nevertheless, the subject of his thesis, "The Waves Produced by a Submarine Earthquake," was inspired by Robinson's own theoretical interests (as well as those of J. A. Jacobs in geophysics). Although Duff took over responsibility as thesis supervisor, Ross duly thanked Robinson at the beginning of his dissertation for having worked with him on the thesis and "suggesting improvements" [Ross 1955].

Robinson had long been interested in the subject of elastic waves and later suggested the topic to Ross for his dissertation, thanks in part to the proximity of the geophysics group on St. George Street. Because the geophysics section of the physics department was directly next door (where the facilities were a bit more spacious and comfortable than they were at number 47), he sometimes went over for tea in the afternoon, where there was always the chance to talk with Jacobs [Ross I–1990].

One afternoon, Duff found Robinson at work in his office, reading Lamb's paper of 1904 on the earthquake problem. "You know," said Robinson, "there is nothing in this paper so very deep or outstanding in itself. All the methods and ideas were well known. It is only when it has all been put together, from the beginning to the end, that you have something really fine" [Duff I–1990]. This might just as well have been said about Robinson's own work, for he often took familiar subjects or results, but by approaching them in a novel way, with fresh combinations in mind, he would obtain far-reaching and sometimes fundamental conclusions.

As Ross began working on his dissertation, Robinson insisted on seeing him regularly. This was important not only to keep Robinson informed, but also to make sure that Ross kept working at a regular pace. This meant that they saw each other at least once a week.

In the classroom, Ross had not found Robinson especially impressive as a lecturer:

> He was well-organized, but even so he was hard to follow. Perhaps he assumed too much. There were also gaps in what he was doing, and you had

to fill them in later yourself. For advanced students, or ones willing to do the extra work, this wasn't probably so bad, but it had obvious drawbacks for others. [Ross I-1990]

Usually, Robinson lectured from notes, "but sometimes it seemed as if he was trying to recall something he had done before, but then forgotten." Doing so many things at once, Ross conjectures that he did not always recall exactly what he needed at a given point in the middle of a lecture [32]. When it came to his dissertation, however, Robinson proved an inspiring and encouraging advisor.

What Ross set out to consider mathematically was a challenging physical problem, namely an underwater earthquake and its resulting effect on the ocean's surface. Representing the earthquake as a radial line source, the earth as a semi-infinite elastic solid, and the sea as a layer of incompressible fluid, he obtained an exact expression for the velocity potential on the surface of the incompressible fluid (under gravity). From this exact solution he went on to give an asymptotic solution valid for large values of time and for the case in which the line impulse was located on the liquid–solid interface. Ross found that his analysis revealed the wavelike effects of the impulse, consisting of disturbances on the fluid surface that were of two types: the one caused by pressure, shear, and Rayleigh waves and the other being (roughly) a gravity-type wave.

11. ROBINSON'S LAST PAPERS AT TORONTO
ON APPLIED MATHEMATICS

In 1956 Robinson published a paper on the motion of small particles in a potential field of flow, relevant both to the problem of aircraft icing and to silting up processes in rivers and estuaries. The approach taken most often to such problems was to compute trajectories of individual particles. Instead, Robinson considered a continuous distribution of small particles and the overall field of flow they would produce [33].

Later that year he published another paper, in which he considered a wave possessing a velocity discontinuity, a problem he approached for wave propagation in an elastic medium with variable properties [Robinson 1957c] [34]. In this paper he showed that longitudinal waves do not transform into transverse waves and vice versa. A year later he went on to examine the transient stresses in a beam of variable characteristics subjected to an implosive or concentrated load and investigated the associated variation in the discontinuity in bending moment that can occur across the front of a shear wave [Robinson 1957a].

By the mid-1950s, however, Robinson's interests in applied mathematics had clearly begun to wane. With the completion of his book with Laurmann on *Wing Theory*, he turned his creative energies almost exclusively to mathematical logic, where he was just beginning to develop exciting new ideas, especially in model theory. His colleague at Toronto in applied mathematics, A. J. Coleman, describes one conversation he had with Robinson at about this time:

> It was in his office, we were seated on either side of his old desk (all desks in the Department of Mathematics were at least 20 and mostly 50 years old!). It was then that I glimpsed for the first time the extent and depth of his knowledge of fluid mechanics, but I left the conversation with the clear impression that this topic was now of only peripheral interest for him. He was deep into logic...and I thought that I had perceived that he had become the purest of pure mathematicians who regarded his mastery of fluid dynamics as a dull necessity forced upon him by the exigency of war work. [But] his heart was now totally caught up with logic... [35].

12. MATHEMATICAL LOGIC AT TORONTO: A. H. LIGHTSTONE'S THESIS

Although Robinson's teaching at Toronto had been exclusively devoted to introductory courses and applied mathematics (with the single exception of a course on foundations of mathematics in his final year), he nevertheless managed to attract the attention of a small group interested in logic [Ross I–1990] [36]. The first of these was A. H. Lightstone, who completed his dissertation with Robinson in 1955. Meanwhile, Paul Gilmore had come to Toronto as a postgraduate student to work with Robinson, and Elias Zakon from the Technion in Haifa arrived to spend a year's sabbatical with Robinson during his last year in Canada.

Lightstone was born in Ottawa in 1926 and received his B.Sc. from Carleton College in the United States and his M.A. at the University of New Brunswick, Canada. It was Robinson, of course, who suggested the subject of Lightstone's dissertation: "Contributions to the Theory of Quantification." In the "Preface" to his thesis, Lightstone was pleased to thank Robinson "for many stimulating hours spent together during the course of the investigation" [Lightstone 1955].

The "investigation" was prompted by a noticeable asymmetry in the restricted predicate calculus, which, as Lightstone observed, permitted only one kind of variable to be quantified. In the Hilbert–Ackermann version, there were three basic variables: sentential, individual, and predicate. Lightstone, in his thesis, introduced a variant calculus designed to include all of the various classes of variables on which the calculus was based. Consequently, the axioms and rules of formation (of well-formed formulae, WFF) did not have to distinguish between relations and individuals. However, a distinction was drawn between types of variables in determining WFFs obtained from a finite sequence of integers.

With slight variations, Lightstone's "Symmetric Predicate Calculus" (SPC) was applicable to the restricted predicate calculus, the extended predicate calculus, and even to the many-sorted calculus. Lightstone also showed that SPC was consistent and complete "by the usual methods." As an application of SPC, Lightstone considered the axioms of incidence in projective geometry, "with a view to proving the principle of duality in a formal fashion." In fact, Lightstone may well have been echoing Robinson in declaring that the "usual" proof of duality "takes no account of the underlying logical calculus."

The second part of Lightstone's thesis was devoted to "Syntactical Transforms." Using his SPC notation, he gave various sets of rules—"syntactical transforms" (mappings of WFFs into themselves; Skolem's normal form is an example of a syntactical transform). Lightstone's aim was the development of transforms under which provability was invariant. The thesis was especially interested in various kinds of continuity—continuity at zero, uniform continuity, and a weakened form of topological continuity at a point. Lightstone observed that the notation of SPC suggested yet another kind of transform, which he called "duplication" (i.e., each variable or functor appearing in a formula was replaced by two symbols of the same type). Not only was provability invariant under this transform, but duplication was also useful in developing the transform for uniform continuity.

Lightstone brought his thesis to a close with a number of applications, all of which were very Robinsonian in character. Among these, he showed that if a WFF holds in every abelian group, then a transform of the formula (expressing the uniform continuity of any functor appearing in the formula) holds in every ordered, completely divisible, abelian group. For a system of equations with integral coefficients and with parameters from an algebraically closed integral domain of characteristic zero with a solution in every such integral domain (including those without a unit element), the solution is continuous in the parameters at zero (i.e., the solution can be made arbitrarily small by choosing suitably small parameters [Lightstone and Robinson 1957a]) [37].

13. HILBERT'S IRREDUCIBILITY THEOREM

In addition to his work with Lightstone, Robinson was also pleased to be working with Paul Gilmore, who had come to Toronto as a postdoctoral student to work with Robinson on mathematical logic in 1953. Gilmore was a Canadian who graduated in 1949 from the University of British Columbia with honors in mathematics and physics after having served a three-year term in the Royal Canadian Air Force. As a graduate student at Cambridge University, on a two-year scholarship from St. John's College, he became interested in logic. After completing his M.A. at St. John's in 1951, he went to Holland to continue his studies in logic with E.W. Beth and A. Heyting. Two years later he was awarded a Ph.D. in mathematics from the University of Amsterdam. It was in his second year, supported by a scholarship from the National Research Council of Canada, that he heard about the availability of NRC postdoctoral fellowships:

> After four years away I wanted to return to Canada. It was probably Professor Beth who suggested Abby's name to me [most friends called him "Abby" Robinson]. When I applied for the NRC fellowship, I was aware of his work in logic and algebra. My application for the fellowship stated that I wanted to use my knowledge of intuitionistic mathematics to further Professor Robinson's work. One of my first memories of Abby is his response to my application when we met in Toronto to discuss what I would do during my two year fellowship. He spoke with some amusement about how my proposed research was more suitable for the purposes of the application than for a serious program of research. Such honesty and openness made his compliments all the more valuable [38].

From 1953 to 1955 Gilmore was a University Research Associate in mathematics at the University of Toronto, under NRC sponsorship. He then taught briefly at Pennsylvania State University before going on to the T. J. Watson Research Center in Yorktown Heights, where he worked for IBM as a mathematician. After managing the group working on combinatorial mathematics for a time, he returned to Canada in 1977 to head the Department of Computer Science at the University of British Columbia.

Working together at Toronto, Gilmore and Robinson succeeded in providing an elegant model-theoretic proof of Hilbert's irreducibility theorem, namely that if a polynomial $F(X_1, \ldots, X_n)$ in n variables over an algebraic number field K is irreducible, then there is an irreducible polynomial in $X_1, \ldots, X_m$ $(0 < m < n)$, obtained from $F(X_1, \ldots, X_n)$ by specializing $X_{m+1}, \ldots, X_n$ to a suitable set of values in K [Hilbert 1892; Gilmore and Robinson 1955].

Their joint result arose from a paper Robinson had given to Gilmore for his comments:

> Abby gave me a draft of a paper he had written and asked me to read it and suggest any corrections or improvements. I could not find any errors in the paper, but I was able to generalize his main theorem. When I described my generalization to Abby, he became very excited and enthusiastic. He rushed us to the library to consult *Math Reviews* and other references which would suggest applications of the extended theorem. Abby listed the papers he wanted me to read and suggested the results we could obtain, and the collaboration was started. He left the writing of the paper pretty much to me, although he suggested changes and improvements to my drafts [39].

Drawing on earlier papers dealing with Hilbert's irreducibility theorem by K. Dörge and W. Franz, Gilmore and Robinson extended a metamathematical result Robinson had already published (in a paper presented at the first CNRS Colloquium in Paris in 1950), from which not only Hilbert's irreducibility theorem followed, but also two other theorems, one related to results established by Dörge and the other related to a new result on irreducibility due to Gilmore and Robinson for fields K with a valuation φ in an ordered field W [Gilmore and Robinson 1955, 487; Robinson 1979, 1, 352].

Again, the main significance of the Gilmore–Robinson paper was the way it showed how algebraic results of substantial and recognized content could be obtained from symbolic logic thanks to Robinson's new methods. The success of an "elegant model-theoretic proof of Hilbert's irreducibility theorem" was itself an important achievement. In Simon Kochen's words, "The possibility now seriously arose that model theory could be a new method of attack on purely algebraic questions. This hope has been amply borne out in a number of cases where the only proof of algebraic results to date is the model-theoretic one" [Kochen 1976, 314].

14. BETH'S THEOREM, 1955

At the end of the year Robinson attended the annual meeting of the Association for Symbolic Logic at Boston University. Following a joint session with the American Philosophical Association in the morning, the

ASL held a separate meeting of its own in the afternoon. Robinson presented a paper "On Beth's Test in the Theory of Definitions," the purpose of which was to demonstrate once again the ease and directness of model theoretic arguments. Given $X = Y(F, G_1, \ldots, G_m)$ in the lower predicate calculus with atomic predicates $F = F(x_1, \ldots, x_n)$ and $G_1, \ldots, G_m$, he took X to define F *implicitly* (semantically) in terms of $G_1, \ldots, G_m$, if the equivalence

$$(x_1) \ldots (x_n)[F(x_1, \ldots, x_n) \equiv F'(x_1, \ldots, x_n)]$$

was deducible from the conjunction

$$Y(F, G_1, \ldots, G_m) \wedge Y(F', G_1, \ldots, G_m).$$

In this case, according to Beth's test, X also defines F *explicitly* (syntactically); i.e. there is a formula $Q(x_1, \ldots, x_n)$ containing only $G_1, \ldots, G_m$ such that

$$(x_1) \ldots (x_n)[F(x_1, \ldots, x_n) \equiv Q(x_1, \ldots, x_n)]$$

is deducible from X [Beth 1953] [40].

In order to obtain this result, Beth had originally used a complicated analysis of derivations in the style of Gentzen. Robinson simplified everything with a model-theoretic lemma:

> Let K be a complete set of sentences (of the lower predicate calculus) and let X_1 and X_2 be two sentences such that all individual constants and atomic predicates which are common to X_1 and X_2 occur also in K. Then if X_1/X_2 is deducible from K, either X_1 or X_2 must be deducible from K [41].

Robinson was candid in admitting, however, that proof of this lemma was not as straightforward as one might suppose. Nevertheless, Beth's Theorem followed as a nearly direct consequence in only a few steps.

Robinson published a more complete version of his model-theoretic proof of Beth's Theorem in a paper Arend Heyting had already communicated in October to the Dutch Academy of Sciences, giving "A Result on Consistency and Its Application to the Theory of Definition." This paper contains what has come to be well known as Robinson's "Consistency Theorem," which he used to establish the general consistency result from which Beth's Theorem followed.

At about this same time, William Craig was also interested in proving Beth's Theorem on definability, although his work on the subject was virtually independent of what Robinson had done:

> Our exchanges about his joint-consistency theorem and my interpolation theorem were not very extended. Apparently each of us originally was mainly interested in providing a proof of Beth's theorem on definability. It took both of us some time, I believe, to realize that the respective theorems that we used in the proof of Beth's results were of intrinsic interest. (Lyndon's strengthening of the interpolation theorem to obtain results on closure under homomorphisms may have been the main factor). The equivalence of Robinson's theorem and mine, for first-order logic, which only then became an issue, also did not come out until some time after our papers had been published [42].

The relations between Beth's Theorem, Robinson's Consistency Theorem, and Craig's Interpolation Theorem are now standard topics in any introduction to metamathematics or model theory. In fact, the search for extensions satisfying one or another of these has been a driving force behind a good deal of work in abstract model theory [43]. In bringing his paper on definability to a close, Robinson also presented an analogous (and simpler) argument to show that the two notions of semantic and syntactic definability coincide for the sentential calculus, along with a generalization of Beth's Theorem for n predicates $F_1, \ldots, F_n$ [Robinson 1956b].

15. PERSISTENCE AND COMPLETENESS, 1956

Considering what Robinson had accomplished in 1955, it is hard to believe that he could have done even more in the following year. In January he visited the University of Montreal, where he presented two lectures meant to provide an "Aperçu Metamathématique sur les Nombres Réels" [44]. Later, he also published two papers that echoed ideas he had pioneered in 1951, inspired this time by a paper of Leon Henkin's.

In "Two Concepts from the Theory of Models," Henkin posed the problem of characterizing syntactically the statements X that are persistent with respect to a given set K (of sentences of the predicate calculus). The idea of persistence was due to Robinson, who introduced it in his first book, *On the Metamathematics of Algebra*. Henkin noted that "Robinson's theorem [on the finite persistence property] can be expressed by saying that the class of abelian groups and the class of commutative fields both have the finite persistence property" [Henkin 1956, 28]. Henkin went on to determine what was required for a statement X to be absolutely persistent. Robinson, in response to Henkin's paper, was able to prove a more general theorem:

> In order that the statement X be persistent with respect to the set K it is necessary and sufficient that $X \equiv Y$ be deducible from K for some existential statement Y (i.e. it is in prenex normal form and contains no universal quantifiers). [Robinson 1956c]

Robinson had submitted his note on Henkin's paper to the *Journal of Symbolic Logic* in April of 1955, but continued to think about the problem. In less than a year he had come up with a longer paper, "Completeness and Persistence in the Theory of Models," which was published in 1956. Here the concept of persistence was extended from its earlier meaning in terms of statements in the lower predicate calculus to include predicates as well. Robinson also added two new concepts of completeness, precompleteness and n-completeness. The paper itself took up various themes, all of them related to completeness, including 1-completeness and restricted completeness, as well as persistence. An entire section of the paper was devoted to tests for completeness, including various necessary and sufficient criteria. For example:

> THEOREM: In order that a set of sentences K be model-complete, it is necessary and sufficient that for every model M of K, the set $K \cup N$ is 1-complete, where N is the diagram of M. [Robinson 1956d, p. 24] [45]

This theorem led directly, Robinson noted, to the model completeness test he had presented at the Amsterdam Colloquium in 1954, but which he was now able to prove using very different methods [Robinson 1955a]. All of this, in fact, looked ahead to the many examples he had systematically worked out as applications of the test for his next major publication, *Complete Theories*.

16.   COMPLETE THEORIES, 1956

Robinson spent much of 1956 seeing *Complete Theories* through press. This volume gathered together many of his recent results, most of which were less than 12 months old [Robinson 1956e, "Preface"]. The new book was built around the idea of model completeness, which provided an ingenious approach for determining the conditions under which various theories were complete and decidable. He also gave a new proof of Tarski's classic result that the field of real numbers is decidable (a problem to which he returned again in 1971, after he had developed the concept of Robinson forcing).

Robinson sent *Complete Theories* to North-Holland in 1955, again for the Dutch series, *Studies in Logic and the Foundations of Mathematics* [46]. Having heard some of his new ideas during the Logic Symposium held in conjunction with the International Congress the year before, the editors of the Studies series invited Robinson to elaborate, and he did [47].

Upon publication, *Complete Theories* was hailed as "a milestone in the development of model theoretic algebra" [Keisler 1977, vi.]. In addition to model completeness, Robinson developed the related concept of model completion (a model completion of K is a theory $K' \subset K$ such that every model of K is embeddable in a model of K', and the union of K' with the diagram of any model of K is complete). The theory of algebraically closed fields, for example, is a model completion of the theory of integral domains. Not all theories, however, are model complete—neither the theory of differential fields nor that of formally real fields has a model completion. But if a model completion does exist for a given theory, then it is unique [48].

One of the leitmotifs of Robinson's career, by now familiar, was his continuing examination of the abstract properties of algebraically closed fields. Although various approaches may be taken to this subject (including the question of categoricity in power of algebraically closed fields of fixed characteristic), Robinson's approach improved upon Tarski's method of quantifier elimination, which thanks to Robinson's generalization of model completion could now be avoided in favor of subtler and more elegant tools.

The basic idea behind *Complete Theories* is comparable to an abstract version of Hilbert's *Nullstellensatz*, one of the most fundamental results in the theory of algebraically closed fields [49]. Once the nature of Robinson's insight is understood, the idea can be widely applied to a number of other important examples. What Robinson did was to see how a general theory that applied to all model-complete theories could be developed. In the course of doing so, he was able to give a short, elegant proof of the completeness of real-closed fields, as well as other completeness results that had been obtained previously using various other methods.

The book begins, after expected preliminaries, with a discussion of model completeness (Chapter 2) [50]. After introducing the idea of persistence, Robinson used this to establish his well-known test for model completion, namely the important Theorem 2.3.1:

> In order that a non-empty consistent set of statements K be model-complete, it is necessary and sufficient that for every pair of models of K, M and

M', such that M' is an extension of M, any *primitive* statement Y which is defined in M can hold in M' only if it holds in M. [Robinson 1956e, 16] [51]

Robinson then considered various examples (in Chapter 3) of model-complete groups and fields. He began with simple examples of certain types of abelian groups (e.g., completely divisible torsion-free abelian groups with at least two elements) and then went on to other examples, including algebraically closed fields, real-closed fields, fields with valuation, integral domains with valuation, and modules (groups with operators).

In the next chapter he took up the problem of finding conditions under which model completeness entailed completeness. Here he introduced another new concept, that of a prime model, which led naturally to the "Prime Model Test":

> PRIME MODEL: A structure M_0 is said to be a prime model of a set of statements K if M_0 is a model of K and if every model M of K contains a partial structure M' (i.e. M is an extension of M'), such that M' is isomorphic to M_0. If K contains any constants then these shall correspond to themselves in the isomorphism. [Robinson 1956e, 72] [52]

> THE PRIME MODEL TEST: Let K be a model-complete set of statements which possess a prime model M_0. Then K is complete. [Robinson 1956e, 74]

The Prime Model Test, combined with his results on model-completeness, provided, as Robinson said, "a very effective tool for the investigation of concrete cases" [Robinson 1956e, 74] [53].

Robinson used the Prime Model Test to establish the completeness of a wide variety of different groups and fields (in Section 4.3 of *Complete Theories*). Although it may be fairly simple to show that K has a prime model, it is usually not so simple to show that it is model complete. Here Robinson's success reflected his considerable facility, not only in his handling of metamathematical arguments, but also in his thorough understanding of the basic algebraic concepts involved [54].

In Chapter 5, a number of direct applications were given, beginning with an example Robinson had discussed at length with the topologist Beno Eckmann. Using topological methods, Eckmann had proved a variation of this theorem for polynomials with complex coefficients:

THEOREM (5.5.1): Let
5.2.2

$$\begin{cases} p_j(x_1, \ldots, x_n) = 0 \quad j = 1, \ldots, n, \, n \text{ odd}, \\ x_1 p_1(x_1, \ldots, x_n) + x_2 p_2(x_1, \ldots, x_n) + \cdots + x_n p_n(x_1, \ldots, x_n) = 0 \end{cases}$$

be a system of polynomial equations with coefficients in a field of characteristic zero. Then the system possesses at least one solution other than

$x_1 = x_2 = \cdots = x_n = 0$ (within some extension of the field) [Eckmann 1942] [55].

Translating the basic idea of this theorem into a statement X of the lower predicate calculus, it follows from the fact that the elementary theory of algebraically closed fields of *specified* characteristic is complete (Robinson's Theorem 4.3.8), that X holds for *all* other algebraically closed fields of characteristic 0. This proves Eckmann's theorem directly, since every field of characteristic 0 can be embedded in an algebraically closed field of characteristic 0.

This very elegant, simple example demonstrates nicely the power and great appeal of the new kind of results Robinson was able to achieve in *Complete Theories*. It also led him to the subject of transfer principles, which were to assume an increasingly important role in virtually all of his future work.

Robinson explained transfer principles as follows:

> By a *transfer principle* we mean a metamathematical theorem which asserts that any statement of a specified type which is true for one particular structure or class, is true also for some other structure or class of structures. Thus the proposition that a particular set of axioms K is complete may be expressed in the form of a transfer theorem since it amounts to the assertion that any statement which is defined in K and which holds in one particular model of K, holds also in all other models of K. [Robinson 1956e, 92]

This notion of transfer principles also harkened back to his thesis and other ideas put forward in *On the Metamathematics of Algebra* [Robinson 1950a, and Robinson 1951b]. A good example was Eckmann's Theorem. Robinson was able to prove this based upon the metamathematics of transferrability: a result that holds in one case for an algebraically closed field of characteristic 0 must then hold in all such cases, including that for which Eckmann's Theorem was formulated.

Robinson went on in this chapter to provide a stronger description for algebraic varieties than he had given on this subject in his earlier paper, "On Predicates in Algebraically Closed Fields" [Robinson 1954b]. But now he added the condition that dimension V_i > dimension V_{i+1}. This was a significant improvement and showed that the length of the chain of varieties in question cannot exceed $n + 2$ (i.e. $2k < n$) [Robinson 1956e, 106].

The final chapter of *Complete Theories* (Chapter 6) was devoted to "syntactical transforms," a subject familiar from the paper Robinson had written with his graduate student Harold Lightstone on this same subject [Lightstone and Robinson 1957a]. After defining syntactical transforms,

Robinson explained how they permitted passage from a phrase like "for all x..." to "for all x which belong to R..." or from "there exists an x..." to "there exists an x which belongs to R...." In mathematics, Robinson observed that "assertions on the *continuity* or boundedness of the solution of certain problems can be obtained from the assertion of the mere existence of a solution by this kind of modification" [Robinson 1956e, 108].

Details for the continuity transform were given in the paper with Lightstone (and indeed constituted a significant part of Lightstone's thesis). In *Complete Theories*, Robinson instead took up the syntactical transform concerned with *boundedness*. Among other examples, he then applied this to results of the following sort: If a statement X that asserts the existence of elements with a certain property holds in all completely divisible torsion-free abelian groups (with at least two elements), then a correlated statement asserting the existence of such elements as bounded functions of the parameters involved holds in all ordered groups of the same type [Robinson 1956e, 125].

Complete Theories was an important book for Robinson—and for model theory. It was also a thorough embodiment of his own characteristic approach to the metamathematics of algebra. *Complete Theories* shows how adept Robinson had become at taking an idea from algebra, interpreting it from a model-theoretic point of view, and then finding the significant generalizations, the important metamathematical properties of the theories themselves that could be used, in turn, to shed new light on various parts of algebra. This was a method Robinson had honed to perfection.

17. THE CANADIAN SUMMER RESEARCH INSTITUTE: KINGSTON, 1956

In the summer of 1956 Robinson was among the group of mathematicians taking part in the annual Canadian Summer Research Institute at Queens University in Kingston, Ontario [56]. Robinson spent most of the summer working on several papers, one of which was later entitled "Some Problems of Definability in the Lower Predicate Calculus" [Robinson 1957b]. Actually, this reflected more work on Hilbert's 17th Problem, applying the idea of model completeness to field extensions. Robinson was again interested in whether or not there was a uniform bound to the number of squares required to express a totally positive element of a finite algebraic extension of an ordered field as a sum of squares of elements of the extension. The answer Robinson gave was "yes," and, in working out the details of his proof, he was led, as he later said, to a relativization of model completeness [57].

Another product of Robinson's concentration that summer was a joint paper written with his just-graduated student Harold Lightstone. Together they worked on the metamathematics of algebraically closed fields, specifically, "On the Representation of Herbrand Functions in Algebraically Closed Fields" [Lightstone and Robinson 1957b].

18. ELIAS ZAKON

Back in Toronto after the Summer Institute, Robinson began a new collaboration with a colleague who had come to Canada specifically to study mathematical logic with him. This proved to be the beginning of a long and productive academic friendship. Some months earlier, Elias Zakon had taken (as he admitted) the bold step of writing to Robinson:

> I have to apologize in advance for the liberty I am taking in writing this letter. I hope, however, that you will find it possible to consider my request favorably.
>
> I am a senior lecturer of mathematics at the "Technion," Israel Institute of Technology. My scientific interests are in the field of set theory, logic and foundations of mathematics. So far I have published several papers, mainly on transfinite numbers. Two of my results, modest as they are, are cited in H. Bachmann's book *Transfinite Zahlen*, Zürich, 1955, p. 52 and 98, footnote (see also *Mathematical Reviews*, September 1953 and May 1954).
>
> In the next academic year I shall have my "sabbatical leave," and I intend to use it for advanced study and research in the field that interests me. The Institute for which I work ("The Technion") is willing to help me by allocating an appropriate stipend, so as to enable me to study abroad, at a university or institute where I could obtain proper facilities and scientific guidance. On my part, I should like very much to spend the period of my leave at Toronto, so as to have the opportunity and the privilege of your most competent guidance, and I should be very grateful to you for agreeing to supervise my work. I wish to add that it is not my intention to become a candidate for a degree, and I should prefer to study as a guest or a visiting research fellow. Of course, I understand that my intended research work will not place any financial obligations on the University.
>
> If my request can be granted, kindly let me have an official confirmation, as such a letter is required for the visa formalities. I am hoping for a favorable reply.
>
> Yours, very truly, E. Zakon [58]

At the time, Zakon was 48, having studied at the Friedrich Wilhelm University in Berlin (1926–1930) and at the Stephan Batory University in Vilna, Lithuania (1930–1934). In 1950 he obtained a position at the Haifa Institute of Technology and by 1954 had been advanced to the position of Senior Lecturer of Mathematics. Among his publications Zakon mentioned two that had appeared in the *Mathematical Quarterly* (*Riveon Lematematika*), published in Jerusalem, "Left Side Distributive Law of the Multiplication of Transinfinite Numbers" and "On the Relation of 'Similarity' between Ordinal Numbers" [Zakon 1953, 1954a]. He had also published a monograph, "On Fractions of Ordinal Numbers" [Zakon 1955], and a three-volume *Collection of Exercises in Higher Mathematics* that was used as a basic textbook in the engineering departments at the Technion [Zakon 1954b]. Meanwhile, he reported that he was preparing another textbook for publication on differential and integral calculus, also intended for use at various institutes of technology in Israel [59].

Robinson was suitably impressed and agreed to invite Zakon to Toronto. He even sent a memorandum to Pounder, urging that Zakon be offered "something better than the formal standing of 'occasional student' (e.g., the title of 'research fellow'). It is understood that would not involve the University of Toronto in any financial obligation" [60].

Indeed, over the next decade and more, Robinson and Zakon worked together on several papers which they jointly published, the first on "Elementary Properties of Ordered Abelian Groups" and later "A Set-Theoretical Characterization of Enlargements" [Zakon and Robinson 1960, 1967]. Zakon, in fact, remained in Canada and soon found a position at Essex College, Assumption University of Windsor, Ontario.

19. PREPARING TO LEAVE TORONTO

Late in 1956 Robinson received an inquiry from Hebrew University, a tempting invitation to accept the chair in mathematics held by his former teacher, Abraham Fraenkel, at the Einstein Institute. The opportunity to return to Jerusalem, and above all to devote his energies full-time to pure mathematics and logic, proved irresistible. On January 9, 1957, Robinson wrote to Pounder about his decision:

Dear Professor Pounder,

Several months ago I was asked by a senior member of the Faculty of Science at the Hebrew University, Jersualem, whether I was interested in an appointment at that University. I replied that I intended to give serious consideration to any offer and, at the same time and subsequently, kept you informed of the gradual development of this matter.

I have now received an official letter from the President of the Hebrew University and have decided to accept his offer of a professorship in Mathematics. Accordingly, I herewith give notice that I wish to resign my position at the University of Toronto at the end of the present academic year.

I may add that this has not been an easy decision for me. I regret the fact that I shall have to sever my connections with the University of Toronto, and I am also sorry that I have to impose on you the task of finding somebody to take my place. However, I have no doubt that in view of its great academic reputation the University of Toronto will have no difficulty in choosing a man who will carry out my duties as well as (or better than) I. [61]

Robinson's last semester at Toronto was a busy one, during the course of which he accepted invitations to participate in two major conferences, a Summer Institute in Logic at Cornell University and the Sixth Mathematics Seminar organized by the Canadian Mathematical Congress (scheduled for August in Edmonton, Alberta). Robinson was asked to give a series of lectures on mathematical logic, meant to introduce the subject in a serious way throughout the country. There could have been no more fitting an end to the five years he had spent in Canada than this.

Before setting off for the summer, however, Robinson must have been surprised to receive a letter from Sidney Smith. Still President of the University of Toronto but about to depart for an official position in the Canadian government as Secretary of State for External Affairs [62], Smith wrote in June with the happy news that Robinson had been promoted to the rank of Professor, to take effect as of July 1, 1957. Apparently, he knew nothing of the letter Robinson had sent to Pounder in January, resigning his position. Smith's letter was followed within a week by even better news: simultaneously, his salary would increase to $8,000 annually [63]. Ironically, as Robinson later told R. A. Ross, had the promotion and salary increase come just a little sooner, he could never have *afforded* to leave Toronto [Ross I–1990].

20. THE EDMONTON SUMMER SCHOOL: ROBINSON'S COURSE ON LOGIC

Abby and Renée arrived in Edmonton just in time for the 15 lectures Robinson had promised to give as an introductory course on mathematical logic for the Sixth Mathematics Seminar, sponsored by the Canadian Mathematical Congress (August 12–31, 1957). The success of the meeting, which was held at the University of Alberta in conjunction with the first seminar of the Canadian Association of Physicists (the Theoretical Physics Divison), was reflected in the number of participants, approximately 190 (including wives and children). Almost everyone was housed in student dormitories.

The seminar opened officially on Monday afternoon, August 12, with welcoming remarks from the President of the University, the Mayor of Edmonton, and the Alberta Minister of Education. In addition to conferring honorary degrees upon Professors H. S. M. Coxeter and Eugene Wigner, a garden party "in glorious weather" was held on the lawn of the President's home on Saturday afternoon, where "the colorful saris of the Indian ladies present provided a picturesque touch" [64]. In addition to other social gatherings, receptions, and dinners, the staff of the Rutherford Library also prepared "an exhibit of old books, pictures and other items of historical and mathematical interest."

Due to the relatively short amount of time at his disposal for an introductory course on mathematical logic, Robinson limited his lectures primarily to the propositional and predicate calculi (first order). The text he used followed basically this same plan, namely the Hilbert–Ackermann *Principles of Mathematical Logic* (1950). Theorems were proven selectively, but with enough detail to give everyone a sense of what the methods and approach were like [65]. When Robinson reached the end of his short-course, with the Löwenheim-Skolem theorem, he admitted that he had reached only "the end of the beginning" [Robinson 1958b, 41].

21. LEAVING TORONTO

After their summer of institutes, seminars, summer schools, and travels, the Robinsons returned briefly to Toronto, to pack and say their goodbyes. Renée recalls that just before they left, Dean Beatty said to her, "This isn't one man, but two." This was doubtless more than a reference to the long hours Robinson spent in his office and the steady, regular, tenacious manner in which he worked. Indeed, while he had been at Toronto, Robinson had excelled in two important areas, both applied and pure mathematics. The Department gave a big going-away party for them at which Beatty, genuinely sorry to lose Robinson, made a "wonderful speech" [Renée Robinson, I–1980].

Of the many expressions of gratitude Robinson received as he was about to leave Toronto, one was a warm letter from Mary Cooper, Corresponding Secretary of the University's organization for "Friendly Relations with Overseas Students" (F.R.O.S.). This group's mission was to welcome foreign students to Canada and (according to its letterhead) "provide opportunities for mutual understanding and appreciation" [66]. Robinson was an active member of the Toronto Committee and under its Chairman Professor, James Ham, served several terms as Vice Chairman of the Executive Committee. As Mary Cooper explained:

> At our Annual Meeting it was the unanimous wish of the Committee that I should express to you our warm appreciation of your interest and support while you have served on the Toronto Community Committee of F.R.O.S. We would like to send warmest good wishes to you and Mrs. Robinson as you take up your new post in Jerusalem. We shall miss you in Toronto, but shall always count you among the good friends of F.R.O.S. wherever you are. [67]

Among the photographs the Robinsons took after five happy and productive years in their little house on Winnett Avenue were several prompted by a few last parties or gatherings of friends. One was of a small group of students, including Vichien Verapanish and Nizam Ahmed Qazi, who gave Robinson a beautiful wooden box, dated August 6, 1957, with greetings carved on the lid in both Hindi and Arabic.

Wim Luxemburg and Donald Coxeter were also invited, along with their wives, for a final farewell before Abby and Renée sailed for Europe. A picture taken on that occasion provides a fitting group portrait, a reminder that, as Robinson prepared to leave Toronto, the decision to do so had not been easy. In giving up his position at the University, he was giving up more than a job, but congenial colleagues and good friends as well. What awaited them in Israel, a country in a state of war with most of its neighbors following the Suez Crisis of 1956, was anybody's guess [68].

NOTES

[1] Samuel Beatty (1881–1970) also served as Dean of the Faculty of Arts from 1936 until his retirement in 1952, when he was elected Chancellor of the University (1953–1957). As Gilbert Robinson has said, "in a very real sense he guided Canadian mathematics from the isolation of the 19th century to a significant role in the 20th century" [G. Robinson 1971, 489].

[2] In 1943 Synge left Toronto for Ohio State University, where he was to be Chairman of the Mathematics Department. In 1948 he returned to Ireland

and took a position as Senior Professor at the Dublin Institute for Advanced Studies [Synge 1972, 255]. From then on, Beatty headed both departments, the Department of Applied Mathematics becoming once again part of the Department of Mathematics [G. Robinson 1979, 52].

[3] In addition to his own autobiography [Infeld 1978], see the obituary in *The New York Times* (January 17, 1968), p. 51, column 2, as well as Infeld [1970] and Trautman [1973].

[4] Igor Gouzenko had been a code clerk at the Soviet Embassy in Canada. Based upon secret reports, he released the names of a Russian "spy ring" comprising of 16 people, most of them intellectuals, operating in Canada. Further details are given in [Infeld 1978, 28–30].

[5] After a visit to Einstein in 1950, Infeld recalled after seeing "the lovely, well-cared-for houses, in such good condition, I pictured America to myself as a bag bursting with gold. And how distasteful I found it! This was something I had never felt before, a proof of the profound impression Poland had made on me" [Infeld 1978, 41].

[6] More inflamatory newspaper and magazine articles were to follow. As for *The Ensign*, the Ottawa Evening *Citizen* said it was "hysterical over the cold war" [Infeld 1978, 51].

[7] Others saw Smith as "buoyant, exuberant, face aglow with good feeling … [a] basic Cape Breton gaelic." Some faculty, however, were critical of Smith's public manner, "which, in its mixture of homeliness and geniality, violated the academic canons of decorum; his critics talked of 'Rotarianism' and suggested that he had a fatal zest for platitudes" [Bissell 1974, 19–20].

[8] Infeld actually put it more sarcastically, the threat from Beatty meaning that "I would lose my wonderful job" [Infeld 1978, 55–56]. To complicate his feelings about Toronto, Infeld was apparently offered a visiting professorship at Princeton for the second half of the academic year in 1949. "It was not possible to arrange this, and the disappointment contributed to a growing feeling of restlessness in the University" [G. Robinson 1971, 302]. Meanwhile, as Gilbert Robinson noticed, Infeld was becoming "hypersensitive to criticism of any kind."

[9] Infeld's political problems with Canada continued, and his two children's Canadian citizenships were revoked. Infeld died in Warsaw on January 15, 1968. Remarkably, Trautman [1973] makes no mention of the tremendous political difficulties Infeld faced in Canada due to anticommunism there.

[10] Later Infeld reiterated that "Fortunately for me, I was not aware of the grimness of the Stalinist times through which I lived" [Infeld 1978, 56]. "My eyes were opened to what was happening only when, in the last months of Stalin's life, it was reported that over a dozen Jewish doctors had been jailed in the Soviet Union …. Then, for the first and only time in my life, I regretted that I had returned to Poland. But soon the Stalin era ended." For details, see Infeld [1978, 84].

[11] Robinson is wrong about A. Robinson having been appointed in 1952; A. Robinson arrived in Toronto the year before, in September of 1951.

[12] For general introductions to Canadian history, see Bothwell [1989] and West [1967].

[13] Tim Rooney also recalls pleasant evenings at the Robinson's home, "the funniest little house, it couldn't have been more than twelve feet wide. Downstairs there was a combined living and dining room, with a little kitchen; upstairs were two bedrooms and a bathroom," [Rooney I–1990].

[14] Similarly, A. J. Coleman recalls the building at 47 St. George Street: "It was a comfortable old building and there were many loud arguments up and down the broken-down stair well." Coleman in a letter to JWD, December 3, 1990.

[15] "Robinson was well appreciated, especially for his versatility, his profundity, his style" [Duff I–1990].

[16] Coxeter, too, remembers that as he was working on his introduction to geometry, Robinson was helpful in suggesting some problems to be included [Coxeter I–1990].

[17] Note that Hart is not listed as one of Robinson's graduate students at Toronto in the list of his former students given in Robinson [1979, vol. 1, 694].

[18] The relatively large values for the spin effects were due to "large coefficients of kinematic origin and appreciable radial parameters." Thus Hart concluded that the spin effects were comparable in magnitude to the electrostatic ones and could not be neglected. This was in turn helpful in explaining deviations from the normal electron level distributions in the atom as predicted by individual electron–nucleus spin interactions. Hart ascribed such anomalies to interconfiguration perturbations.

[19] Later, rarefied gas flow, blast phenomena, spacecraft orbital mechanics, acoustics, wind loads on buildings, etc., were also added to the areas taught at the Institute. For details, see Green [1970].

[20] Fowell is also not mentioned in the list of Robinson's graduate students given in Robinson [1979, vol. 1, 694], nor is he included in Gilbert Robinson's list of Ph.D. students in the Department of Mathematics at Toronto in Robinson [1979, 96–99]. But the dissertation on file in the University archives lists Abraham Robinson along with G. N. Patterson as major advisors and thanks them both in the "Acknowledgement" at the beginning of the thesis "for their guidance and counsel during the course of this project." See Fowell [1955].

[21] L. L. Campbell, in a letter to JWD, October 30, 1990.

[22] L. L. Campbell, in a letter to JWD, October 30, 1990. See as well Campbell [1955].

[23] Campbell's dissertation dealt only with real functions and variables. See Campbell [1955].

[24] One widely used method for solving the initial value problem for an equation of arbitrary order was to solve an equivalent problem for a system of

first-order equations. This was a method also familiar in solving ordinary differential equations, but introducing the various derivatives of z as new variables does not take advantage of any of the properties of the characteristic. Instead, the reduction in this case to a first order system usually introduces characteristic curves that are *not* characteristic curves of the original equation. Robinson's method, by reducing the problem to one of integrating a succession of first-order systems, overcame both of these difficulties. The resulting first-order systems were in characteristic form and introduced no new characteristic curves. For further details, see Campbell [1955, 93–94].

[25] Actually, it was also intended to reach those "in industry, at the universities, or at research establishments, who are interested in airfoil theory for either practical or theoretical reasons." The major mathematical prerequisites were a "sound knowledge" of calculus and the theory of functions of a complex variable. Those parts of general hydrodynamics needed for the development of airfoil theory were included in Chapter 1 [Laurmann and Robinson 1956, Introduction, v].

[26] According to royalty statements from Cambridge University Press, the book continued to sell throughout the 1960s. Robinson would periodically receive a check every few years for amounts of £ 10–20 (averaging $30–50). By 1970, sales had dwindled considerably; only one copy was sold—for a payment to Robinson of 4 shillings. Figures based on royalty statements in Robinson's files, part of the Robinson Papers, Department of Manuscripts and Archives, Yale University Library, Box 1603A, Yale Station, New Haven, Connecticut 06520, referred to hereafter as RP/YUL.

[27] The Introduction, oddly enough, is dated Toronto, 1953 (December). See Laurmann and Robinson [1956, vi].

[28] John A. Laurmann to JWD, December 28, 1990. Laurmann received his M.Sc. degree from Cranfield in 1951. He went on to earn a Ph.D. in Engineering at the University of California, Berkeley, in 1958. Since then, he has worked as a research scientist or consultant to NASA, Lockheed (Missile and Space Division), the Institute for Defense Analysis, and the National Academy of Sciences, among other institutions. From 1976 to 1988 he was also Senior Research Associate in the Department of Mechanical Engineering, Stanford University.

[29] The book also permitted Robinson to correct some errors in previous publications, especially one formula containing an arithmetical mistake in Robinson [1948b], noted in Ward [1952, 446].

[30] For transsonic speeds, the results proved by linearized theory are definitely not reliable. In fact, analyses of shock waves and the particular problem of shock wave boundary layer interaction are well beyond the scope of linearized theory, although these phenomena are very important in determining lift and drag in the transsonic range.

[31] J.A. Laurmann to JWD, December 28, 1990.

[32] This was a tradition in England, Ross points out, where one did not pay too much attention to lectures. "Instead, you made it up as you went along" [Ross I–1990].

[33] Robinson used Stokes' law to determine the motion of an individual particle, assuming that the fluid exerted a drag force on the particle proportional to the vector difference between the velocity of the particle and the velocity of the fluid in question [Robinson 1956g]. The paper also calculated the total mass of the particles striking a segment of the contour of an obstacle placed in the field of flow of an incompressible fluid.

[34] He noted that H. Jeffreys had already studied problems of reflections and refractions of waves between interfaces of homogeneous layers of an elastic medium and the loss of energy (negligible) from transverse to longitudinal waves or vice versa. Robinson approached the problem by a more direct method. Assuming the wave possessed a discontinuity across the wave front, he analyzed the system of ordinary differential equations satisfied by such a discontinuity along the rays. Noting that his conclusions were more precise than Jeffreys', Robinson concluded that it was impossible to create sharp longitudinal waves by a gradual transformation from a sharp transverse wave and vice versa. Robinson suggested that his conclusions might well apply to seismic shocks.

[35] A.J. Coleman to JWD, December 3, 1990.

[36] Because he was already finished with his requirements, Ross did not take the course.

[37] In a note at the beginning of the paper, the authors indicated that the results reported in the article constituted a part of Lightstone's thesis.

[38] Paul C. Gilmore in a letter to JWD, February 8, 1991.

[39] Gilmore to JWD, February 8, 1991.

[40] He also discusses his theorem, as well as proofs by Robinson, Craig and Lyndon in Beth [1959, 293].

[41] See the report of Robinson's paper in *Journal of Symbolic Logic* **21** (1956), 220–221.

[42] William Craig to JWD, December 18, 1990. See also Craig [1957] and Lyndon [1959].

[43] See, for example, Keisler [1971] and the articles by Makowski and Shelah [1979].

[44] See item 41 in the "Bibliography" of Robinson's works [Robinson 1979, **1**, 577].

[45] Robinson gave several other necessary and sufficient conditions for K to be model complete, namely that, for every model M of K, $K \cup N$ is precomplete (Theorem 4.7) and that K be precomplete (Theorem 4.8). Consequently, the notions of model-completeness and pre-completeness are equivalent. But, alternatively, Robinson also gave examples to show that 1-completeness and

precompleteness did not imply each other and that completeness did not imply precompleteness). Finally, Robinson brought his paper to an end with a necessary and sufficient condition for modelcompleteness stated in terms of Henkin's related concept of Γ-completeness.

[46] Robinson signed the contract for *Complete Theories*, to appear in a North-Holland's series, *Studies in Logic*, on May 2, 1955; RP/YUL.

[47] The paper "Ordered Structures and Related Concepts" was presented during the Symposium on the Mathematical Interpretation of Formal Systems, held in conjunction with the International Congress of Mathematicians meeting in Amsterdam in 1954; RP/YUL.

[48] In 1969 Eli Bers (who Keisler has called Robinson's "fourpartite colleague") improved upon the concept of model completion with the idea of model companion. This is a more general concept in that it concerns a theory, $K' \subset K$ such that every model of K is embeddable in a model of K' where K' is model-complete. This is an important generalization of Robinson's idea because (for example) the theories of formally real fields and differential fields do have model companions (but not model completions).

[49] For further evaluation of this approach, see [Keisler 1979, xxxiii].

[50] This chapter also introduces related notions of "partial completeness" and "partial model-completeness," along with corresponding versions of the model-completeness and prime model tests. R. L. Vaught has pointed out, however, that the only example Robinson gives here for partial completeness is wrong. See [Vaught 1960, 176].

[51] A WFF Y is *primitive* if it is of the form:
$$Y = [(\exists y_1) \ldots (\exists y_n) Z(y_1, \ldots, y_n)],$$
where Z is a conjunction of atomic formulae (or negations of same).

[52] For example, the field of rational numbers is a prime model of the set of axioms K for the concept of a commutative field of characteristic zero. Note that no prime model exists if the characteristic of the field is not specified.

[53] Robinson also gave a corresponding test for partially complete sets.

[54] At the Cornell Summer Institute in 1957, Tarski and Vaught presented a paper later published in *Compositio Mathematica*, in which they also considered model-completeness. Their paper showed that the idea could be formulated conveniently in terms of arithmetical extensionality, and that the model-completeness of the theories of algebraically closed and of real closed fields could both be established, of course, by elimination of quantifiers, a persistent theme in Tarski's work.

[55] See also Robinson [1956e, 90].

[56] The inspiration for such meetings had come from Ralph Jeffrey, and the Summer Institutes proved to be one of the Canadian Mathematical Society's "most successful projects," according to Robinson [1971, 490]. The idea was to provide a stipend for everyone who participated, supporting three months of summer research at Queens University. The Institutes, in fact, were so successful that they continued for some 25 years [Duff I–1990].

[57] This idea had already been presented in his paper on ordered structures at the Amsterdam symposium in 1954, and was included in his book on *Complete Theories* [Robinson 1956e]. Robinson acknowledged suggestions that Harold Lightstone had made to help improve the presentation. Moreover, just as the paper was about to appear, Robinson learned that Kreisel had published a notice of work that he was doing on bounds of degrees of polynomials in the *Bulletin of the American Mathematical Society* **63** (1957), 100. Robinson therefore added a reference to Kreisel's notes in his own paper [Robinson 1957b, 395].

[58] E. Zakon to A. Robinson, May 14, 1956, in Robinson's administrative file, Department of Mathematics, University of Toronto.

[59] E. Zakon to A. Robinson, June 19, 1956, in Robinson's administrative file, Department of Mathematics, University of Toronto.

[60] A. Robinson to I. R. Pounder, July 1, 1956, in Robinson's administrative file, Department of Mathematics, University of Toronto.

[61] A. Robinson to I. R. Pounder, January 9, 1957, in Robinson's administrative file, Department of Mathematics, University of Toronto.

[62] When Smith left the University for this new position in the fall of 1957, Robinson sent a warm note of congratulations from Jerusalem. Smith replied on October 15, 1957, thanking Robinson for his congratulatory letter "on my assuming the portfolio for External Affairs," and hoping their paths would cross again soon. Sidney Smith to A. Robinson, Ottawa, October 14, 1957; RP/YUL.

[63] C. E. Higginbottom, Bursar and Secretary to the Board of Governors, University of Toronto, letter to A. Robinson of June 27, 1957; RP/YUL.

[64] "The Canadian Mathematical Congress Seminar in Edmonton" *Canadian Mathematical Bulletin* **1** (1958), 62.

[65] Robinson published his presentation in four parts, as an "Outline of an Introduction to Mathematical Logic" [Robinson 1958b].

[66] Cook 1961. In 1965 F.R.O.S. changed its name to the "International Student Centre of the University of Toronto." See *The Varsity* (student newspaper, Univerity of Toronto), July 1, 1965.

[67] Mary Cooper to Abraham Robinson, May 27, 1957; RP/YUL.

[68] For discussion of the years Robinson spent at the Hebrew University in Jerusalem (1957–1962), see Chapter 7 of Dauben [1995, 243–304].

REFERENCES

Beth, E. W. 1953. On Padoa's method in the theory of definition. *Koninklijke Nederlandse Akademie van Wetenschappen, Proceedings* **56**; *Indagationes Mathematicae* **15**, 330–339.

— 1959. *The Foundations of Mathematics. A Study in the Philosophy of Science.* Amsterdam: North-Holland.

Bissell, C. 1974. *Halfway Up Parnassus. A Personal Account of the University of Toronto, 1932–1971.* Toronto: University of Toronto Press.

Bothwell, R., Drummond, I., and English, J. 1989. *Canada since 1945: Power, Politics, and Provincialism.* Toronto: University of Toronto Press, 1981. Revised ed. 1989.

Campbell, L. L. 1955. *Mixed boundary value problems for hyperbolic differential equations.* Ph.D. Thesis, University of Toronto.

Cohen, P. J. 1971. Comments on the foundations of set theory. *Proceedings of Symposia in Pure Mathematics* **13**, Part I, 9–15.

Cook, M. 1961. Foreign student's haven. *The Varsity* (Student Newspaper, University of Toronto), December 13, 1961, p. 5.

Coxeter I–1990: H. S. M. Coxeter, interviewed in his office at the University of Toronto, November 12, 1990.

Coxeter, H. S. M., and Escher, M. C., *et al.*, Eds., *M. C. Escher, Art and Science: Proceedings of the International Congress on M. C. Escher* (Rome, Italy, 26–28 March, 1985), Amsterdam: North-Holland.

Coxeter, H. S. M., Emmer, M., Penrose, R., and Teuber, M. L. Eds. 1986. *M.C. Escher: Art and Science.* Amsterdam: North-Holland.

Craig, W. 1957. Three uses of the Herbrand-Genzen theorem in relating model theory and proof theory. *Journal of Symbolic Logic* **22**, 269–285.

Dauben, J. W. 1995. *Abraham Robinson. The Creation of Nonstandard Analysis, A Personal and Mathematical Odyssey*, Princeton: Princeton University Press.

Duff I–1990: George F. D. Duff, interviewed in his office at the University of Toronto, November 13, 1990.

Duff, G. F. D. 1969. Arthur Francis Chesterfield Stevenson. *Proceedings and Transactions of the Royal Society of Canada* **7**, 103–107.

Eckmann, B. 1942. Systeme von Richtungsfeldern in Sphären und stetige Lösungen komplexer linearer Gleichungen. *Commentarii Mathematici Helvetici*, **15** (1942–43), pp. 1–26.

Fowell, L. R. 1955. An exact theory of supersonic flow around a delta wing. *Ph.D. Thesis*, University of Toronto.

Gilmore, P. C. and Robinson, A. 1955. Metamathematical considerations on the relative irreducibility of polynomials. *Canadian Journal of Mathematics* **7**, 483–489; In [Robinson 1979, **1**, 348–354].

Green, J. J. 1970. *Aeronautics, Highway to the Future. A Study of Aeronautical Research and Development in Canada* (Background Study for the Science Council of Canada, Special Study No. 12). Ottawa: Queens Printer for Canada.

Hart, J. F. 1953. Theory of spin effects in the quantum mechanics of a many-electron atom with special reference to Pb^{III}. *Ph.D. Dissertation*, University of Toronto.

Henkin, L. 1956. Two concepts from the theory of models. *Journal of Symbolic Logic* **21**, pp. 28–32.

Hilbert, D. 1892. Ueber die irreducibilität ganzer rationaler functionen mit ganzzahligen coefficienten. *Journal für die reine und angewandte Mathematik* **110**, pp. 104–129.

Infeld, E. 1970. Leopold Infeld bibliography. *General Relativity and Gravitation* **1**, 191–208.

Infeld, L. 1978. Canada. In *Why I left Canada. Reflections on Science and Politics* (H. Infeld, Trans.) Montreal: McGill-Queen's University Press.

Keisler, J. 1971. *Model Theory for Infinitary Logic.* Amsterdam: North-Holland.

— 1977. Introduction to A. Robinson. *Complete Theories*, 2nd ed. Amsterdam: North-Holland.

— 1979. Introduction to A. Robinson. *Selected Papers of Abraham Robinson*, H. J. Keisler, S. Körner, W. A. J. Luxemburg and A. D. Young, Eds. New Haven: Yale University Press **1**, pp. xxxiii–xxxvii.

Kochen, S. 1976. The pure mathematician. On Abraham Robinson's work in mathematical logic. In [Young 1976, 312–315].

Laurmann, J., and Robinson, A. 1956. *Wing Theory.* Cambridge, UK: Cambridge University Press.

Lighthill, M. J. 1957. Review of [Robinson and Laurmann 1956]. *Mathematical Reviews* **18**, p 529.

Lightstone, A. H. 1955. Contributions to the theory of quantification. *Ph.D. Dissertation*, University of Toronto.

Lightstone, A. H. and Robinson, A. 1957a. Syntactical transforms. *Transactions of the American Mathematical Society* **86**, pp. 220–245; in Robinson 1979 **1**, pp. 120–145.

— 1957b. On the representation of Herbrand functions in algebraically closed fields. *Journal of Symbolic Logic* **22**, pp. 187–204; in Robinson 1977 **1**, pp. 396–413.

Lyndon, R. 1959. An interpolation theorem in the predicate calculus. *Pacific Journal of Mathematics* **9**, 129–142.

Makowski, J. A., and Shelah S. 1979. The theorems of Beth and Craig in abstract model theory. *Transactions of the American Mathematical Society* **256**, 213–139; Parts (II) and (III) published subsequently.

Pyenson, L. 1978. "Introduction to [Infeld 1978].

Robinson Interviews: I–1980 Renée Robinson, interviewed by JWD at Yale University in New Haven, Connecticut, June 23, 1980. I–1983 Renée Robinson, interviewed by JWD at her home in Hamden, Connecticut, May 10, 1983. I–1990 Renée Robinson, interviewed by JWD at her home in Hamden, Connecticut, October 10, 1990.

Robinson, A. 1945. Shock transmission in beams. *Aeronautical Research Council, Reports and Memoranda*, No. 2265 (8769, 9306 and 9344). London: Ministry of Supply (1950).

— 1947. The effect of the sweepback of Delta wings on the performance of an aircraft at supersonic speeds. *Aeronautical Research Council Reports and Memoranda*, (No. 2476, March, 1947). London: Ministry of Supply (Her Majesty's Stationery Office), 1951. This paper is not reproduced in [Robinson 1979].

— 1948a. On some problems of unsteady supersonic aerofoil theory. *Proceedings of the 7th International Congress for Applied Mechanics, London* **2**, 500–514.

— 1948b. On source and vortex distributions in the linearized theory of steady supersonic flow. *Quarterly Journal of Mechanics and Applied Mathematics* **1**, 408–432.

— 1950a. On the application of symbolic logic to algebra. *Proceedings of the International Congress of Mathematicians, Cambridge, Massachussets, 1950*, Vol. 1, pp. 686–694. Providence, RI: American Mathematical Society, 1951 (in [Robinson 1979, **1**, 3–11]).

— 1950b. On the integration of hyperbolic differential equations. *Journal of the London Mathematical Society* **25**, pp. 209–217.

— 1951a. On the foundations of dimensional analysis, dated RMS Franconia, August/September 1951. This is a 25-page manuscript cited as Robinson MS 1951 (RP/YUL).

— 1951b. *On the Metamathematics of Algebra*. Amsterdam: North-Holland.

— 1953a. Non-uniform supersonic flow. *Quarterly of Applied Mathematics* **10**, pp. 307–319 (in [Robinson 1979, **3**, 183–195]).

— 1953b. Flow around compound lifting units. *Symposium on High Speed Aerodynamics, Ottawa, Canada 1953*, pp. 26–29. Ottawa: High Speed Aerodynamics Laboratory, National Aeronautics Establishment.

— 1954a. On some problems of unsteady wing theory. *Second Canadian Symposium on Aerodynamics, Toronto, Canada*, (publication details), pp. 106–122; Robinson 1979 **3**, pp. 196–211.

— 1954b. On predicates and algebraically closed fields. *Journal of Symbolic Logic* **19**, pp. 103–114; in Robinson 1979 **1**, pp. 332–343.

— 1955a. Ordered structures and related concepts. *Mathematical Interpretation of Formal Systems*, pp. 51–56. Amsterdam: North-Holland (in [Robinson 1979, 1, 99–104]).

— 1955b. Théorie métamathématique des idéaux. Paris: Gauthier-Villars.

— 1956a. Further remarks on ordered fields and definite functions. *Mathematische Annalen* **130**, 405–409 (in [Robinson 1979, **1**, 370–374]).

— 1956b. A result on consistency and its application to the theory of definition. *Koninklijke Nederlandse Akademie van Wetenschappen, Proceedings* **59**; *Indagationes Mathematicae* **18**, 47–58 (in [Robinson 1979, **1**, 87–98]).

— 1956c. Note on a problem of Leon Henkin. *Journal of Symbolic Logic* **21**, 33–35 (in [Robinson 1979, **1**, 105–107]).

— 1956d. Completeness and persistence in the theory of models. *Zeitschrift für mathematische Logik und Grundlagen der Mathematik* **2**, 15–26 (in [Robinson 1979, **1**, 108–119]).

— 1956e. *Complete Theories* Amsterdam: North-Holland.

— 1956f. Solution of a problem by Erdös-Gillman-Henriksen. *Proceedings of the American Mathematical Society* **7**, 908–909 (in [Robinson 1979, **1**, 559–560]).

— 1956g. On the motion of small particles in a potential field of flow. *Communications of Pure and Applied Mathematics* **9**, 69–84 (in [Robinson 1979, **3**, 212–227]).

— 1957a. Transient stresses in beams of variable characteristics. *Quarterly Journal of Mechanics and Applied Mathematics* **10**, 148–159 (in [Robinson 1979, **3**, 241–252]).

— 1957b. Some problems of definability in the lower predicate calculus. *Fundamenta Mathematica* **44**, 309–329 (in [Robinson 1979, **1**, 375–395]).

— 1957c. Wave propagation in a heterogeneous elastic medium. *Journal of Mathematics and Physics* **36**, 210–222 (in [Robinson 1979, **3**, 228–240]).

— 1958a. Relative model-completeness and the elimination of quantifiers. *Dialectica* **12** (1958), pp. 394–407; Robinson 1977 **1**, pp. 146–159.

— 1958b. Outline of an introduction to mathematical logic. *Canadian Mathematical Bulletin* **1** (1958), pp. 41–54, 113–127, 193–208; and **2** (1959), pp. 33–42.

— 1958c. On the concept of a differentially closed field. *Bulletin of the Research Council of Israel* **8** (1958), pp. 113–128.

— 1979. *Selected Papers of Abraham Robinson*. H. J. Keisler, S. Körner, W. A. J. Luxemburg and A. D. Young, eds. New Haven: Yale University Press (in three volumes).

Robinson, G. de B. 1968. Leopold Infeld. *Proceedings and Transactions of the Royal Society of Canada* **6**, 123–125. Reprinted in *Canadian Mathematical Bulletin* **14**, 301–302, (1971).

— 1971. Samuel Beatty. *Proceedings and Transactions of the Royal Society of Canada* **9**, 31–34. Reprinted in *Canadian Mathematical Bulletin* **14**, 489–490, (1971).

— 1979. *The Mathematics Department in the University of Toronto, 1827–1978*. Toronto: Department of Mathematics, University of Toronto.

Rooney I–1990: P. G. Rooney, interviewed in his office at the University of Toronto, November 12/13.

Ross I–1990: A. R. Ross, interviewed in his office at the University of Toronto, November 14.

Ross, R. A. 1955. The waves produced by a submarine earthquake. *Ph.D. Thesis*, University of Toronto.

Steketee R–1976: "Notes on A. Robinson in Toronto," sent by J. A. Steketee to George Seligman, August 17, 1976, Robinson Papers, Yale University Library.

Steketee, J. A. 1955. Some problems in boundary-layer transition. *Ph.D. Thesis*, University of Toronto.

Synge, J. L. 1972. Curriculum vitae of J. L. Synge. In *General Relativity (Papers in Honor of J. L. Synge)*. Oxford: Clarendon Press.

Trautman, A. 1973. Leopold Infeld. *Dictionary of Scientific Biography*, Vol. 7, pp. 10–11. New York: Scribner's.

Vaught, R. L. 1960. Review of Robinson 1956e, *Complete Theories*, in *Journal of Symbolic Logic* **25**, pp. 174–176.

Ward, G. N. 1952. On the integration of some vector differential equations. *Quarterly Journal of Mechanics and Applied Mathematics* **5**, p. 446.

West, B. 1967. *Toronto*. Toronto: Doubleday Canada Limited.

Young R–1975: Alec Young, reminiscences of Abraham Robinson, in RP/YUL.

Young, A. D. 1976. The applied mathematician, in Young *et al.*, "Abraham Robinson." *Bulletin of the London Mathematical Society* **8**, pp. 307–323.

Zakon, E. 1953. Left side distributive law of the multiplication of transifnite numbers. *Riveon Lematematika* **6**, pp. 28–32 (in Hebrew).

— 1954a. On the relation of 'similarity' between ordinal numbers. *Riveon Lematematika* **7**, pp. 44–49 (in Hebrew).

— 1954b. *Collection of Exercises in Higher Mathematics*. Haifa: Technion Edition, 1954/55.

— 1955. On fractions of ordinal numbers. *Scientific Publications of the Hiafa Institute of Technology* **6** (1955).

— 1960. Elementary properties of ordered Abelian groups. *Transactions of the American Mathematical Society* **96**, 222–236 (in [Robinson 1979, 1, 456–470]).

— 1967. A set-theoretical characterization of enlargements; in W. A. J. Luxemburg, Ed. *Applications of Model Theory to Algebra, Analysis and Probability*. New York: Holt, Rinehart and Winston, 109–122; in Robinson 1979 **2**, 206–219.

N. N. Luzin and the Affair of the "National Fascist Center"

Sergei S. Demidov

Department of the History of Mathematics,
Institute of the History of Natural Science and Technology,
Russian Academy of Sciences, Staropansky 1/5, 103012 Moscow, Russia

and

Charles E. Ford

Department of Mathematics and Computer Science,
Saint Louis University, St. Louis, Missouri 63103-2097

The "affair of Academician Luzin" was a major event in the history of Soviet mathematics. It involved the renowned mathematician N. N. Luzin (1883–1950) who was publicly attacked in July 1936, a clear prelude to his arrest and destruction. Quite suddenly the campaign came to an abrupt halt, without serious consequences for him. This unexpected outcome has never been adequately explained. A recent article together with new historical analyses make possible the formulation of new hypotheses about this outcome. The article revealed new information about the arrest and execution of the philosopher, theologian, and scientist P. A. Florensky (1882–1937), a long-time friend of Luzin. When Florensky was arrested in 1933, he was forced to implicate Luzin in a fictitious "nationalist–fascist" plot involving Hitler. Recent historical analyses of the Soviet political situation in the 1930s have shown that by the summer of 1936 Stalin was strongly convinced that he could soon conclude a nonaggression pact with Hitler. We offer the hypothesis that the attack on Luzin was a prelude to the unveiling of this "nationalist–fascist" plot. This was halted by Stalin who, hopeful of a Nazi–Soviet pact, did not wish to risk alienating Hitler by "uncovering" such a plot at this time.

1. THE "AFFAIR OF ACADEMICIAN LUZIN"

Interest in the affair of Academician Luzin has greatly increased [Levin 1990, Yushkevich 1989, Demidov 1993]. This interest is easily explained and completely justified. Although not well known like the great purges of the Stalin terror of the 1930s, it, nonetheless, affected for many years the arrangement of forces in the Soviet mathematical community and the behavior of the members of this community. The memory of this affair lived on for a long time in the minds of Soviet mathematicians, influencing not only their behavior, but also even at times the choice of their research topics.

We will recount the main events of this affair. Academician N. N. Luzin (1883–1950) was, by the 1930s, a world-renowned scientist. He was the head of a scientific school that was known throughout the world and represented by such names as D. Ye. Menshov, A. Ya. Khinchin, M. Ya. Suslin, P. S. Aleksandrov, P. S. Uryson, L. G. Shnirelman, L. A. Lyusternik, A. N. Kolmogorov, and P. S. Novikov. By the 1930s this school had in fact already disintegrated into many successor schools. Very difficult, even antagonistic, relationships had developed between the teacher and his former students. This resulted from both objective factors (the departure of the former students from the topics of the teacher and their transformation into independent researchers, sometimes into the chiefs of their own schools) and subjective factors (above all the complexity of the creative personality of Luzin himself). By the middle of the 1930s the influence of Luzin at the Academy of Sciences had greatly increased. He was head of its mathematical group, and the politics of the Academy in the field of mathematics depended, in very large measure, on him. As a result, his estrangement from some of his students, Aleksandrov and others, was further exacerbated.

Such situations are hardly unique. They are quite natural in the development of scientific communities. Old schools disintegrate and new schools take their place. This does not proceed smoothly—people are people. However, in the specific atmosphere of the Stalinist terror of the 1930s, this natural process became extremely distorted. In this case it became a public political affair. The initiative came from below. "Red professors" and "revolutionary students" in their search for "wrecking" among the professorate could hardly ignore the figure of Luzin. After the arrest, exile, and death of his teacher D. F. Egorov [Ford 1991], Luzin was in line to be the next target. He was a philosopher with a non-Marxist orientation. He was a religious believer who did not accept the new regime. Although, unlike Egorov, he neither publicly identified himself with the Church, nor openly criticized the regime, Luzin was, nonetheless, an excellent target for the Marxist ideologues.

The most active figure in these attacks was E. Kolman—the "black angel" of Moscow mathematics, indeed of the whole scientific community in Moscow. His attacks were the more dangerous because he held very high positions in the Moscow party organizations. Luzin was especially vulnerable because of his philosophical idealism, which his attackers, quite correctly, connected with the worldview of the prerevolutionary mathematical–philosophical school at Moscow University. This school was denounced in the Soviet press of this period as being nothing more than "monarchist" and "black hundred." (The Black Hundred was a prerevolutionary antisemitic, antirevolutionary organization.) Kolman attacked Luzin's views on the foundations of mathematics and connected his works with "reactionary" currents in modern Western philosophy [Kolman 1931], labeling his philosophical position as "pro-fascist" [Kolman 1936, 290].

The epithet "fascist" was utilized in the Soviet press of that time in a broad sense. The western social democrats, for example, were routinely decorated with it. In the case of Luzin, however, this term was utilized in its narrow sense. This can be seen from the accusations of V. N. Molodshii, who compared Luzin's supposed fascist views with the ideas of the German mathematician and Nazi-sympathizer L. Bieberbach about "Aryan" and "non-Aryan" mathematics [Molodshii 1936, 17].

Kolman was then chief of the Department of Science of the Moscow Committee of the Party. It is clear that he played a major role in the organization and preparation of the campaign against Luzin. These preparations were carefully planned. Luzin was invited to view the examination process in one of the Moscow secondary schools and was then asked to give his impressions in the newspaper *Izvestya*. The organizers of the campaign were counting on Luzin's tendency to give inflated evaluations of the work of others. Luzin, as expected, grossly exaggerated the achievements of the pupils in *Izvestya*, on June 27, 1936. The very same issue of *Izvestya* contained an article by the indignant director of the school, who insisted that the pupils were not nearly as good as Luzin had described. Soviet schools need not exaggerated praise, but constructive criticism. "Was it not your goal," said the organizers of the campaign through the mouth of the director, "to whitewash our shortcomings and thus damage our school?" Thus the charge of sabotage was introduced. This charge was highlighted in a dramatic article on the front page of *Pravda* on July 3, 1936, entitled "On Enemies behind a Soviet Mask." There is every reason to suppose, as A. E. Levin has argued, that the author of this article was none other than Kolman himself [Levin 1990, 99–100].

The campaign to destroy Luzin was underway. A flood of denunciations appeared in the central and provincial press. At meetings in universities and institutions the "hostile activities of the Academician" were held up to condemnation. Many of Luzin's former students were assigned a special role in this chorus of denunciation and rose to the task by leading the attack. The well-planned events developed with astonishing speed. Already on July 7 a special Commission of the Presidium of the Academy of Sciences was formed, headed by the Vice President of the Academy, G. M. Krzhizhanovsky. In the meetings of this Commission, attacks on Luzin were leveled by his former students, above all by Aleksandrov.

The charges against Luzin boiled down to the following: (1) he systematically gave undeservedly generous reviews of the works of other mathematicians; (2) he allegedly idolized the West, publishing his best works abroad, while publishing only minor papers in the USSR; and (3) he misappropriated the results of his own students. This was all construed as outright sabotage.

As an analysis of the stenographic notes of the Commission [Demidov 1993, 49] reveals, its work was proceeding swiftly to its predetermined conclusion. The Commission was preparing to condemn the activity of Luzin as wrecking and recommend to the General Assembly of the Academy his expulsion. Then, after this essentially *pro forma* procedure of expulsion, the affair would be transferred to the OGPU (a predecessor of the KGB).

Such, however, was not to be. Somebody at the top (Stalin himself?) totally unexpectedly stopped the campaign. This occurred sometime between the Commission meeting of July 11 and that of July 13. The tone changed dramatically between these two meetings, as the stenographic notes clearly reveal. The result was that, on the basis of the conclusions of the Commission, the Presidium of the Academy of Sciences offered the following resolution. "It is sufficient to reprimand and to warn Luzin that in case he does not change his line of conduct the Presidium of the Academy will have to raise the question of excluding N. N. Luzin from the Academy" [Demidov 1993, 49].

The natural question arises: why and by whom was the "Luzin affair" terminated? In an article [Demidov 1993, 49], written in 1989, one could only agree with the opinion of Yushkevich [1989, 113] in pointing to a very important reason for its termination. At that time, the preparations for the trial of Zinoviev and Kamenev, which began in August, were well under way. Compared with this formidable trial of Stalin's major opponents and rivals, the case of Luzin seemed unimportant.

It was not possible to say anything more at that time. However, in 1990 some materials from the archives of the KGB, which opened up a new approach to this question, were published.

2. THE AFFAIR OF THE "NATIONAL FASCIST CENTER"

In 1990 the article "The Fate of Greatness" by Vitalii Shentalinskii was published in the popular, mass-circulation journal *Ogonyok* [Shentalinskii 1990]. This article, based on documents in the archives of the KGB, is devoted to the circumstances of the arrest and execution of the philosopher, theologian, and scientist P. A. Florensky (1882–1937).

Details about the life and work of this exceptional person can be found in Slesinski [1984] and Florensky [1987]. Here we shall give only a brief sketch of his life. P. A. Florensky graduated in 1904 from the Mathematics Department of Moscow University and, in spite of an invitation to remain at the University "for preparation for the professorate," he entered the Moscow Theological Academy. After graduating in 1908, he was appointed to the chair of the History of Philosophy and taught there until the closing of the Academy after the Bolshevik revolution in 1917. In 1911 he became a priest. His treatise "The Pillar and Foundation of Truth" [Florensky 1914] is one of the finest works of Orthodox philosophy in the 20th century. Educated as a mathematician, Florensky remained interested in mathematics his entire life and never abandoned his use of mathematical concepts. His place in the history of mathematics is discussed in Demidov [1988]. His activities during his student years had an influence on the process of the birth of the Moscow school of the theory of functions of a real variable. Florensky was held in high esteem by such important members of the mathematics faculty as Egorov and N. Ye. Zhukovsky. Florensky and Luzin developed a close friendship during their student years together at Moscow University, a friendship that endured. Eighteen years of correspondence between them was published in 1989 [Demidov, Parshin, and Polovinkin 1989].

Florensky was a person with very broad interests: from linguistics, folklore, and the history of art to a variety of questions in science and technology. He had a special interest in electrotechnical materials, on which was one of the most competent specialists in the entire country. This became a major focus for his research after the closing of the Theological Academy. For several years he worked as the head of the Department of Material Science of the State Electro-Technical Institute and also as an editor of the Technical Encyclopedia. In 1928 he was arrested for his support of a group of clergymen and former nobility who lived in the village of Sergiev Posad. Florensky himself lived in this village, located not far from Moscow. It was connected with the Troitsky–Sergiev Lavra, in which the then-closed Moscow Theological Academy was located. After an investigation, Florensky was banished to Nizhny Novgorod. Fortunately, he was allowed to return after only a few months.

Florensky was a significant figure, connected on the one hand with the church and on the other with the scientific community, who had been previously arrested. He made an excellent target for the fabricated "construction" of counterrevolutionary organizations by the OGPU. These "constructions" were modeled on the "brilliant success" of the "Industrial Party" affair of 1928–1930. Among the important defendants accused of wrecking in this affair was the well-known engineer L. K. Ramzin. Under interrogation he was coerced into actively assisting the OGPU in "constructing" a counterrevolutionary organization, with many branches and connections to foreign countries. After this the OGPU set to work on the "construction" of similar affairs.

The documents from the KGB archive published by Shentalinskii [1990] show us one such construction—the affair of the counterrevolutionary organization "Party for the Rebirth of Russia." According to the conception of the OGPU this new party was to be born from the remnants of the earlier "monarchist" organization the "All-Peoples' Union for the Struggle for the Rebirth of Russia." This fictitious party had been "liquidated" by the OGPU in 1930. Among its "leaders" were the well-known Academicians and historians S. F. Platanov and Ye. V. Tarle. The nucleus of the new organization was to be organized from scientists with a supposed pro-German tendency united by reactionary convictions.

The new affair apparently began with the arrest in January 1933 of P. V. Gidulyanov, a professor of jurisprudence and a research worker at the Central Antireligious Museum. On February 25, 1933, Florensky was arrested. The certificate for the arrest contains the following information about Florensky [Shentalinskii 1990, 24].

> A member of the center of the counterrevolutionary organization the "Party for the Rebirth of Russia." Denounced by the evidence of the accused professors Gidulyanov, Ostroukhov and the member of the counterrevolutionary organization Zhirikhin.

Then began the long, arduous interrogation employing means, imaginable and unimaginable, by which the accused was reduced to giving the testimony desired by the OGPU.

The first to "confess" was Gidulyanov. We give an excerpt from his own letter to the prosecutor, written from exile in Alma Ata and preserved in the archives of the KGB. This letter requested that the Office of the Prosecutor investigate his case, fabricated by the OGPU, and change the sentence. Written in the apparent hope of obtaining justice, the letter was simply added to his OGPU dossier.

Gidulyanov explains that, after prolonged interrogation by Shuleiko, an interrogator for the OGPU, and after receiving the promise that at the end of the investigation he would be freed and allowed to resume the study of science, he practically dictated the story the interrogators sought.

> I gave myself entirely to the power of the...OGPU and became the producer and first tragic actor in the frame-up process of the nationalists, turned by the will of the OGPU into the national fascists....I declared myself to be the organizer of the Committee for National Organization which, after a series of attempts inside the walls of the OGPU, was christened "the national center." Moreover the members of this mythical Committee, whose names were dictated to me, were already in the hands of the OGPU, my colleagues Chaplygin, Luzin, and Florensky.

The Chaplygin referred to here is the well-known specialist in mechanics and mathematics, professor at Moscow State University, Academician S. A. Chaplygin. Neither Chaplygin nor Luzin had been arrested. The news of their arrest was given to Gidulyanov as disinformation by the interrogator. At first Gidulyanov was to construct a nationalist organization. However, between the time of his arrest and that of Florensky came the appointment of Hitler on January 30, 1933, as Chancellor of Germany. Thus by the time of Florensky's arrest, the label "fascist" had been added to the title of the organization.

> To manifest true repentence the main role had to be taken by me. I presumably made the connection with Florensky...and through him the connection with Chaplygin and Luzin. Thus was created a mythical committee! The Chairman was Chaplygin, I was secretary, Florensky was ideologue, and Luzin was for contacts abroad.

In another part of his long letter, Gidulyanov wrote that he had never met Chaplygin or Luzin and that he had made the acquaintance of Florensky only when confronted with him during the interrogation [Shentalinskii 1990, 25].

> The platform of the party of nationalists I fabricated myself with the gracious help of...Radzivilovsky, who with his own hand had written down my "detailed evidence."
> The party of nationalists is to begin its activities after the taking of Moscow and the military occupation of Russia by the Germans. Moreover,

the foundation of its platform was to be the principle of "Soviets without communists" under the cover of the bourgeois order.

On March 3, 4, and 5 Florensky began to give testimony. He named himself as the head of the organization. And in his own hand he drew the "scheme of the structure of the national fascist center." His immediate deputies were Luzin and Gidulyanov. The connection with Chaplygin was through Luzin. In this organizational center, constructed by the OGPU, Luzin had a special role—the connection with foreign countries.

On June 30 this "investigation" reached its conclusion, and a document of 30 pages was issued. It contains the following two points [Shentalinskii 1990, 26].

> The Moscow district of the OGPU uncovered and liquidated a counterrevolutionary national fascist organization which named itself "Party for the Rebirth of Russia."

The OGPU forgot to add that it had also created this organization.

> The organization was headed by the leadership center consisting of Professors Florensky, Gidulyanov, and Academicians Chaplygin and Luzin. It grew, in fact, out of the survivors of the defeated remnants of the monarchist organization the "All-Peoples' Union for Struggle for the Rebirth of Russia," headed by Academician Platanov and others, which had been liquidated by the OGPU in 1930. Connection had been made with the émigré White Guards and a confidential meeting with Hitler had been arranged...

Responsibility for having arranged the meeting with Hitler presumably fell to Luzin, who was in charge of foreign contacts.

In this summary, 12 persons were named, among them Florensky, Gidulyanov, and Ostroukhov. The material about Chaplygin and Luzin was set aside from the case, presumably for later use, with the notation: "Orientation: counterrevolutionary intelligentsia" [Shentalinskii 1990, 26].

On the foundation of this conclusion the Special Troika of the Moscow district of the OGPU on July 26, 1933, sentenced the participants of the "center." Gidulyanov was exiled. Florensky was convicted and sentenced on November 10, 1933, to 10 years in corrective labor camps.

Although Academicians Luzin and Chaplygin were not touched at this time, the OGPU preserved this material about their supposed membership in the leadership of the National Fascist Center. This material could have remained unused in the archive of OGPU forever. On the other hand, it may have been used, as we believe, in the affair of Luzin in 1936.

Supporting this hypothesis are the following two points:

1. The Luzin affair was carefully prepared. It is likely that the OGPU was involved.
2. The attacks of Kolman and Molodshii on Luzin include the accusations of fascism. To the objection that this accusation was a common theme in the Soviet press of this period, we have argued above that their accusations against Luzin were not about fascism in the broad sense, but about Nazi fascism in the narrow sense.

If our hypothesis is correct, then the campaign against Luzin could have been the preliminary stage of a much larger affair involving the National Fascist Center. The attack on Luzin may have been started by local initiative with the involvement of the OGPU. A campaign against the National Fascist Center, however, would touch on the Soviet relationship with Germany. Indeed, as constructed, it would directly implicate Hitler. Permission to unleash this campaign could be given only by Stalin himself. Therefore, it seems certain that the campaign against Luzin was halted by Stalin. The question remains—why?

The suggestion was advanced that perhaps the orientation of Stalin toward Germany had already began to change, anticipating the Molotov–Ribbentrop nonaggression pact between the Soviet Union and Germany concluded in 1939. Did the great leader think, perhaps, that an attack against internal "fascists" would work against the new political orientation then being developed?

This idea was criticized on the grounds that the change of direction in Soviet politics came only later. Still to come in 1936 were the antifascist intervention in the Spanish Civil War; the trial of Pyatakov in 1937, in which anti-German charges were very prominent; the attack against Tukhachevsky and the leadership of the Soviet military in 1937, in which they were accused of spying for Germany; and other such events. However, new revelations in the historical literature of the 1980s offer evidence in favor of our hypothesis.

3. THE GERMAN POLITICS OF STALIN, 1934–1938, AND THE AFFAIR OF ACADEMICIAN LUZIN

The Soviet diplomat and journalist Ye. Gnedin, who worked in the years 1935–1936 in the Soviet embassy in Berlin, wrote:

I remember that we members of the Berlin embassy staff were rather taken aback when Eliava, the deputy commissar for foreign trade, who was passing through Berlin (in 1936, as I recall) and who had access to Stalin because of long-standing personal ties, gave us to understand that "at the top" Hitler was viewed "differently" than he was in the Soviet press or by the Soviet embassy staff in Berlin [Heller 1985, 326].

In fact, as a number of authoritative historians have noted, the change in Stalin's attitude toward Hitler and Nazi fascism had actually begun much earlier (see Heller [1985, 322f], Tolstoy [1981, 83f], and Conquest [1990, 195f]). It goes back at least to the infamous massacre organized by Hitler on June 30, 1934, directed primarily against Ernst Roehm and the leadership of the Nazi "Brownshirts." In response to this, Stalin announced to the Politboro: "The events in Germany do not at all indicate the collapse of the Nazi regime. On the contrary, they are bound to lead to the consolidation of that regime and the strengthing of Hitler himself" [Tolstoy 1981, 83–84]. Hitler's action had earned him the sympathy and respect of Stalin. Within a few weeks Stalin began preparations for his own first bloody purge and the murder of S. M. Kirov.

From this moment on, convergence with Hitler became the *idée fixe* in the mind of Stalin. In 1935–1936 he was particularily active in preparations for achieving good relations with Berlin. To this end, he sent his personal confidant David Kandelaki to Berlin as a Soviet trade representative to pursue secret negotiations for a nonaggression pact. As a result of the efforts of Kandelaki, a trade agreement with Germany in which the Soviet Union received very good credit was concluded in May 1936. This convinced Stalin that talks with Hitler would lead to a successful agreement. According to the chief of the Soviet military intelligence network in Western Europe, W. Krivitsky, Stalin told Yezhov, "In the immediate future we shall consummate an agreement with Germany." This certitude, however, was ill founded. The talks, successful from the Soviet perspective, were cancelled by the German side. On February 11, 1937, the German Minister of Foreign Affairs, von Neurath, announced that the Soviet proposals had been rejected by Germany [Heller 1985, 327]. Hitler continued to rebuff Stalin.

As noted above, there were two trials in 1937 that had strong anti-Nazi overtones. Even though these came after Hitler's rejection of Stalin's overtures, the Soviets were careful to signal to the Nazis that they were not the target. In the Pyatakov trial, the German Military Attaché who was supposedly implicated was not even declared *persona non grata* by the Soviet Union [Conquest 1990, 197]. The Tukhachevsky trial actually involved cooperation among the Soviet military intelligence, the NKVD, and

Reinhardt Heydrich's SD [Heller 1985, 305, Conquest 1990, 198]. It was clear to the Nazis that these trials were not directed against them. Anti-Nazi accusations were being used merely for domestic political purposes.

Certainly Stalin would have been out of character not to continue talks on two fronts. He continued his maneuvers with Britain and France to create an antifascist coalition. For him, however, the union with Germany was by far the more preferable. It is not his fault that he could reach agreement only in 1939. In the summer of 1936 Stalin was sure that an agreement with Germany was near. He could certainly have regarded as badly timed a major affair that would serve no other purpose than anti-Nazi propaganda. Even worse, the accusations were aimed at Hitler himself. This, together with preparations for the much more important trial against Kamenev and Zinoviev that began in August, could have caused Stalin to abort the campaign against Luzin.

Thus concluded the affair of the National Fascist Center. It had touched only Luzin, who never knew the precipice before which he stood. It never touched the "fascist" Academician Chaplygin, who never learned of the materials against him contained in the files of the OGPU. It ended tragically for Gidulyanov: in 1937 he was rearrested and shot.

Before it was transformed into an antifascist campaign, this affair had originally been conceived of as antireligious, a continuation of one involving the "All Peoples' Union for Struggle for the Rebirth of Russia." This was the affair of Academician Platonov in 1930 in which about 100 leading scholars were arrested. Most had been practicing religious believers, including some clergymen [Pospielovsky 1988, 43]. Florensky was a priest and, unlike Luzin, continued to openly identify himself with the Church. The years 1936–1939 saw the liquidation of most of the remaining bishops, clergy, and active lay believers [Pospielovsky 1987, 65–66, 68]. This may explain the tragic fate of Florensky. Following the decision of a Special Troika of the Leningrad district of the UNKVD, Florensky was shot on December 8, 1937.

ACKNOWLEDGMENTS

Research for this paper was supported by a grant from the International Research and Exchanges Board, with funds provided by the U.S. Department of State (Title VIII) and the National Endowment for the Humanities. None of these organizations is responsible for the views expressed.

REFERENCES

Conquest, R. 1990. *The Great Terror: A Reassessment.* New York: Oxford University Press.

Demidov, S. S. 1988. On an early history of the Moscow School of theory of functions. *Philosophia Mathematica* **3**, 29–35.

— 1993. The Moscow school of the theory of functions. In *Golden Years of Moscow Mathematics*, S. Zdravkovska and P. L. Duren, Eds., History of Mathematics, Vol. 6, pp. 35–53. New York: American Mathematical Society.

—, Parshin, A. N., and Polovinkin, S. M., Eds. 1989. The correspondence of N. N. Luzin with P. A. Florensky. *Istoriko-Matematicheskiye Issledovaniya* **31**, 116–191 (in Russian).

Florensky, P. A. 1914. *The Pillar and Foundation of Truth.* Moscow (B. Jakim, Trans.). Princeton, NJ: Princeton University Press (to appear).

— 1987. *Salt of the Earth* (R. Betts, Trans.). Platina, CA: St. Herman of Alaska Brotherhood.

Ford, C. E. 1991. Dmitrii Egorov: Mathematics and religion in Moscow. *The Mathematical Intelligencer* **13** (2), 24–30.

Heller, M., and Nekrich, A. 1985. *Utopia in Power.* New York: Summit Books.

Kolman, E. 1931. Preface to: *Struggle for a Materialistic Dialectics in Mathematics*, pp. 5–9. Moscow-Leningrad: Gosudarstvennoye Nauchno-Tekhnicheskoye Izdatel'stvo (in Russian).

— 1936. *The Subject and the Method of Contemporary Mathematics.* Moscow: Sotsekgiz (in Russian).

Levin, A. E. 1990. Anatomy of a public campaign: "Academician Luzin's case" in Soviet political history. *Slavic Review* **49** (1) (Spring), 90–108.

Molodshii, V. N. 1936. On a scientific enemy behind a Soviet mask. *Pod Znamenem Marksizma* No. 9, 8–18 (in Russian).

Pospielovsky, D. V. 1987. *A History of Soviet Atheism in Theory and Practice, and the Believer.* Vol. 1: *A History of Marxist-Leninist Atheism and Soviet Anti-Religious Policies.* New York: St. Martin's Press.

— 1988. *A History of Soviet Atheism in Theory and Practice, and the Believer.* Vol. 2: *Soviet Anti-Religious Campaigns and Persecutions.* New York: St. Martin's Press.

Shentalinskii, V. 1990. The fate of greatness, *Ogonyok* No. 45, 23–27 (in Russian).

Slesinski, R. 1984. *Pavel Florensky: A Metaphysics of Love.* Crestwood, NY: St. Vladimir Seminary Press.

Tolstoy, N. 1981. *Stalin's Secret War.* London: Jonathan Cape.

Yushkevich, A. P. 1989. The case of Academician N. N. Luzin. *Vestnik Akademiyi Nauk SSSR* No. 4, 102–113 (in Russian).

Johannes Praetorius (1537–1616) — ein bedeutender Mathematiker und Astronom des 16. Jahrhunderts [*]

Menso Folkerts

Institut für Geschichte der Naturwissenschaften, Universität München,
D-80306 München, Germany

After Johannes Praetorius (1537–1616) studied in Wittenberg, he was later an instrument maker in Nürnberg (1562–1569) and taught mathematics at the universities in Wittenberg (1571–1576) and then in Altdorf (1576–1616). The major part of his *Nachlass*, which is to be found primarily in the Municipal Library in Schweinfurt and in the University Library in Erlangen, was exhibited in 1993 on the occasion of the 500th anniversary of the appearance of Copernicus' great work. Based substantially on an examination of the manuscripts of Praetorius, his accomplishments as an astronomer, geodetic surveyor, and mathematician are described, as well as his importance within the circle of humanists.

Im Jahre 1543 erschien in Nürnberg das Hauptwerk von Nicolaus Copernicus, seine "Sechs Bücher über die Umläufe der Himmelssphären" (*De revolutionibus orbium coelestium libri sex*). Bekanntlich hatte Copernicus lange gezögert, das Buch herauszubringen. Es war vor allem Georg Joachim Rheticus (1514–1574), der sich von 1539 bis 1541 bei Copernicus in Frauenburg aufhielt und ihn schließlich überredete, dem Druck zuzustimmen. Rheticus, seit 1536 Professor für Mathematik an der Universität Wittenberg, kümmerte sich um die Drucklegung der Schrift, die schließlich im Mai 1543 in Nürnberg erschien. Er konnte nicht verhindern, daß Andreas Osiander, einer der führenden Theologen der Reformation, eigenmächtig dem Buch eine selbstverfaßte Vorrede beifügte, in der die heliozentrische Lehre als Hypothese bezeichnet wurde, deren Wahrheit nicht behauptet werden sollte.

Als das Hauptwerk des Copernicus im Druck erschien, hat man die Schrift und seine heliozentrische Lehre zunächst relativ sachlich betrachtet. Zu polemischen Auseinandersetzungen kam es eigentlich erst im 17. Jahrhundert; es ist bekannt, daß ihr prominentestes Opfer Galilei wurde. In den Jahrzehnten nach Copernicus' Tod hat gerade die katholische Kirche seine Lehre nicht massiv verurteilt. Der Jesuitenpater Christoph Clavius, einer der bedeutendsten Astronomen seiner Zeit, lobte in seinem Kommentar zur *Sphaera* des Sacrobosco, der 1570 erschien, Copernicus als einen bedeutenden Astronomen, bezeichnete seine Lehre allerdings als falsch. Die wissenschaftliche Arbeit des Copernicus wurde auch dadurch gewürdigt, daß bei der Kalenderreform durch Papst Gregor XIII. (1582) für die Umlaufzeit des Mondes die Zahlenangaben des Copernicus benutzt wurden. Doch auch dies bedeutete natürlich keine Anerkennung seiner Lehre. In den letzten beiden Jahrzehnten des 16. Jahrhunderts versuchte der dänische Astronom Tycho Brahe auf seiner Sternwarte im Großen Sund, mit Hilfe von genauen astronomischen Beobachtungen zu klären, ob das ptolemaeische oder das copernicanische System besser sei. Obwohl seine Beobachtungen genauer waren als die aller anderen Zeitgenossen, ließen sie keine Entscheidung zu. Vermutlich aus religiösen Gründen hielt Brahe an der Zentralstellung der Erde fest und stellte 1583 als Kompromißlösung ein eigenes System auf, nach dem die Erde, die sich im Zentrum des Universums befindet, von Mond und Sonne umlaufen wird, während die übrigen Planeten nicht die Erde, sondern die Sonne umkreisen. Sein System fand nur vorübergehend einige Anhänger.

In protestantischen Kreisen wurde die copernicanische Lehre in der ersten Zeit viel heftiger kritisiert. Luther äußerte sich völlig ablehnend dazu. Auch Michael Maestlin, seit 1582 Professor für Mathematik in Tübingen, sprach sich zunächst gegen das heliozentrische System aus. Er hat es später aber in seinen Vorlesungen trotzdem erwähnt, und dies führte dazu, daß sein Schüler Kepler darauf aufmerksam wurde.

Unter dem Einfluß der Kritik Luthers und Melanchthons konnte sich das copernicanische System anfangs an der Universität Wittenberg nicht durchsetzen. Doch die beiden Professoren der Mathematik, Rheticus und Erasmus Reinhold (1511–1553), interessierten sich dafür. Besonders Reinhold unternahm viele Berechnungen nach der neuen Lehre und begann auch mit der Aufstellung von Tafeln, die es ermöglichen sollten, die Bewegungen der Planeten zu berechnen. 1551 publizierte Reinhold diese Tafeln unter dem Titel "Prutenische Tafeln". Sie waren berechnet nach den Daten von *De revolutionibus*, konnten aber auch unter der ptolemäischen Hypothese benutzt werden.

Durch Rheticus und Reinhold, die das geozentrische und das heliozentrische System in gleicher Weise kannten, bot in der 2. Hälfte des 16.

Jahrhunderts die Universität Wittenberg wahrscheinlich besser als jede andere Universitätsstadt die Möglichkeit, Vor- und Nachteile der beiden Lehren kennenzulernen. Ein Beispiel hierfür ist Johannes Praetorius, der in Wittenberg studierte und später für kurze Zeit dort Professor war, bevor er eine Professur in Altdorf übernahm.

Es ist ein Glücksumstand, daß man die Bibliothek des Praetorius recht gut kennt, weil sich nicht nur ein Verzeichnis seines Nachlasses erhalten hat (vermehrt um die Bände, die sein Nachfolger Saxonius hinzuerwarb), sondern auch die Mehrzahl der Handschriften und Bücher, die Praetorius bzw. Saxonius besaßen, noch vorhanden sind. Die meisten dieser Bücher und einige Handschriften befinden sich heute in der Stadtbibliothek Schweinfurt. Die einzigartige Sammlung des Praetorius/Saxonius macht den besonderen Wert dieser Bibliothek aus. In Verbindung mit einer Ausstellung zum 450. Jahrestag des Erscheinens von *De revolutionibus* wurde die Sammlung des Praetorius/Saxonius erstmals fast komplett präsentiert und durch einen Katalog erschlossen [Müller 1993].

Wie kam es dazu, daß die Bibliothek des Praetorius und auch seine Arbeiten so wenig beachtet wurden? Da Praetorius, abgesehen von Kalendern, nur zwei Arbeiten veröffentlichte, ist es verständlich, daß er in der Folgezeit beinahe in Vergessenheit geriet. Fast alles, was man in der Literatur über ihn liest, geht auf die Informationen zurück, die Johann Gabriel Doppelmayr im 18. Jahrhundert veröffentlichte [Doppelmayr 1730]. Nach Doppelmayr hat nur noch Ernst Zinner neue Erkenntnisse hinzugefügt. Zinner gab den Katalog der Bibliothek von Praetorius heraus und räumte ihm einen gebührenden Platz in seinem grundlegenden Werk über die Entstehung und Ausbreitung der copernicanischen Lehre [Zinner 1943] ein. Bei der Vorbereitung der Schweinfurter Ausstellung konnten die bisherigen Kenntnisse über Praetorius in manchen Punkten erweitert werden. Aus vielen Mosaiksteinchen formte sich das Bild eines Mannes, der in vieler Hinsicht bemerkenswert war: Er war zu seiner Zeit einer der bedeutendsten Astronomen und Mathematiker in Deutschland. Darüber hinaus pflegte er intensive wissenschaftliche Kontakte mit Gelehrten im In- und Ausland. Ich möchte dies im folgenden näher ausführen. Nach einem kurzen Abriß über sein Leben möchte ich auf seine Leistungen im Bereich der Astronomie, der Geodäsie und der Mathematik eingehen und skizzieren, welche Bedeutung er innerhalb des Kreises der Humanisten besaß. Dabei werde ich von den Büchern und Handschriften in seinem Nachlaß ausgehen, aber auch seinen Briefwechsel einbeziehen. Es versteht sich von selbst, daß das Thema in diesem Beitrag nicht erschöpfend behandelt werden kann.

1. PRAETORIUS' LEBEN [1]

Johannes Praetorius wurde 1537 in Joachimsthal in Böhmen geboren. Im Sommersemester 1557 immatrikulierte er sich an der Universität Wittenberg, die durch Luther und Melanchthon zur bedeutendsten protestantischen Universität geworden war [2]. An der 1502 gegründeten Universität war das Fach Mathematik zunächst nicht vorgesehen. Erst ab 1514 wurde auch Mathematik gelehrt und für die Abschlüsse (Baccalaureat und Magister) verbindlich gemacht. 1536 wurden zwei Lehrstühle errichtet und mit Georg Joachim Rheticus (1514–1574) und Erasmus Reinhold (1511–1553) hervorragend besetzt. Als Praetorius sein Studium begann, war Reinhold schon tot und Rheticus nicht mehr in Wittenberg (er übernahm 1542 eine Professur in Leipzig), aber durch Caspar Peucer (1525–1602), den Nachfolger Reinholds und Schwiegersohn Melanchthons, konnte er mit dem Denken dieser beiden bedeutenden Naturwissenschaftler vertraut werden. Zu Praetorius' Lehrern dürfte auch Sebastian Dietrich (Theodoricus) von Winsheim (†1574) gehört haben, der ebenfalls in Wittenberg Mathematik unterrichtete und ein später weit verbreitetes Buch über Astronomie schrieb [Müller 1993, 241–242, Nr. 82]. 1562 erwarb Praetorius den Grad eines Magisters an der untersten, der artistischen, Fakultät [3].

Von 1562 bis 1569 lebte Praetorius als Mechaniker in Nürnberg. Dort fertigte er mathematische und astronomische Instrumente an. Einige von ihm hergestellte Globen, Sonnenuhren, Astrolabien und andere astronomische Instrumente sind heute noch vorhanden, die meisten davon im Germanischen Nationalmuseum in Nürnberg (siehe [Zinner 1967, 471–472]).

Von besonderer Bedeutung für Praetorius wurde seine Begegnung mit dem Humanisten und kaiserlichen Rat Andreas Dudith (1533–1589), den Praetorius im Jahre 1569 in Krakau traf [4] und mit dem er in den folgenden drei Jahren mehrere Reisen unternahm, u.a. nach Prag und Wien. In Krakau traf Praetorius auch mit Rheticus zusammen, der ihm bereitwillig Einsicht in seine eigenen trigonometrischen Arbeiten und in Schriften von verstorbenen Mathematikern und Astronomen des süddeutschen Raumes gewährte, die er besaß. Auf diese Weise lernte Praetorius Arbeiten von Johannes Regiomontanus und von Johannes Werner kennen, die damals noch nicht gedruckt waren. Doppelmayr berichtet, daß Praetorius in dieser Zeit auch Kaiser Maximilian II. in Mathematik unterrichtet hat – vermutlich durch Vermittlung von Dudith

[Doppelmayr 1730, 84]. Praetorius besuchte Dudith noch einmal im Jahre 1575 in Krakau; im folgenden Jahr traf er auf Bitten Dudiths mit dem Kaiser in Regensburg zusammen [5].

1571 wurde Praetorius als Professor der höheren Mathematik nach Wittenberg berufen [6]. Er übernahm dort die Professur von Sebastian Dietrich, der zur medizinischen Fakultät übergewechselt war. In den fünf Jahren, die Praetorius in Wittenberg wirkte, führte er die Tradition von Peucer fort; so las er 1572 über Peucers *Hypotheses astronomicae*, die im Jahr zuvor gedruckt worden waren [7]. Mit Wolfgang Schuler, der zur gleichen Zeit wie Praetorius in Wittenberg Mathematik lehrte, beobachtete Praetorius den 1572 neu erschienenen Stern in der Cassiopeia. Dazu benutzte er zunächst einen von Reinhold stammenden Quadranten und dann einen von ihm selbst angefertigten Dreistab [Zinner 1967, 471]. Praetorius veröffentlichte über die Supernova von 1572 im Jahre 1578 eine lateinische Schrift, die noch im gleichen Jahr ebenfalls in deutscher Sprache herauskam [8]. In dieser Schrift bemerkte Praetorius, weil der Stern wieder verschwunden sei, müsse es doch wohl ein "Meteoron" gewesen sein, vielleicht von etwas reinerer Konsistenz als die Kometen. Tycho Brahe hat die Beobachtungen des Praetorius eingehend gewürdigt und ihn als einen der führenden *mathematici* bezeichnet [9].

1576 verließ Praetorius Wittenberg und wurde Professor an der neuge-gründeten Akademie in Altdorf [10]. Sie steht in der Tradition der huma-nistischen Schulgründungen in den Reichsstädten Straßburg, Nürnberg und Augsburg. Der Nürnberger Rat richtete 1526 im ehemaligen Kloster St. Egidien die "Obere Schule" ein, wobei Melanchthon mitwirkte und seine Vorstellungen einbrachte (siehe hierzu [Müller 1984, 101–102; Schindling 1984, 109–110]). Sie sollte ein Propädeutikum zu den auswärtigen Universitäten anbieten. Da aber die Abgrenzung zu den vier elementaren Lateinschulen der Stadt nicht geklärt war, ergaben sich Probleme, die dazu führten, daß die Schule 1575 nach Altdorf verlegt wurde. Nach dem Vorbild der kursächsischen Fürstenschulen und der Straßburger Schule, die vor allem durch die didaktische Methode ihres Leiters Johann Sturm geprägt war, wurde das Altdorfer Gymnasium hochschulmäßig ausgebaut. Es erhielt durch den Kaiser 1578 das Privileg einer Seminar-Universität mit dem Recht, die Titel eines Baccalaureus und eines Magister artium zu verleihen. Promotionen waren erst seit 1622 möglich, als die Akademie ein volles Universitätsprivileg erhielt. Trotz dieser Einschränkung zählte Altdorf schon bald nach seiner Gründung zu den größeren deutschen Hochschulen: Zur Zeit des Praetorius gab es dort etwa 200 Studenten (1582) und etwa 14 Lehrstühle (um 1600), die sich auf die vier Fakultäten verteilten. In der artistischen Fakultät, in der Praetorius tätig war, hatte er die Stelle des *mathematicus* inne; daneben gab es noch vier weitere Professoren, und zwar den *orator*, den *logicus*, den *ethicus* und den *historicus* [Schindling 1978, 165, 175]. Die Altdorfer Universität wurde 1809 aufgelöst; ihre Sammlungen wurden der 1743 gegründeten Universität Erlangen einverleibt.

Schon kurz nach seinem Ruf nach Altdorf hatte Praetorius vom hessischen Landgrafen das Angebot erhalten, nach Kassel zu kommen und die dortige Sternwarte zu übernehmen. Er hatte es aber vorgezogen, in Altdorf zu bleiben, und so lehrte er dort 40 Jahre lang, bis zu seinem Tode am 27. Oktober 1616, Mathematik und Astronomie. Praetorius bekleidete in dieser Zeit viermal das Amt des Rektors [11].

In der langen Zeit, die Praetorius in Altdorf verbrachte, hat er fast nichts publiziert: Abgesehen von einer größeren Anzahl Prognostiken und Kalendern, die zwischen 1578 und 1614 überwiegend in Nürnberg herausgegeben wurden [12], sowie der bereits genannten Kometenschrift, hat er nur eine mathematische Arbeit über Sehnenvierecke drucken lassen, die 1598 erschien [13]. Ist die Zahl seiner Veröffentlichungen auch gering, so war er doch in Wirklichkeit ein außerordentlich produktiver Wissenschaftler, der sich für die Mathematik und viele naturwissenschaftliche Disziplinen interessierte. Um seine wahre Bedeutung zu ermessen, muß man auch die Handschriften des Praetorius heranziehen, die sich heute in Erlangen, Schweinfurt und München befinden. Meine folgenden Bemerkungen beruhen wesentlich auf seinem handschriftlichen Nachlaß.

2. PRAETORIUS ALS ASTRONOM

Praetorius war mit der Astronomie seiner Zeit bestens vertraut. Betrachtet man die Bibliothek des Saxonius, deren Kern die Bücher des Praetorius bilden, so wird klar, daß sie die wichtigsten astronomischen und mathematischen Schriften des 16. Jahrhunderts enthielt. Zinners Bemerkung trifft zu, daß in Altdorf eine Bibliothek entstand, die in ihrem Umfang damals wohl nur von wenigen Privatbibliotheken übertroffen wurde [Zinner 1943, 425]. Vorhanden waren und sind heute noch Ausgaben aller wichtigen antiken und mittelalterlichen Autoren zur Astronomie, Astrologie und Mathematik, einschließlich der "moderneren" Autoren wie Regiomontanus, Copernicus, Tycho Brahe, Clavius und Maestlin. Eine Reihe von

Büchern wurden im Ausland, vor allem in Paris und Venedig, gedruckt; es gibt auch Texte in italienischer, französischer und holländischer Sprache. Die Sammlung besitzt ebenfalls eine Reihe von Jahrbüchern. Viele Bände enthalten Bemerkungen von Praetorius' Hand, die zeigen, daß er sie teilweise intensiv studierte [14]. Erwähnenswert ist, daß Praetorius zwei Exemplare von Copernicus' Hauptwerk besaß. Eines davon, das ihm um 1560 der Wittenberger Professor Paul Eber schenkte, befindet sich noch in Schweinfurt (siehe [Müller 1993, 226–228, Nr. 66]). Das zweite Exemplar, das heute in der Bibliothek der Yale University in New Haven liegt, stammt aus dem Besitz des Leipziger Professors Johannes Hommel (siehe O. Gingerich, in [Müller 1993, 20]). In beiden Exemplaren finden wir zahlreiche, teils umfangreiche handschriftliche Bemerkungen von den Vorbesitzern Eber und Hommel und auch von Praetorius selbst.

Durch sein Studium in Wittenberg vertraut mit dem geozentrischen System des Ptolemaeus, dem heliozentrischen des Copernicus und dem vermittelnden System von Brahe, hielt Praetorius in Altdorf oft Vorlesungen über Planetentheorie; zahlreiche handschriftliche Entwürfe dazu, meist unter dem Titel *Hypotheses astronomicae* oder *Theoriae planetarum*, sind im Nachlaß vorhanden. In diesen Vorlesungen stellte Praetorius des öfteren die beiden hauptsächlichen Systeme einander gegenüber. Auch wenn diese Handschriften noch nicht im einzelnen ausgewertet sind [15], kann man sagen, daß Praetorius kein bedingungsloser Anhänger des heliozentrischen Systems des Copernicus war. Vielmehr vertrat er – ähnlich wie Rheticus und Reinhold – den Standpunkt, daß das copernicanische System gegenüber dem ptolemäischen theoretische und praktische Vorteile aufwies. Es erschien ihm sinnvoll, das Werk des Copernicus zu studieren, um damit Ungenauigkeiten im ptolemäischen System zu verbessern. Ob Praetorius eigene Beobachtungen anstellte, kann heute nicht mehr geklärt werden; vielmehr versuchte er, mit Hilfe der bekannten Beobachtungswerte Modelle durchzurechnen, die entweder auf der ptolemäischen oder auf der copernicanischen Basis beruhten. Am Ende seiner Betrachtungen zog er jedoch das geozentrische Schema vor. In einer Handschrift aus dem Jahre 1605 bemerkt Praetorius ausdrücklich, es stehe dem Astronomen frei, Kreise, Epizykel und ähnliches zu erfinden, auch wenn es derlei Dinge in der Realität nicht gebe. Wer über die Natur urteilen wolle, der betreibe Physik und nicht Astronomie und könne überhaupt nichts Sicheres aussagen. Man könne durchaus die Positionen nach Ptolemaeus beibehalten und dennoch die in manchen Dingen leistungsfähigere Theorie des Copernicus verwenden (W. Kokott, in [Müller 1993, 357]). Praetorius war also ein Pragmatiker; seine Entscheidung für ein System war weniger durch philosophische Überlegungen als durch technische bestimmt.

Unter diesen Umständen überrascht es nicht, daß Praetorius sich skeptisch zu Keplers Arbeiten äußerte. Als er 1598 durch Vermittlung des bayerischen Kanzlers Herwart von Hohenburg (1553–1622) Keplers *Mysterium cosmographicum* erhielt, in dem Kepler in pythagoreisch-platonischer Art die Bewegung der Planeten durch ein Modell verdeutlichte, bei der die platonischen Körper mit Hilfe von Kugelschalen ineinander verschachtelt sind, ging sein ursprüngliches Interesse in Frustration über, weil in seinen Augen Keplers Darstellung auf Spekulation beruhte und daher physikalisch und nicht astronomisch war [16]. Von daher ist es verständlich, daß Kepler in Praetorius' Bibliothek nur mit einigen weniger bedeutenden Werken vertreten war; Keplers Hauptschrift, die *Astronomia nova* (1609), fehlt völlig.

Praetorius war wahrscheinlich besser als seine Zeitgenossen über die Geschichte des Druckes von Copernicus' Hauptwerk informiert. Er wußte, daß die anonyme Vorrede von Osiander hinzugefügt worden war, ohne daß Copernicus davon erfahren hatte [17]. Es ist sehr wahrscheinlich, daß Praetorius diese Information erhalten hatte, als er Rheticus besuchte. Durch Rheticus lernte er auch bisher ungedruckte Schriften bzw. Briefe von Regiomontanus, Werner und Copernicus kennen, darunter den sogenannten Wapowski-Brief, den Copernicus an den Krakauer Domherrn Wapowski richtete. In dieser Schrift, die in Briefform gekleidet ist, kritisierte Copernicus die Erklärung, die Werner für die Präzessionsbewegung gegeben hatte, scharf.

Bemerkenswert ist auch, daß Praetorius von einem Brief des Regiomontanus wußte, in dem dieser geschrieben hatte, es sei notwendig, die Bewegung der Sterne ein wenig zu ändern wegen der Erdbewegung [18]. Dies könnte ein Hinweis auf die aus dem Arabischen kommende Trepidationslehre sein; es ist aber auch daraus vermutet worden, daß Regiomontanus schon etwa 70 Jahre vor Copernicus von der Bewegung der Erde um die Sonne ausgegangen ist.

In seiner Kometenschrift von 1578 geht Praetorius auch kurz auf das Wesen der Kometen ein [19]. Ganz im Sinne der überkommenen Lehre bezeichnet er sie als "moles irdischer dünst"; er ordnet sie also dem irdischen und nicht dem himmlischen Bereich zu. Die Deutung des Kometen erfolgt nach Ptolemaeus. Der Schweif des Kometen von 1577 ist nach Praetorius' Beobachtungen eher als "Durchleuchtung" denn als Feuer zu betrachten, doch sollten dies die "Physici" beurteilen. Praetorius beklagt, daß er in Ermangelung geeigneter Instrumente keine Positions- und

Parallaxenbestimmungen durchführen konnte. Er fügt ein Verzeichnis historischer Kometen bei und beruft sich dabei auf seinen "lieben Herrn Praeceptor" Camerarius (hierzu siehe Abschnitt 5). Es liegt in der Zeit, daß sich Praetorius auch für Kalenderfragen interessierte. Er erklärte 1596 den neu eingeführten Gregorianischen Kalender [Müller 1993, 355–356, Nr. 172] und besaß auch eine Handschrift des Leipziger Professors J. Hommel über dieses Thema [Müller 1993, 350–351, Nr. 168]. Praetorius verfaßte 1583 eine Arbeit über astronomische Chronologie (jetzt Erlangen, Ms. 815; siehe M. Folkerts in [Müller 1993, 35]) und vermittelte den Druck eines Werkes über die Einteilung der seit Erschaffung der Welt verstrichenen Zeit, das der Theologe Paul Crusius verfaßt hatte [Müller 1993, 346–347, Nr. 163].

Praetorius beschäftigte sich in Altdorf, wie schon in seinen jungen Jahren in Nürnberg, auch mit der Herstellung von Instrumenten für astronomische und mathematische Zwecke. Zwei Astrolabien und eine Armillarsphäre aus seinem Besitz, die allerdings nicht von ihm hergestellt wurden, befinden sich in Schweinfurt [Müller 1993, 361–381]. Seine besondere Vorliebe galt den Sonnenuhren. Ausgehend von einer Würfelsonnenuhr, die außer den gewöhnlichen Stunden auch die Planetenstunden und die Zeichen zeigte, konstruierte er eine vielflächige Sonnenuhr mit ebenen und krummen Flächen. Vermutlich hat er auch die kunstvollen Sonnenuhren an der Universität Altdorf entworfen [Zinner 1967, 83–84, 471–472]. Im Nachlaß gibt es mehrere Handschriften über Sonnenuhren; über dieses Thema hielt er auch Vorlesungen. Praetorius hinterließ auch mehrere Arbeiten und Entwürfe über Herstellung und Gebrauch des Quadranten, jedoch nicht als Autographen [20].

Zuletzt sei noch ein anderer wichtiger Bereich angeschnitten: die Astrologie. Anders als die meisten Astronomen seiner Zeit, war Praetorius ein heftiger Gegner der Astrologie, wie schon die Inschrift zeigt, die sich auf seinem Grabmal befunden haben soll: *Astronomus insignis, vanitatis astrologicae osor acerrimus* [Will 1795, 102]. In einem Brief, den Praetorius 1581 an den Nürnberger Hieronymus Paumgartner schrieb, bezeichnet er die Astrologie als Wahrsagekunst (*divinatrix ars*), die jetzt, nachdem alle früheren Befürworter gestorben seien, die durch Gelehrsamkeit und Wissen herausragten, keinen Fürsprecher mehr habe und sich auf keine Grundlagen stützen könne. Dies habe auch Konsequenzen für die Kalender, aus denen man die astrologischen Prognostiken entfernen müsse [21]. Ein kleines Kuriosum am Rande: In einem Brief, den Dudith an Praetorius am 21. Februar 1589 schrieb, notiert er am Ende, die Sonnenfinsternis vom 15. des Monats sei im Zeichen des Wassermanns aufgetreten, in seinem Tierkreiszeichen, und wenn die Astrologie wahr sei, so kündige das seinen baldigen Tod oder eine schwere Krankheit an [22]. Tatsächlich starb Dudith zwei Tage später. Offenbar hat dieses Ereignis Praetorius nicht veranlaßt, seine Ansicht über die Astrologie zu ändern.

3. PRAETORIUS' BEDEUTUNG FÜR DIE GEODÄSIE; DER MESSTISCH

Neben der Astronomie interessierte sich Praetorius auch für geodätische Fragen. Mit der Vermessung beschäftigen sich zahlreiche Handschriften in seinem Nachlaß. Es gibt Texte darüber, wie man den Abstand zweier Orte bestimmt, die Fläche von Landstücken berechnet und Karten des vermessenen Gebiets anlegt. In einer anderen Arbeit gibt Praetorius eine Methode an, gezeichnete Figuren mechanisch zu verkleinern oder zu vergrößern. Eine Handschrift hat die Fortifikation zum Thema, die mit der Feldmessung eng verbunden ist [23].

Zur Erleichterung der geodätischen Arbeiten ersann Praetorius Hilfsmittel oder verbesserte die gebräuchlichen Instrumente, z.B. den Jakobstab. Mit Hilfe der (schon bekannten) Wasserwaage ließ er eine Wasserleitung nach Altdorf anlegen. Will berichtet darüber [Will 1795, 277–278]: "Der berühmte Prof. Praetorius leitete bald, mit Hülfe seiner Wasserwaage, aus dem benachbarten Pühlheim lebendiges Wasser durch Röhren herein, an welchem es dem Städtgen gänzlich fehlte, so daß Altdorf, welches in trockner Sommerszeit noch immer einigen Mangel an Wasser hat, bei öfters entstandener Feuersgefahr schon längstens würde abgebrannt seyn." Es wird auch berichtet, daß mit Hilfe des Praetorius eine neue Straße von Altdorf nach Nürnberg angelegt worden ist, die sehr viel kürzer als der alte Weg war.

Zur Erleichterung der Messungen im Gelände und der Kartographie ersann Praetorius den Meßtisch, der nach ihm *mensula Praetoriana* genannt wird. Dabei handelt es sich um eine quadratische Holzplatte, die auf einem Stativ befestigt ist. Auf der Platte liegt ein Zeichenblatt. Ein Lineal kann mit Hilfe eines Gleitschuhs an der Kante des Meßtisches entlanggeführt und fest angelegt werden. An seinen Enden befindet sich eine Visiereinrichtung, mit der Punkte im Gelände angepeilt werden können. Man zeichnet auf dem Papier eine Standlinie. Von ihren Endpunkten werden die Zielstrahlen nach Geländepunkten eingezeichnet. Die Schnittpunkte sind die Abbilder der Geländepunkte auf der Karte. So kann man auf dem Meßtisch mit Hilfe des sog. Einschneideverfahrens eine

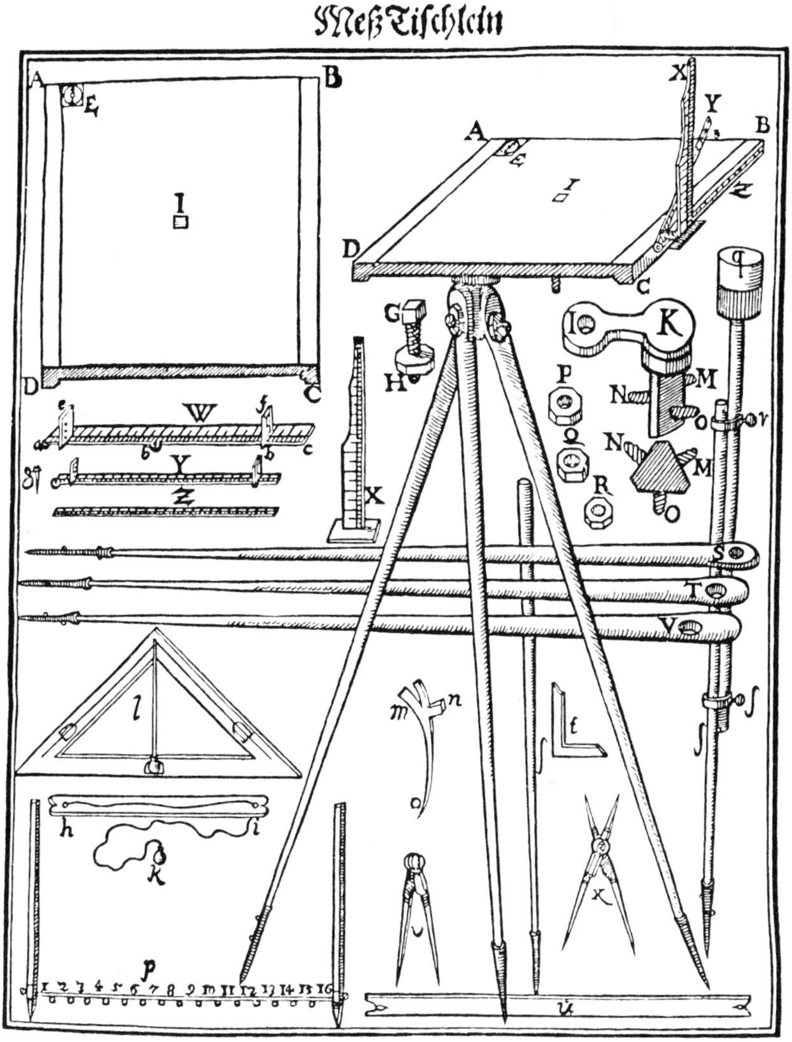

Meß Tischlein

Abbildung 1. Meßtisch von Praetorius (aus Schwenter 1623–1625).

maßstabsgetreue Karte der Landschaft anfertigen, ohne Winkelmessungen durchführen zu müssen. Dieses Gerät, das Praetorius um 1590 erfunden haben soll [Doppelmayr 1730, 86], wurde vor allem durch den Bericht von Daniel Schwenter (1585–1636) bekannt. Schwenter, seit 1602 Schüler des Praetorius, wurde 1608 Professor für Hebräisch in Altdorf und erhielt 1628 als Nachfolger von Praetorius und Saxonius die Professur für Mathematik. 1619 veröffentlichte er eine Schrift über Praetorius' Meßtisch, in der auch die verschiedenen Verwendungsmöglichkeiten angegeben sind [Schwenter 1623–1626]. Das Prinzip des Meßtischs setzte sich schnell durch und ist bis in die Gegenwart in Gebrauch.

4. PRAETORIUS ALS MATHEMATIKER

Neben Astronomie und Geodäsie war ein weiteres großes Gebiet in Praetorius' Schaffen die Mathematik. Im Nachlaß gibt es Handschriften zur Arithmetik, Algebra, Geometrie und Trigonometrie. Sie erlauben Aussagen über die Themen, die Praetorius besonders interessierten [24].

Praetorius schrieb elementare Darstellungen über das Rechnen mit ganzen Zahlen und mit Brüchen. Das Wort *logistica*, das er dafür gebraucht, weist auf das praktische Rechnen der Griechen, die "Logistik", hin, die Praetorius aus byzantinischen Texten vertraut war. Um komplizierte Rechnungen zu erleichtern, die vor allem in der Astronomie nötig waren, benutzte Praetorius Multiplikations- und Divisionstafeln, darunter auch solche, die der bayerische Kanzler Herwart von Hohenburg erstellt hatte und die es ermöglichten, Produkte von je zwei dreistelligen Zahlen durch einfaches Aufschlagen zu ermitteln – die Logarithmen waren ja noch nicht bekannt [25].

In der Tradition der Griechen stehen auch Praetorius' Arbeiten zur Proportionenlehre und zu den verschiedenen Mitteln. Er interessierte sich insbesondere für die Bestimmung einer oder zweier mittlerer Proportionalen zu zwei gegebenen Größen. Im ersten Fall geht es um die Berechnung des geometrischen Mittels, im zweiten um die Ermittlung zweier Werte, die insbesondere zur Lösung des klassischen Problems der Würfelverdopplung herangezogen werden können. Es paßt zu Praetorius, daß er sich vor allem für mechanische Lösungen dieses Problems interessierte, das auch zu seiner Zeit sehr beliebt war. Praetorius befaßte sich auch mit magischen Quadraten, u.a. mit der Herstellung eines Quadrates mit 15 × 15 Feldern nach der Methode von Johannes Faulhaber (1580–1635). Aus seinen Bemerkungen wird klar, daß er auch mit den Arbeiten Michael Stifels (1487?–1567) zu diesem Thema vertraut war.

Praetorius interessierte sich für die Proportionenlehre nicht nur in Verbindung mit den natürlichen Zahlen, sondern insbesondere auch mit ihrer Anwendung auf allgemeine Größen. Er war mit der Theorie der Proportionen vertraut, die Eudoxos entwickelt hatte und die in Buch 5 von Euklids *Elementen* dargelegt wird. Praetorius besaß mehrere Exemplare von Euklids Hauptwerk. Wie andere Mathematiker seiner Zeit inter-

essierte sich Praetorius auch für die Anwendungen dieser Proportionenlehre auf die Theorie der Irrationalitäten, die Euklid in Buch 10 der
Elemente darstellt und die für die Behandlung der regulären Körper in
Buch 13 benutzt wird. Mehrere umfangreiche Schriften in seinem Nachlaß
über das Rechnen mit irrationalen Größen dürften mit Vorlesungen
zusammenhängen, die Praetorius in Altdorf gehalten hat.

Ein wesentliches Teilgebiet der antiken Zahlentheorie, die auf die
Pythagoreer zurückgeht und die z.B. in den Arbeiten von Nikomachos und
Boethius niedergelegt ist, bilden die figurierten Zahlen, bei denen
Steinchen in geometrischer Form dargestellt und Zusammenhänge zwi-
schen den durch sie repräsentierten Zahlen abgeleitet werden. In einer
längeren Abhandlung beschäftigt sich Praetorius vor allem mit den Polygo-
nalzahlen und den drei Mitteln. Er kannte auch die Abhandlung des
Diophant über die Polygonalzahlen, von der er eine lateinische
Übersetzung besaß. Praetorius hatte vor Februar 1572 eine griechische
Handschrift des Diophant durch seinen Freund Dudith erhalten, die an W.
Holtzmann (Xylander) weitergeleitet werden sollte [Allard 1985]. Dieser
wollte den griechischen Text edieren; durch seinen frühen Tod erschien
aber nur eine lateinische Übersetzung (Basel 1575). Durch die Zusendung
der Diophant-Handschrift kannte Praetorius das wichtigste griechische
Werk zur Zahlentheorie und Algebra und war auch mit Methoden vertraut, unbestimmte Gleichungen zu lösen. Praetorius hat eigenhändig die
Sätze der griechisch erhaltenen sechs Bücher der *Arithmetica* des Diophant und Auszüge aus Kommentaren zu diesem Werk abgeschrieben
sowie Sätze behandelt, die mit Diophants Schriften zusammenhängen. Er
war ebenfalls vertraut mit Jordanus Nemorarius' *Arithmetica*, einer wichtigen Arbeit aus dem 13. Jahrhundert, die gleichfalls zahlentheoretische
Themen behandelt, dabei aber oft über die griechischen Vorbilder hinaus-
geht.

Im Praetorius-Saxonius-Nachlaß sind außer den Arbeiten zu Diophant
einige andere Texte vorhanden, die sich explizit mit Algebra beschäftigen.
Praetorius kannte die Algebra der süddeutschen Cossisten und benutzte
ihre Darstellung der Unbekannten und ihrer Potenzen. Eine größere
algebraische Schrift des Praetorius, die auf den Methoden von C. Clavius
beruhte, wurde von Schwenter an die Universität Altdorf gegeben, ist aber
heute nicht mehr nachweisbar [Doppelmayr 1730, 86]. Vorhanden ist eine
Übersetzung der *Algebra* von Petrus Nonius (Nuñez) (Antwerpen 1567)
aus dem Spanischen ins Lateinische, die Praetorius im Jahre 1615 erstellte.
Nonius gehört zu den bevorzugten mathematischen Autoren des Praetorius: er zitiert oft aus seinen Werken. Schließlich befindet sich im Nachlaß

noch ein längerer algebraischer Traktat, der sich mit den Symbolen der
Cossisten, dem Rechnen mit Binomen aus Wurzeln und mit der Lösung
der kubischen Gleichung beschäftigt, wobei auch Cardano erwähnt wird.
Dieser Text wurde allerdings nicht von Praetorius geschrieben.

Wenden wir uns nun der Geometrie zu. Praetorius hatte Interesse an
den regelmäßigen Polygonen und an den regulären Körpern. Aus dem
Jahre 1612 stammt ein Text über das in einen Kreis einbeschriebene
Fünfeck. Besonders wichtig ist Praetorius' 1598 gedruckte Schrift über
Sehnenvierecke, in der er das Problem behandelt, aus vier gegebenen
Seiten ein Viereck zu konstruieren, das einem Kreise einbeschrieben
werden kann. Er wurde dabei durch Arbeiten von Regiomontanus, Simon
Jacob und François Viète angeregt, führte ihre Ansätze aber originell
weiter [26]. Praetorius schrieb auch Luca Paciolis Abhandlung über die
fünf regulären Körper ab, die 1509 erschienen war. Dreiecksberechnungen
scheint Praetorius besonders geliebt zu haben; oft gibt er Lösungen an, die
er bei anderen Autoren (Geber, Regiomontanus, Copernicus, Cardano,
Tartaglia, Stifel, Nonius) gefunden hat. Praetorius beschäftigte sich auch
intensiv mit den Schriften des Archimedes. Kommentare zu seiner
Kreismessung, zur *Sandzahl* und zu den Kegelschnitten, die im Nachlaß
vorhanden waren, sind nicht erhalten. Es gibt aber Auszüge aus
Archimedes' *De sphaera et cylindro*, Aufgaben zur Volumenberechnung
der Kugel und zur Teilung einer Kugel in drei volumengleiche Abschnitte.
Schließlich sind auch Arbeiten zur Bestimmung des Inhalts von Fässern
bzw. zur Herstellung einer dazu geeigneten Visierrute erhalten.

Offenbar hat das Problem der Kreisquadratur Praetorius besonders
fasziniert. In einem dicken Faszikel sind fremde Entwürfe zeitgenössischer
Mathematiker zur Kreismessung zusammen mit Bemerkungen des Praetorius gesammelt [27]. Die Schriften des Leidener Philologen Scaliger, der
behauptete, das Problem der Kreisquadratur ebenso gelöst zu haben wie
die Würfelverdopplung, hat Praetorius intensiv studiert; seine Bemerkungen (hierzu siehe [Müller 1993, 190–193, Nr. 38–40]) zeigen, mit welcher
Leidenschaft er sich gegen die falschen Behauptungen Scaligers wandte—
ganz in Übereinstimmung mit den anderen bedeutenden Mathematikern
seiner Zeit.

Demgegenüber treten Arbeiten über Kegelschnitte zurück. Abgesehen
von den Bemerkungen zu Archimedes, ist nur ein kurzer Text über die
Parabel erhalten, jedoch soll Praetorius auch das wichtigste antike Werk
über die Kegelschnitte, die *Conica* des Apollonios, exzerpiert und
Archimedes' Schrift *De conoidibus et sphaeroidibus* kommentiert haben.

Die Trigonometrie hatte, da sie für astronomische Rechnungen grundlegend war, auch für Praetorius besondere Wichtigkeit. So verwundert es nicht, daß trigonometrische Arbeiten ebenfalls in großer Zahl im handschriftlichen Nachlaß vorhanden sind. Selbstverständlich kannte Praetorius die wichtigsten antiken Schriften zur Kugelgeometrie: die *Sphaerica* des Theodosios und des Menelaos. Über beide Autoren hielt er 1582/1583 in Altdorf Vorlesungen. Die Dreieckslehre des J. Werner war ihm durch Rheticus bekannt. Praetorius besaß auch Abschriften der trigonometrischen Arbeiten und der Tafeln des Rheticus, lange bevor diese gedruckt wurden; Zusätze von Praetorius' Hand zeigen, daß er sie genau gelesen hat. Rheticus' Schriften dürften Praetorius zu eigenen Arbeiten angeregt haben. Im Nachlaß gibt es Texte zur Berechnung rechtwinkliger sphärischer und allgemeiner Dreiecke. Im Jahre 1612 berechnete Praetorius eigene trigonometrische Tafeln, die die Werte der sechs Winkelfunktionen im Sexagesimalsystem angeben [Müller 1993, 360, Nr. 176].

5. PRAETORIUS ALS HUMANIST

Man würde Praetorius nicht gerecht, würde man nur seine Leistungen in der Astronomie, Mathematik und Geodäsie erwähnen. Bedeutend war er auch dadurch, daß er sich intensiv darum bemühte, Texte herbeizuschaffen, die noch nicht gedruckt waren, um sie selbst auszuwerten oder an Freunde weiterzugeben. In diesem Punkt ist Praetorius ein typischer Humanist. Sein Wohnsitz in der Nähe des großen Handelszentrums Nürnberg bot gute Voraussetzungen, an Bücher zu kommen und Gelehrte zu treffen. Praetorius hat von dieser Möglichkeit Gebrauch gemacht. Er hatte Kontakt zu Politikern, vor allem zu dem bayerischen Kanzler Herwart von Hohenburg, der auch an Mathematik und Astronomie interessiert war [28], und zu dem kaiserlichen Rat Dudith, der wohl sein bester Freund war [29], und er nutzte diese Kontakte, um an ungedruckte Bücher zu kommen. Herwart erbat von Praetorius des öfteren Stellungnahmen zu neu erschienenen Büchern, die er zu diesem Zweck Praetorius zukommen ließ, und besorgte ihm auch Abschriften lateinischer und griechischer Texte aus den Bibliotheken in München und Wien.

Der umfangreiche Briefwechsel des Praetorius – er befindet sich heute insbesondere in den Bibliotheken in Berlin, Breslau, München, Nürnberg, Oxford und Paris – läßt erkennen, daß neben Herwart und Dudith vor allem Joachim Camerarius d.J. und Henry Savile zu Praetorius' Freunden zählten.

Der Vater des jüngeren Camerarius, Joachim Camerarius d.Ä. (1500–1574; siehe [Pfeiffer 1977]), wurde durch Melanchthon im Jahre 1526 der erste Leiter der höheren Schule in Nürnberg. 1535 erhielt er eine Professur für Gräzistik in Tübingen und ging 1541 an die Universität Leipzig. 1565 verfaßte er ein Gutachten über die Verlegung des Nürnberger Egidiengymnasiums, das zehn Jahre später bei der Einrichtung der Altdorfer Akademie zum Tragen kam. Sein Sohn, Joachim d.J. (1534–1598), war wie der Vater ein echter Humanist. Er ließ sich 1564 als Arzt in Nürnberg nieder und hatte auch als Botaniker internationalen Ruf; in einer Ausstellung, die die Universitätsbibliothek Erlangen für 1995 plant, soll das Wirken von Joachim d.J. auf dem Gebiet der Botanik gewürdigt werden [Wickert 1993]. Der jüngere Joachim führte eine umfangreiche Korrespondenz mit Naturwissenschaftlern in ganz Europa. Für Praetorius war er die wichtigste Kontaktperson in Nürnberg; 24 Briefe an Camerarius aus den Jahren 1577–1597 sind erhalten [30].

Praetorius war auch mit Henry Savile (1549–1622) befreundet [31]. Savile erwarb im Jahre 1570 am Merton College in Oxford den Grad eines M.A. und bewies bei dieser Gelegenheit seine guten Kenntnisse in der griechischen Sprache und in der Astronomie durch Vorlesungen über Ptolemaeus' *Almagest*. Hierzu benutzte er eine lateinische Übersetzung von Ptolemaeus' Schrift und von Theons Kommentar dazu, die er 1568 selbst angefertigt hatte; sie ist heute noch in Oxford erhalten [32]. Im Jahre 1578 unternahm er eine Reise auf den Kontinent und traf dort mit zahlreichen Gelehrten zusammen, vermutlich auch mit Praetorius. In der Folgezeit beschäftigte er sich hauptsächlich damit, Handschriften griechischer Autoren zu besorgen. Einige Codices kopierte er selbst, darunter auch Schriften antiker Mathematiker und Astronomen [33]. Savile war später Vorsteher am Merton College und in Eton. In den Jahren nach 1604 widmete er sich vor allem der griechischen Literatur und edierte Schriften des Chrysostomos und Xenophon, war aber weiterhin auch den mathematischen Wissenschaften verbunden. 1621 erschienen seine Vorlesungen über Euklids *Elemente*. Von seinen engen Beziehungen zu Praetorius zeugen dessen Briefe aus den Jahren 1580–1595, die sich in der Bodleian Library befinden [34]. Praetorius besorgte für Savile die Verbindungen zu anderen Gelehrten auf dem Kontinent; er spielte eine Rolle bei der griechischen Erstausgabe des Geminos [35] und übergab Savile auch eine lateinische Handschrift mit optischen Texten [36].

Meist durch Vermittlung der genannten Personen, erhielt Praetorius Drucke oder Abschriften von schwer zugänglichen mathematischen oder astronomischen Texten. Zu den mathematischen Schriften, die Praetorius selbst kopierte, gehören die mathematischen Werke von François Viète, den er hoch schätzte, und die Schrift des P. Nonius über die Irrtümer von Orontius Finaeus. In der Astronomie schrieb Praetorius J. Aurias Übersetzungen der Werke des Autolykos und Theodosios ab. 1572 kopierte er in Wittenberg Auszüge aus Fracastoros Werk über die homozentrischen Sphären. Schließlich besaß Praetorius Abschriften bzw. Exzerpte von Maurolicos Kosmographie und von seinen mathematischen Schriften, unter denen sich auch Maurolicos Übersetzungen von Werken des Theodosios, Menelaos, Autolykos und Euklid befanden. Als sich Praetorius 1570 in Krakau bei Rheticus befand, schrieb er 19 Sätze aus Maurolicos *Sphärik* ab [37].

6. SCHLUSSBEMERKUNGEN

Die vorstehenden Ausführungen dürften einen Eindruck von dem breiten Spektrum der Arbeiten des Praetorius vermittelt und klargemacht haben, daß er zu den bedeutenden Mathematikern und Astronomen des 16. Jahrhunderts gerechnet werden muß. Es ist an der Zeit, daß sich die Wissenschaftshistoriker mit dem Inhalt des umfangreichen Nachlasses, mit seinen Handschriften und seinen Bucheintragungen näher beschäftigen und seine Rolle innerhalb des Kreises der Humanisten herausarbeiten. Das Material dazu ist bereitgestellt.

ANMERKUNGEN

[*] Überarbeitete Fassung eines Vortrags, der am 22. November 1993 in Schweinfurt anläßlich der Eröffung der Copernicus-Ausstellung gehalten wurde.

[1] Hierzu siehe außer [Doppelmayr 1730, 83–90] und [Apinus 1728, 11–31] etwa [Günther 1888; Zinner 1967, 471–472]; Georg Drescher, in [Müller 1993, 142–143].

[2] Der Eintrag in der Matrikel lautet [Foerstemann 1841, 329]: *Johannes Praetorius Vallensis, Mathesii affinis*. Die grundlegende Darstellung zur Geschichte der Universität Wittenberg ist [Friedensburg 1917].

[3] Im "Philosophischen Dekanatsbuch" wird unter XIII. Cal. Mai. 1562 (= 19.4.1562) als Magister "Iohannes Praetorius Hertzbergensis" genannt (Universitätsarchiv Halle-Wittenberg, Rep. 1, Nr. XXXXV, 1, Bd. 2, S. 175). Ein "Ioannes Maior Ioachimus, ex valle Ioachimica", der am 27.2.1556 Magister wurde (ebda., S. 143), ist nicht mit Johannes Praetorius identisch.

[4] So in einem Brief von Praetorius an G. M. Lingelshemius (Paris, BN, Dupuy 348, f.156r): *Non diu post has nuptias Cracoviam veni an. 1569 et ei* (= Dudith) *in ocio degenti triennium fere inservivi.*

[5] So Praetorius in dem in Anm. 4 erwähnten Brief.

[6] Im "Philosophischen Dekanatsbuch" (siehe Anm. 3) wird unter den Magistern, *quibus tributus est locus in collegio*, auch Praetorius erwähnt: *Decano M. Alberto Lemeigero Hamburgensi* (= 1571/1572) *collegio Philosophico adscripti sunt, privatim, M. Johannes Praetorius Joachimus postulantibus et iubentibus coeteris collegis, quod ipsi mandato illustrissimi Electoris Augusti, professio Mathematum publica, demandata esset.* ... (S. 757). Unter dem 4.3.1572 ist verzeichnet: *Iohannes Praetorius Ioachimus Mathematum professor publicus* (S. 249).

[7] Ein Bruchstück dieser Vorlesung befindet sich in Wien, ÖNB, Cod. 10641, f.1r–11r.

[8] *De cometis, qui antea visi sunt, et de eo qui novissime mense Novembri apparuit, narratio,* Nürnberg 1578; deutsche Fassung: *Narratio oder historische Erzelung dern Cometen so vor dieser Zeit sind gesehen worden vnd dann auch dessen, der jüngst erschienen ist,* Nürnberg 1578. [Zinner 1941, Nr. 2845 und 2846.]

[9] Brahe geht ausführlich auf Praetorius' Schrift über die Erscheinung von 1577 ein [Brahe 1916, III, 153–157, 1918, IV.1, 356–358] und bezeichnet Praetorius im Zusammenhang damit als *clarissimus mathematicus, excellens mathematicus, eruditissimus, doctissimus mathematicus* und als *clarissimus et eruditissimus mathematicus.*

[10] Zur Akademie in Altdorf siehe [Will 1795; Recktenwald 1966; Schindling 1978, 159–161, 1984, 115–116].

[11] 1584/1585, 1592/1593, 1596/1597, 1604/1604; siehe [Will 1795, 36–38; Recktenwald 1966, 34].

[12] Zinner erwähnt Drucke für die Jahre 1578–1581, 1587, 1588, 1590, 1593–1598, 1606, 1608, 1614 [Zinner 1941, Nr. 2847, 2897, 2954, 3006, 3271, 3311, 3398, 3549, 3586, 3630, 3677, 3737, 3789, 4116, 4201, 4481].

[13] *Problema, quod iubet ex quatuor rectis lineis datis quadrilaterum fieri, quod sit in Circulo. Aliquot modis explicatum a Johan. Praetorio Ioachimico.* Nürnberg: Fuhrmann, 1598.

[14] Näheres dazu in den Beschreibungen in [Müller 1993].

[15] Die einzige ausführliche Arbeit ist [Westman 1975, 289–305].

[16] Brief an Herwart; abgedruckt in [Kepler 1945, 205–206].

[17] Den Sachverhalt drückt er sehr deutlich in einem Brief an Herwart aus dem Jahre 1609 aus (teilweise abgedruckt in [Zinner 1943, 454]).

[18] *necesse est motum stellarum paululum variari propter motum terrae.* Praetorius zitiert diese Bemerkung Regiomontans in einer Vorlesung über die Planetentheorie, die er 1613 hielt (siehe [Zinner 1968, 280–281]).

[19] Die folgenden Informationen verdanke ich Herrn Dr. Wolfgang Kokott.

[20] Siehe hierzu die zusammenfassenden Bemerkungen von M. Folkerts in [Müller 1993, 30] mit Hinweis auf die Beschreibungen in [Müller 1993, 34–50].

[21] Nürnberg, Stadtbibliothek, Autogr. 1959.

[22] *Eclip⟨sis⟩ ⟨solis⟩ 15. huius mensis, incidit in ⟨Aquarium⟩, qui meus horoscop⟨us⟩ est, Si vera est Astrologia, mortem mihi affert aut graviss⟨imum⟩ aliquem morbum, tu quid censes?* Der Brief befindet sich in Paris, BN, Dupuy 348, f.154r-v.

[23] Näheres hierzu bei M. Folkerts in [Müller 1993, 30–31] mit Hinweis auf die Beschreibungen in [Müller 1993, 34–50].

[24] Der folgende Abschnitt entspricht M. Folkerts in [Müller 1993, 31–33]; dort werden die einzelnen Aussagen durch Verweis auf die Texte in Praetorius' Handschriften [Müller 1993, 34–50] belegt.

[25] Das Werk hatte ein Format von 52 × 27 cm und war über 10 cm dick; siehe [Cantor 1900, 721–722].

[26] Zu dieser Schrift, deren Titel in Anm. 13 angegeben ist, siehe M. Folkerts in [Müller 1993, 196–197, Nr. 43].

[27] Folgende Autoren sind vertreten: Ludolph van Keulen, Christoph Hutzler, Jacob Falco, Thomas Gephirandus, Joseph Scaliger, Simon a Quercu (Duchesne).

[28] Fünf Briefe von Praetorius an Herwart befinden sich in der UB München (Ms. 2° 692, Nr. 47, 119, 128, 130, 166), zwei weitere in der BSB München (Clm 1607, f.94r–95v; Clm 1608, p. 799–806).

[29] Einige Briefe Dudiths an Praetorius sind ediert bei [Apinus 1728, 25, 26–30] (siehe [Doppelmayr 1730, 84]). In der Universitätsbibliothek Breslau befindet sich ein Konvolut von Praetorius' Briefen an Dudith. Zu Dudiths Bedeutung für die Mathematik siehe [Allard 1985].

[30] Die meisten Briefe an Camerarius befinden sich in der BSB München (Clm 10362, 10365). Einzelne Briefe, die ursprünglich dieser Sammlung angehörten, liegen heute in Nürnberg, Germanisches Nationalmuseum (Archiv, Autographen, K 34: 2 Briefe) bzw. Stadtbibliothek (Autogr. 1721: 1 Brief) und in Berlin (Staatsbibliothek, Slg. Darmstaedter, F 2a 1570 Praetorius: 2 Briefe).

[31] Zu Savile siehe allgemein [Dictionary 1897, 367–370]. Über seine Leistungen als Mathematiker informiert [Mogenet 1985, 85].

[32] Bodleian Library, MS. Savile 26–28. Siehe [Madan and Craster and Denholm-Young 1937, 1108–1109].

[33] Heute Oxford, MS. Savile 5, 6, 9–11. Die Handschriften, die Savile 1581 kopierte, enthalten Texte von Geminos, Ptolemaeus, Diophant mit Scholien des Planudes, Argyros, Barlaam, Theon von Smyrna, Theodosios, Euklid, Pappos, Autolykos, Aristarch, Hypsikles, Apollonios, Triklinios und Philoponos [Gamillscheg und Harlfinger 1981, 78–79].

[34] MS. Savile 108, f.104–117.

[35] Siehe hierzu M. Folkerts in [Müller 1993, 332–333].

[36] Ptolemaeus, Optik; Alkindi, *De aspectibus*; heute Oxford, BL, Savile 24.

[37] Zu den Handschriften, die Praetorius kopierte, siehe M. Folkerts in [Müller 1993, 33]; dort wird auf die Beschreibungen in [Müller 1993, 34–50] hingewiesen.

LITERATURVERZEICHNIS

Allard, A. 1985. Le manuscrit des arithmétiques de Diophante d'Alexandrie et les lettres d'André Dudith dans le Monacensis lat. 10370. In *Mathemata. Festschrift für Helmuth Gericke*, M. Folkerts und U. Lindgren, Hrsg. S. 297–315. Stuttgart: Franz Steiner.

Apinus, J. 1728. *Vitae professorum philosophiae qui a condita Academia Altorfina ad hunc usque diem claruerunt.* Nürnberg/Altdorf: Johann Daniel Taubers Erben.

Brahe, T. 1916. *Tychonis Brahe Opera omnia*, Tomus III. Kopenhagen: Gyldendal.

Brahe, T. 1918. *Tychonis Brahe Opera omnia*, Tomus IV.1. Kopenhagen: Gyldendal.

Cantor, M. 1900. *Vorlesungen über Geschichte der Mathematik*, 2. Band von 1200–1668. 2. Aufl. Leipzig: Teubner.

Dictionary. 1897. *Dictionary of National Biography* Vol. 50. London: Smith, Elder, & Co.

Doppelmayr, J. G. 1730. *Historische Nachricht von den Nürnbergischen Mathematicis und Künstlern . . .* Nürnberg: Peter Conrad Monath.

Foerstemann, C. E., Hrsg. 1841. *Album academiae Vitebergensis ab a. Ch. MDII usque ad a. MDLX.* Leipzig: Carl Tauchnitz.

Friedensburg, W. 1917. *Geschichte der Universität Wittenberg.* Halle: Niemeyer.

Gamillscheg, E., und Harlfinger, D. 1981. *Repertorium der griechischen Kopisten 800–1600.* 1. Teil: *Handschriften aus Bibliotheken Großbritanniens, A: Verzeichnis der Kopisten.* Wien: Verlag der Österreichischen Akademie der Wissenschaften.

Günther, S. 1888. Johannes Praetorius. *Allgemeine deutsche Biographie* **26**, 519f.

Kepler, J. 1945. *Gesammelte Werke.* Bd. 13: *Briefe 1590–1599*, M. Caspar, Hrsg. München: C. H. Beck.

Madan, F., Craster, H. H. E., and Denholm-Young, N. 1937. *A Summary Catalogue of Western Manuscripts in the Bodleian Library at Oxford*, Vol. II, Part II. Oxford: Clarendon Press.

Mogenet, J. 1985. *Le "Grand Commentaire" de Theon d'Alexandrie aux tables faciles de Ptolémée. Livre I. Histoire du texte, édition critique, traduction, revues et complétées par Anne Tihon. Commentaire par Anne Tihon* (= Studi e testi, 315). Città di Vaticano: Biblioteca Apostolica Vaticana.

Müller, G. 1984. Philipp Melanchthon zwischen Pädagogik und Theologie. In *Humanismus im Bildungswesen des 15. und 16. Jahrhunderts*, W. Reinhard, Hrsg.

(Mitteilung XII der Kommission für Humanismusforschung), S. 95–106. Weinheim: Verlag Chemie.

Müller, U. 1993. *450 Jahre Copernicus "De revolutionibus".* *Astronomische und mathematische Bücher aus Schweinfurter Bibliotheken* (Veröffentlichungen des Stadtarchivs Schweinfurt), U. Müller, Hrsg., Nr. 9. Schweinfurt: Stadtarchiv.

Pfeiffer, G. 1977. Joachim Camerarius d.Ä. In *Fränkische Lebensbilder*, G. Pfeiffer und A. Wendehorst, Hrsg., Bd. 7. S. 97–108. Würzburg: Schöningh (in Kommission).

Recktenwald, H. C., Hrsg. 1966. *Gelehrte der Universität Altdorf.* Nürnberg: Lorenz Spindler.

Schindling, A. 1978. Straßburg und Altdorf – Zwei humanistische Hochschulgründungen von evangelischen freien Reichsstädten. In *Beiträge zu Problemen deutscher Universitätsgründungen der frühen Neuzeit*, P. Baumgart und N. Hammerstein, Hrsg. (Wolfenbütteler Forschungen), Bd. 4, S. 149–189. Nendeln: KTO Press.

— 1984. Die humanistische Bildungsreform in den Reichsstädten Straßburg, Nürnberg und Augsburg. In *Humanismus im Bildungswesen des 15. und 16. Jahrhunderts*, Wolfgang Reinhard, Hrsg. (Mitteilung XII der Kommission für Humanismusforschung), S. 107–120. Weinheim: Verlag Chemie.

Schwenter, D. 1623–1626. *Mensula Praetoriana. Beschreibung deß nutzlichen Geometrischen Tischleins / von dem fürtrefflichen und weitberühmten Mathematico M. Johanne Praetorio S. erfunden* (= Daniel Schwenter. *Geometriae practicae novae et auctae tractatus III.*). Nürnberg: Jeremias Dümler. (Erstauflage 1619.)

Westman, R. S. 1975. Three responses to the Copernican Theory: Johannes Praetorius, Tycho Brahe, and Michael Maestlin. In *The Copernican Achievement*. R. S. Westman, Ed., pp. 285–345. Berkeley: University of California Press.

Wickert, K. 1993. *Das Camerarius-Florilegium.* Erlangen: Universitätsbibliothek. (Kulturstiftung der Länder, Patrimonia 61)

Will, G. A. 1795. *Geschichte und Beschreibung der Nürnbergischen Universität Altdorf.* Altdorf: Akademische Monath-Kußlerische Buchhandlung (in Kommission).

Zinner, E. 1941. *Geschichte und Bibliographie der astronomischen Literatur in Deutschland zur Zeit der Renaissance.* Leipzig: Karl W. Hiersemann.

— 1943. *Entstehung und Ausbreitung der copernicanischen Lehre.* Erlangen: Max Mencke (in Kommission) (*Sitzungsberichte der Physikalisch-medizinischen Sozietät zu Erlangen*, **74**), 2. Aufl. 1988. München: C. H. Beck.

— 1967. *Deutsche und niederländische astronomische Instrumente des 11.–18. Jahrhunderts*, 2. Aufl. München: C. H. Beck.

— 1968. *Leben und Wirken des Joh. Müller von Königsberg genannt Regiomontanus*, 2. Aufl. Osnabrück: Otto Zeller.

The Transmission and Evolution of Ideas:
Various Traditions

Einiges über die Handschrift Leiden 399, 1 und die arabisch-lateinische Übersetzung von Gerhard von Cremona

Hubert L. L. Busard

Herungerstraat 123, NL-5911 AK Venlo, The Netherlands

There are two versions of a commentary on the *Elements* of Euclid attributed to one Anaritius. One, in Arabic, is combined with Euclid's *Elements* I–VII.Def.7 in MS Leiden 399.1 (= **L**); the second is known only in the Latin translation by Gerard of Cremona (= **A**). Some differences between the two manuscripts are discussed, but these do not suffice to establish that there were two different authors. Next the question is considered: which version does **L** contain? Some proofs given in **L**, and the same proofs given in the texts of Adelard of Bath and Gerard of Cremona, are compared. From this it follows that **L** is a result of the combination of various elements, and certainly not an unadulterated version of al-Ḥajjāj II, as was previously assumed. In the second part of this paper, the question of who may have been responsible for the alternative proofs in the translation by Gerard of Cremona is discussed.

I

Es gibt zwei Fassungen eines Kommentars zu den *Elementen* Euklids, die einer Person Abu l-ʿAbbās al-Faḍl b. Ḥātim an-Nairīzī, der im Abendland unter der latinisierten Form seines Namens als Anaritius bekannt ist, zugeschrieben werden. Die eine, arabische, Fassung ist zusammen mit Euklids *Elementen* I–VII. Def. 7 enthalten im Manuskript Leiden, Or. 399/1, ff. 1–81 (= **L**) und in einer zweiten, erst kürzlich bekannt gewordenen, Handschrift[1]. Der Anfang des arabischen Textes von **L** wurde mit einer lateinischen Übersetzung zunächst von R. O. Besthorn und J. L. Heiberg veröffentlicht [Besthorn und Heiberg 1893ff., I–III.1]; der Rest wurde von G. Junge, J. Raeder, und W. Thomson ediert [Besthorn und Heiberg 1893ff., III.2]. Die zweite Fassung des Nairīzī-Kommentars ist enthalten in der lateinischen Übersetzung, die Gerhard von Cremona

anfertigte und von der die ersten 10 Bücher erhalten sind; sie wurde von
M. Curtze ediert ([Curtze 1899] = **A**). Bevor ich näher auf den Verfasser
eingehe, möchte ich einige Bemerkungen vorausschicken:

1. In **L** fehlen nach Angabe des Herausgebers (G. Junge, in [Besthorn
   und Heiberg 1893ff., III.2, 208]) 6 Folioblätter, so daß von den
   Definitionen des Buchs I nur Def.1 und der Anfang des Kommentars
   zu Definition 1 erhalten ist. Die Handschrift fährt fort mit dem
   Kommentar zu der letzten Definition, d.h. außer Def.1 fehlen alle
   Definitionen von Buch I [2].
2. Die Edition von Curtze stützt sich auf nur eine Handschrift. Inzwischen sind noch zwei weitere Handschriften bekannt geworden, die
   den Text enthalten: Madrid, Bibl. Nac. 10010, ff.13v−50r, und Vatikan,
   Biblioteca Apostolica, Reg. lat. 1268, ff.144r−207v. Folglich ist die
   Edition von Curtze nicht sehr zuverlässig [3].
3. Die Zahlenbeispiele gehören nicht zum Euklid-Text [Klamroth 1881,
   310, 314, Murdoch 1971, 439, De Young 1991, 662]. Dies ergibt sich
   nicht nur daraus, daß diese Beispiele in den Text von **A** aufgenommen worden sind, sondern auch aus der Tatsache, daß in **L** bei
   V.Def.10 und VI.25, 26, 28, 29 dazu bemerkt wird: *Commentator dixit*
   bzw. *al-Narîzî dixit*. In II.1 steht am Anfang: "Hinzufügung," und in
   II.3−5 steht diese Bemerkung als Interlinearglosse. Nur in II.2 gibt es
   keine Bemerkung. Weitere Zahlenbeispiele sind in **L** nicht vorhanden.

Die beiden Fassungen des Kommentars sind sicher nicht unabhängig
voneinander entstanden. Es gibt aber Unterschiede zwischen **L** und **A**:

1. Vor dem Beginn des Buchs VII lesen wir in **A** [Curtze 1899, 190]:
   *Ipsius tamen elementa non reperi, neque primum theorema, quoniam
   deerat primum folium in exemplari.* **L** enthält von Buch VII nur die
   ersten 7 Definitionen ohne Kommentar und bricht dann ab. Bei
   Definition 1 ist hinzugefügt: *Principium est numerorum.* Dieser
   Zusatz fehlt in allen mir bekannten Texten ([De Young 1981, 4, 285,
   564]; Adelard [Busard 1983a, 196]; Gerhard [Busard 1983b, 165]). Er
   bezieht sich auf eine Bemerkung im Kommentar zu der letzten
   Definition von Buch I in **L** und **A**: *Punctum principium est magnitudinum continuarum, unitas discretarum* (**L**, [Besthorn und Heiberg
   1893ff. I, 13]; siehe auch [Tummers 1984, II, 145]). Da der Kommentar in allen anderen Büchern entweder nach Definition 1 (I, II, III,
   VI) oder nach Definition 2 (IV, V) anfängt, muß man annehmen,

daß der Verfasser von **L** in **A** keinen Kommentar vorgefunden hat und daß die Bemerkung in **A** von Anaritius herrührt. Da **A** sich nur auf eine, **L** aber auf mehrere Fassungen stützt, mag das ein Indiz dafür sein, daß **L** nach **A** geschrieben wurde.

2. Es gibt viele Stellen sowohl im Euklidtext als auch im Kommentar, an denen **A** das Pronomen "ich," **L** aber das Pronomen "wir" verwendet.

3. Ein Abschnitt in **A** [Curtze 1899, II.12, 108 *Yrinus*, bis S. 109 *expansum*] fehlt in **L**. Dieser Text enthält die Erklärung, warum Heron die Umkehrungen der Sätze II.12 und 13 gibt: *Inquit Yrinus: Nos vero in hac figura faciemus, quod Euclides in prima parte (I.48) fecit, et ostendemus istud in hac figura et in figura, que sequitur eam.* Statt in **A**, S. 109: *nos itaque ostendemus, quod* heißt es in **L**, S. 71: *Additamentum. Hero dixit.*

4. Entsprechend der Aussage *Hero dixit: Contactum ante sectionem posuimus quia contactus sectione prior est* (**L**, III.5 [Besthorn und Heiberg 1893ff., II.2, 19]) hat **L** die sonst übliche Reihenfolge der Sätze III.5 und 6 geändert. Diese Bemerkung fehlt in **A**, aber **A** bringt zu III.1−6 Bemerkungen, die in **L** fehlen.

5. Der Satz III.22 (**A** [Curtze 1899, 134]) befindet sich in **L** am Rande und wird dort eingeleitet mit: *al-Narizi dixit* (**L** [Besthorn und Heiberg 1893ff., II.2, 91]).

6. Der letzte Satz des Kommentars zu Buch IV wird in **A** [Curtze 1899, 155] eingeleitet mit: *Dixit Euclides.* Diese Einleitung fehlt in **L**. Der Satz rührt aber nicht von Euklid her, sondern von Heron.

7. Der erste Zusatz zu V.18 in **L** [Besthorn und Heiberg 1893ff., III.2, 73]: *Propositio addita propositioni XVIII libri quinti* fehlt in **A**. **L** enthält zwei Zusätze. Den zweiten Zusatz: *Secunda propositio addita propositioni XVIII* enthält auch **A** [Curtze 1899, 172].

8. Der Zusatz zu V.21 in **L** [Besthorn und Heiberg 1893ff., III.2, 81] fehlt in **A**.

9. Im Kommentar zu V.23 heißt es in **L** u.a. [Besthorn und Heiberg 1893ff., III.2, 87]: *Duae propositiones priores, 20 et 21, probari non possunt, si plures sunt quam tres magnitudines; etiamsi vero fieri posset, non multum proficeretur.* Obwohl **L** im Kommentar zu V.23 mit **A** übereinstimmt, spricht **L** in den Enuntiationen der Sätze V.20−23 in Übereinstimmung mit Isḥāq-Ṯābit von beliebig vielen Größen. Im griechischen Euklid wird nur in V.22 von beliebig vielen Größen gesprochen und bei Adelard an allen Stellen von nur drei Größen.

10. Nach dem Beweis des Satzes VI.23 [Besthorn und Heiberg 1893ff., III.2, 163] wird in **L** noch ein zweiter Beweis gegeben, der eingeleitet wird mit: *Est tamen facilior demonstratio, quae sequitur*. Dieser Beweis ist falsch und unverständlich, denn die Aussage (S. 165): *Et omnes figurae, quarum anguli aequales sunt, latera angulos aequales comprehendentia proportionalia habent, per propositionem 4 libri sexti*, steht im Widerspruch zur Aussage am Ende von VI.20 (S. 155): *al-Narîzî dixit: Triangulus constructus est, quia, si anguli triangulorum aequales sunt, latera proportionalia sunt; quod non sequitur in parallelogrammis*. Hieraus kann man nur schließen, daß der zweite Beweis nicht von Anaritius herrührt. Für diese Vermutung spricht auch, daß der Beweis in **A** fehlt.

11. In **A** heißt es [Curtze 1899, 178]: *Nostra autem additio in hoc est huius⟨modi⟩*; in **L** [Besthorn und Heiberg 1893ff., III.2, 145] dagegen: *Al-Narîzî dixit*.

Nach diesen Bemerkungen stellt sich für mich die Frage: War Anaritius wirklich der Verfasser des Kommentars, der in **L** enthalten ist, oder stammt dieser Kommentar von einer anderen Person, die den Text, der in **A** erhalten ist, kannte und ausgiebig benutzte? Diese Frage kann zur Zeit nicht beantwortet werden, da die arabische Vorlage von **A** nicht bekannt und **A** selbst unzuverlässig ist.

II

Die nächste Frage lautet: Welche Version enthält **L**?

1. Schon Junge hat bemerkt, daß nach der Enuntiation des Satzes VI.25 [Besthorn und Heiberg 1893ff., III.2, 167] in den Text eine Glosse aufgenommen ist: *Sequuntur haec in textu, quod sine dubio glossema est*, und eine andere in VI.28 [Besthorn und Heiberg 1893ff., III.2, 179]: *Quae sequuntur, videntur esse glossema in textum insertum, vel potius duo glossemata*.

2. Einen anderen Zusatz gibt es nach der Enuntiation von III.34 [Besthorn und Heiberg 1893ff., II.2, 121]: *Sex sunt huius sectionis rationes, aut ut sectio utriusque lineae in centro sit, aut ut altera per centrum ducta sit et alteram in duas partes aequales secet et ad angulos rectos, aut ut altera per centrum ducta sit, sed alteram in duas partes aequales non secet, aut ut neutra per centrum ducta sit et altera alteram*

in duas partes aequales secet, aut ut neutra per centrum ducta sit et altera alteram nec in duas partes aequales nec ad angulos rectos secet, aut ut neutra per centrum ducta sit nec altera alteram in duas partes aequales secet, sed inter se ad angulos rectos secent. Ponimus igitur sex deinceps casus difficiles, I, II, III, IV, V, VI, et sex circuli sint, et in singulis A B G D. Auch dieser Zusatz ist meines Erachtens eine Glosse, die nicht zum Text gehört. Dasselbe gilt auch für den Zusatz in der Enuntiation III.35 (S. 135): *Et haec [propositio] in tres casus dividitur, cum recta secans aut per centrum ducitur aut per semicirculum inter centrum rectamque circulum contingentem positum aut per alterum semicirculum.* Im griechischen Euklid kann dieser Text nicht gestanden haben, da dort nur ein Fall behandelt wird.

Wahrscheinlich ist auch der Anfang des Beweises in III.32 [Besthorn und Heiberg 1893ff., II.2, 115] ein Zusatz des Verfassers: *Exemplificatio: Linea A B est linea data, angulus datus rectus angulus G D E, obtusus angulus* HΘK, *acutus angulus* NΞO. *Demonstrare volumus, quo modo in linea* A B *segmentum circuli erigamus, quod capiat angulum angulo* G D E *aequalem, deinde segmentum, quod capiat angulum angulo* HΘK *aequalem, deinde segmentum, quod capiat angulum angulo* NΞO *aequalem.*

3. In die Enuntiation III.15 [Besthorn und Heiberg 1893ff., II.2, 69] ist ein Zusatz aufgenommen, der nicht zu Euklid gehört: *omnesque lineae, quarum positio haec est, circulum contingunt.*

4. Wir wollen die in Buch IV von **L** gegebenen Definitionen mit denen von Gerhard und Adelard vergleichen.

L:

Euclides dixit: Figura in figura descripta dicitur, ubi anguli figurae intrinsecus descriptae latera figurae exterioris tangunt.

Figura vero extra figuram descripta dicitur, ubi singula latera [figurae] extrinsecus descriptae singulos angulos figurae intra descriptae tangunt.

Si figura in figura est, et latera figurae exterioris angulos figurae interioris tangunt, exterior interiorem comprehendere dicitur [4].

Gerhard:

[i] *Figura intra figuram signari dicitur cum unusquisque angulorum ipsius contingit unumquodque ex lateribus figure intra quam ipsa signatur.*

[ii] *Figura circa figuram dicitur signari cum unumquodque ex lateribus eius contingit unumquemque ex angulis figure circa quam ipsa signatur.*

Adelard:

Cum figura intra figuram contenta fuerit angulique contente continentis latera tetigerint, exterior quidem respectu interioris continens dicitur.
Der Vergleich zeigt, daß die beiden ersten Definitionen in **L** mit Gerhard und die dritte mit Adelard übereinstimmen.

5. Am Ende von III.1 [Besthorn und Heiberg 1893ff., II.2, 11] lesen wir: *Hac nostra de centro circuli demonstratione simul demonstratum est, si chorda aliam in duas partes aequales et ad rectos angulos secet, in ea centrum circuli positum esse. Q.n.e.d. Demonstratum est, fieri non posse; ut in circulo chorda chordam in duas partes aequales et ad rectos angulos secans per centrum circuli non transeat.* Im griechischen Euklid steht an dieser Stelle [Busard 1987, 68]: *Ex hoc ergo manifestum, quoniam si in circulo recta aliqua rectam aliquam in duo equalia et ad rectos secat, in secante est centrum circuli.* Gerhard hat: *Et ex hoc manifestum est quod non est possibile ut sint due linee in circulo quarum una aliam in duo media secet super rectos angulos quin linea que aliam in duo media secat super rectos angulos per centrum transeat circuli.* Der Vergleich dieser drei Texte zeigt, daß es zwei Lesarten in **L** für dasselbe Korollar gibt, und zwar stimmt die erste mit dem griechischen Euklid, die zweite mit Gerhard überein.

6. In IV.5 steht (**L**, [Besthorn und Heiberg 1893ff., III.1, 19]): *ex demonstratione propositionis, quam post hanc propositionem afferimus.* Diese Bemerkung verweist auf einen Beweis, der von Anaritius (der in **A**, [Curtze 1899, 142], namentlich genannt wird) herrührt. Dieser Beweis, der nach dem Beweis des ersten Teils hinzugefügt wurde, ist also kein Bestandteil des Euklid-Textes.

7. Die Bemerkung in VI.29 (**L**, [Besthorn und Heiberg 1893ff., III.2, 187]): *Nihil differt, utrum D Z sit figura quadrilatera aequalium laterum an inaequalium; et figura quadrilatera et aequilatera sit, si velis, rectangularis aut habeat angulos inaequales,* stimmt mit dem letzten Satz zu VI.28 in **A** [Curtze 1899, 186] überein. Dasselbe gilt auch für die Bemerkung in VI.30 (**L**, [Besthorn und Heiberg 1893ff., III.2, 191]): *Nihil differt, utrum figura A D rectangularis sit an non, quoniam opus est tantum, ut latera sint aequalia; et hoc est, cur dicat: "Et sit excessus figura similis figurae A D", quoniam "figura" latius patet quam "rectangulum". Si A D rectangularem esse voluisset, excessum quadrilaterum et rectangularem esse oportere dicendum fuit,* die mit **A**, VI.30 [Curtze 1899, 188] übereinstimmt, und ebenso für das Ende von VI.19 (**L**, [Besthorn und Heiberg 1893ff., III.2, 151f.]), das

eingeleitet wird mit: *Triangulus per intersectionem [linearum] effectis Elementa eodem modo utuntur, quo nos utimur in demonstratione nostra, quae sequitur,* das mit **A**, VI.19 [Curtze 1899, 179–181] übereinstimmt. Alle drei Sätze werden nicht wie üblich eingeleitet (in **L**, VI.29, folgt sogar unmittelbar nach dem Satz die Bemerkung: *Al-Narîzî dixit*), dennoch gehören sie zum Kommentar.

8. Der Satz in VI.7 (**L**, [Besthorn und Heiberg 1893ff., III.2, 119]): *et si addimus angulum* G B H, *anguli tres trianguli* G B H *coniuncti maiores sunt duobus rectis* ist eine Interpolation, die erst später eingebracht wurde, denn der darauf folgende Verweis auf I.17 macht die Interpolation völlig überflüssig.

9. Das Ende von V.4 (**L**, [Besthorn und Heiberg 1893ff., III.2, 39]) lautet: *Ac sicut primae et tertiae multiplicia multiplicibus secundae et quartae aut maiora sunt aut minora aut aequalia suo ordine sumpta, si quattuor magnitudines tales sunt, ut prima ad secundam sit ut tertia ad quartam, et primae et tertiae aeque multiplicia sumpta sunt, atque etiam secundae et quartae, ita etiam huius contrarium sequitur: prima est ad secundam ut tertia ad quartam, si quattuor magnitudines tales sunt, ut, cum primae et tertiae multiplicia sumpta sint, atque etiam secundae et quartae, primae et tertiae multiplicia multiplicibus secundae et quartae aut maiora aut minora aut aequalia sint suo ordine sumpta.* Dieser Text ist wahrscheinlich ein Zusatz des Verfassers; er fehlt im griechischen Euklid, bei Adelard und bei Gerhard.

10. In I.34 (**L**, [Besthorn und Heiberg 1893ff., I, 147]) muß der Satz: *Et demonstravimus, esse* AG = BD ersetzt werden durch: *Et demonstravimus, esse angulus* A B D = *angulus* A G D (siehe Gerhard [Busard 1983b, 26]).

11. Der Zusatz zur Enuntiation in I.22 (**L**, [Besthorn und Heiberg 1893ff., I, 99]): *quoniam ratio trianguli ex I.20 ea est, ut duo latera eius coniuncta semper tertia maiora sunt* gehört sicherlich nicht zu Euklid.

12. Ebenso in I.29 (**L**, [Besthorn und Heiberg 1893ff., I, 135]): *Sed ex eo, quod postulavit Euclides, et quod Geminus in propositionibus, quas praemisit, demonstravit, efficitur...*

13. Ebenso in I.44 (**L**, [Besthorn und Heiberg 1893ff., I, 169]): *Itaque ex eo, quod Geminus in demonstratione propositionum propositioni XXIX praemissarum demonstravit, et ex eo, quod Euclides in postulato [5] praemisit, duae lineae* K H, L B *productae concurrunt.*

14. Ebenso in I.45 (**L**, [Besthorn und Heiberg 1893ff., I, 171]): *ut in demonstratione propositionis propositioni XI additae demonstratum est.*

15. Ebenso in I.47 (**L**, [Besthorn und Heiberg 1893ff., I, 189]): *ita ut in demonstratione propositioni 11 addita demonstratum est.*

16. Ebenso in II.1 (**L**, [Besthorn und Heiberg 1893ff., II.1, 7]): *sicut in demonstratione propositionis ad I.11 adiectae explicavimus.*

17. Ebenso in II.10 (**L**, [Besthorn und Heiberg 1893ff., II.1, 55]): *Quare ex demonstratione Gemini in praemissis ad I.29 et ex eo, quod ei a nobis additum est, duae lineae E B, Z D in directum productae concurrent.*

18. Ebenso in III.32 (**L**, [Besthorn und Heiberg 1893ff., II.2, 115]): *Et diametrus circulum in duas partes aequales dividit, ut Simplicius in definitione libri primi demonstravit.*

19. Ebenso in IV.2 (**L**, [Besthorn und Heiberg 1893ff., III.1, 9]): *ut in propositione ad III.16 adiecta demonstravimus.*

20. Ebenso in IV.5 (**L**, [Besthorn und Heiberg 1893ff., III.1, 25]): *Ex demonstratione praemissa propositionis adiecta demonstrabitur.*

21. Ebenso in IV.7 (**L**, [Besthorn und Heiberg 1893ff., III.1, 35]): *Per puncta A, B, G, D ex propositione ad III.16 adiecta.*

22. Ebenso in VI.12 (**L**, [Besthorn und Heiberg 1893ff. III.2, 129]): *ut demonstratum est in demonstratione propositionis additae propositioni 31 libri primi* [5].

Aus all diesen Beispielen, die noch ergänzt werden können, geht hervor, daß wir es in **L** nicht mit einem reinen Text des arabischen Euklid zu tun haben [6].

III

Ich möchte jetzt einige Beweise in **L** und bei Adelard miteinander vergleichen:

L (III.3, secunda pars):

Rursus supponimus, lineam AB lineam GD in puncto E ad angulos rectos secare. Dico, eandem eam in duas partes aequales secare.

Demonstratio. Triangulus GZD aequicrurius est; crus enim ZD cruri ZG aequale, quoniam utrumque a centro ad ambitum ductum est; *quare ex I.5* ∠ZGD = ∠ZDG. *Iam autem demonstravimus, angulum rectum GEZ angulo DEZ aequalem esse; itaque duo anguli ZGE, ZEG duobus angulis ZDE, ZED aequales sunt.* Relinquitur igitur ex I.32∠GZE = ∠DZE. *Et linea ZE communi sumpta* duo latera GZ, ZE duobus lateribus DZ, ZE aequalia erunt. *Iam autem demonstratum est, angulum GZE angulo DZE aequalem esse; itaque ex I.4 basis GE basi DE aequalis erit. Ergo demonstratum est, lineam AB lineam GD in duas partes aequales secare. Q.n.e.d.*

Adelard (III.3, secunda pars):

Item linea a b *lineam* g d *supra angulum rectum dividat. Dico itaque quia eam in duo media dividit.*

Rationis causa: Si enim g z *sicut* z d *fuerit, erit angulus* z g d *sicut angulus* z d g. *Atqui* g h z *et* d h z *recti anguli. Duorum itaque triangulorum* g h z *et* h z d *duo anguli unius scilicet* z g h *et* g h z *duobus angulis alterius* z d h *et* d h z *quisque se respiciens equales. Latus autem duos angulos equales respiciens commune estque* z h. *Duo ergo latera reliqua equalia scilicet* g h *sicut* h d. *Linea igitur* a b *lineam* g d *in duo media dividit. Et hoc est quod demonstrare proposuimus.*

L (VI.3, secunda pars):

Rursus sit ratio lateris BA *ad latus* AG *aequalis rationi partis* BD *ad partem* DG. *Dico igitur angulum* BAG *in duas partes aequales sectum esse.*

Demonstratio. Ducatur linea GE *[lineae]* AD *parallela, et producatur latus* BA, *ut in eam incidat in puncto* E; *convertatur autem prior demonstratio. Dicimus igitur, quoniam pars* BD *sit ad partem* DG *ut latus* BA *ad latus* AG, *atque etiam ut latus* BA *ad lineam* AE, *ea de causa lineam* BA *ad utramque lineam* AE *et* AG *eandem rationem habere. Itaque secundum ea, quae demonstrata sunt in propositione 9 libri quinti, linea* AE *aequalis est lineae* AG. *Atqui anguli, qui sunt super basim trianguli, qui duo latera aequalia habet, aequales sunt secundum demonstrationem 5 libri primi. Itaque angulus* AEG *aequalis est angulo* AGE, *et secundum ea, quae antea demonstrata sunt, angulus* DAG *aequalis est angulo* AGE *et angulus* AEG *aequalis est angulo* BAD, *quoniam posuimus lineam* AD *lineae* GE *parallelam. Ergo angulus* BAD *aequalis est angulo* DAG. *Quod erat demonstrandum.*

Adelard (VI.3, secunda pars):

Item ponatur quanta b d *ad* d g *tanta* a b *ad* a g. *Iungaturque a cum* d. *Dico quia angulus* b a d *sicut angulus* d a g.

Sitque eorum tedebir unum. Quanta itaque b d *ad* d g *tanta* b a *ad* a h. *Quantaque* b d *ad* d g *tanta* b a *ad* a g. *Atque linea* a b *linearum* a g *et* a h *similitudo una. Atqui* a g *sicut* a h. *Angulus itaque* a g h *sicut* a h g. ⟨*Atqui angulus* a g h *sicut* d a g.⟩ *Atqui* a h g *sicut* b a d. *Est igitur* b a d *sicut* d a g. *Et hoc est quod demonstrare intendimus.*

Aus diesen beiden Beispielen geht die Arbeitsweise klar hervor: In **L** wird jeder Schritt erklärt, auch wenn dies schon vorher geschehen ist, wie im ersten Beispiel: *quoniam a centro ad ambitum ductae sunt* (dies wurde schon im ersten Teil des Beweises gesagt). Der Text in **L** ist sehr weitläufig, und für den Fall, daß der Leser noch nicht weiß, warum die Aussage richtig ist, wird noch auf den benutzten Satz hingewiesen. Vielleicht ist dies auch der Grund dafür, daß in **L** der Beweis zu III.3 etwas abgeändert wurde, denn dort wird er mit Hilfe von I.4 geführt statt, wie im griechischen Euklid, sofort aus I.26 zu schließen, daß *GE* = *DE* ist. Auch im zweiten Beispiel ist **L** sehr weitläufig, auch wenn man bedenkt, daß Adelard etwas zu viel gekürzt hat, denn bei ihm ist ausgelassen, daß die Winkel *a g h* und *d a g* gleich groß sind. Auf welche Weise Adelard den Text kürzte, erkennt man sehr gut am Anfang des Beweises VI.3; dort heißt es bei Adelard: *Sitque eorum tedebir unum*, während in **L** die Konstruktion des ersten Teiles wiederholt wurde. Ganz allgemein kann man sagen, daß die Beweise bei Adelard kürzer sind als in **L**; hierzu siehe z.B. V.8. Nun wird auch verständlich, was gemeint ist, wenn es im Prolog von **L** heißt: "left out the superfluities, filled up the gaps, corrected or removed the errors, until he had perfected the book and made it more certain, and had summarized it, as it is found in the present version. This was done for specialists, without changing any of its substance, while he left the first version as it was for the vulgar" ([Besthorn und Heiberg 1893ff., I, 5]; siehe [Murdoch 1971, 439]). Denn für Fachleute braucht man nicht jeden Schritt zu erklären, und auch Verweise auf die angewandten Sätze sind überflüssig. Aus diesem Vergleich ergibt sich, daß es sehr unwahrscheinlich ist, daß **L** eine al-Ḥaǧǧāǧ II-Version enthält, wie man bisher annahm [Sezgin 1974, 104, Murdoch 1971, 453].

Ich habe schon bemerkt, daß **L** keine reine Fassung enthält. Dies möchte ich im folgenden näher ausführen. Nach Angabe des Pseudo-Ṭūsī fehlen bei al-Ḥaǧǧāǧ die Sätze I.45, III.36 und VI.12. Über I.45 wird dort gesagt: Ḥaǧǧāǧ bringt den Satz nicht, wohl aber Ṯābit; und über VI.12 L (= VI.11 bei Gerhard): Ṯābit macht dieses Porisma zur selbständigen Proposition; es steht als solches aber nicht in den griechischen und syrischen Ausgaben, und deshalb hat Ḥaǧǧāǧ es auch in seiner Ausgabe fortgelassen [Thaer 1936, 118, 120]. De Young [1991, 656] sagt über III.36 folgendes: "In each of these Taḥrīr, however, we are informed that this proposition was not included in the translation of al-Ḥajjāj (Pseudo-Ṭūsī adds that it was lacking from the known Greek manuscripts and that is why al-Ḥajjāj omitted it), and that Thābit is responsible for its inclusion here because it is needed in proposition 10 of Book IV." Tatsächlich fehlen alle drei Sätze bei Adelard, während in **L** nur I.45 fehlt. Deshalb stimmt auch die Formulierung von II.14 in **L** überein mit Adelard und nicht mit Isḥāq-Ṯābit (**L**: Ein einem gegebenen *Dreieck* gleiches Quadrat zu errichten; Isḥāq-Ṯābit: Ein einer gegebenen *geradlinigen Figur* gleiches Quadrat zu errichten).

Was III.36 anbetrifft, so ist die Lage nicht so einfach. De Young [1991, 655] sagt darüber: "This time, however, the manuscripts do not explicitly

relate these added cases to the translation of al-Ḥajjāj. They merely say
that there are found in manuscripts other than this one (apparently
referring to the manuscript being copied) three cases, of which Isḥāq
mentioned only a part of one and that all three have been added to the
text of Isḥāq because the proof used is different from Isḥāq's." Im
Manuskript Brügge 529, das die Bücher I-VIII von Adelard enthält, heißt
es nach III.35: *Explicit liber iii. Nota quia in quibusdam libris tres figure
prescriptis similes adduntur ad demonstrationem conversionis predicte propositionis que sic habet: Cum fuerit punctus extra circulum a quo due linee ad
circulum producantur alia incidens alia circumferentie applicata fueritque
quod ex ductu totius linee incidentis in partem extrinsecam sicut quod ex ductu
applicate in seipsam erit linea applicata contingens quod per mutationem
superiorum argumentationum posse probari satis evidens est* [Busard 1983a,
126]. Gerhard gibt zuerst den Beweis für die 3 Fälle und sagt dann: *Hoc
etiam theorema taliter invenitur in alio libro uno solo signatum caractere*, d.h.
er gibt dann Isḥāqs Beweis [Busard 1983b, 95−97], der mit dem griechischen übereinstimmt. Der Unterschied zwischen den beiden Beweisen
besteht darin, daß bei Isḥāq noch eine Tangente von einem Punkt
außerhalb des Kreises gezogen wird und daß aus der Gleichheit von zwei
Winkeln, von denen der eine ein rechter ist, geschlossen wird, daß auch
der andere ein rechter ist, während Gerhards Beweis sich auf I.48 (= I.47
L) stützt. L enthält nur den Beweis für die 3 Fälle, und De Young [1991,
656] sagt: "The claim that this proposition was missing from the translation
of al-Ḥajjāj is also difficult to understand because the proposition is clearly
present in the Leiden ms., as well as in Ibn Sīnā's Kitāb al-Shifā'." Die
Vorlage Gerhards gehört also zu den Handschriften, von denen in der
ersten Aussage von De Young die Rede ist. Es ist möglich, daß der
Verfasser von L den Beweis hinzugefügt hat, wie es auch mit VI.12
(Heiberg) geschehen ist. Die Tatsache, daß der Beweis in L aufgenommen
ist, bedeutet noch nicht, daß er von al-Ḥaǧǧāǧ herrührt (siehe die zweite
Aussage von De Young). Da A(naritius) keine Bemerkung zu III.36
enthält, hilft er in dieser Frage nicht weiter.

Es ist sehr wohl möglich, daß der Verfasser von L die Isḥāq-Ṭābit-Tradi-
tion gekannt und auch einiges daraus entnommen hat, wie z.B. die Definitionen VI.3−5. Diese Definitionen fehlen bei Adelard, stimmen aber mit
denen bei Gerhard überein. Es fällt auf, daß Gerhard die Definition iii mit
"Dixit Thebit" einleitet, während in L kein Name genannt wird und A sagt
[Curtze 1899, 177]: *Duo elementa, que sequuntur, quia bene translata sunt in
Euclide, praetermisi*. Somit nahm Anaritius an, daß die zwei Definitionen
zum Euklidtext gehören. Der Text dieser Definitionen lautet in L bzw. bei

Gerhard folgendermaßen:
L (VI.Def.3–5):

> *Euclides dixit: Altitudo figurae perpendicularis est a vertice ad basim ducta.*
> *Linea recta in mediam atque extremam rationem secari dicitur, si ratio totius*
> *lineae ad maiorem partem aequalis est rationi maioris partis ad minorem.*
> *Ratio dicitur e rationibus composita esse, si quantitates harum rationum per*
> *se ipsas multiplicatae aliquam rationem efficiunt vel aliam.*
> *Alius codex dicit*: *Ratio est divisa in (plures) rationes, si ratio alia per aliam*
> *divisa rationem aliquam efficit vel aliam.*

Gerhard:

> [iv] *Altitudo est perpendicularis cadens a puncto capitis figure cuiuslibet,*
> *quemcumque figura fuerit, super basim.*
> [v] *Linea recta dicitur dividi secundum proportionem habentem medium et*
> *duo extrema, quando fuerit proportio totius linee ad maiorem eius sectionem*
> *sicut proportio maioris divisionis ipsius ad minorem eius sectionem.*
> [iii] *Dixit Thebit*: *In hoc loco inveni in alia scriptura: Dicitur quod proportio*
> *ex proportionibus aggregatur, quando ex multiplicatione quantitatis propor-*
> *tionum, cum multiplicantur in seipsas, provenit proportio aliqua.*
> *Dicitur quod proportio dividitur in proportiones cum ex divisione propor-*
> *tionum, quando alie per alias dividuntur, provenit proportio quelibet.*

Wir wollen jetzt noch den Text von VI.Def.2 in L, A, bei Gerhard und
Adelard vergleichen:

> **L**: *Reciprocae figurae sunt, quarum latera proportionalia sunt ad anteceden-*
> *tiam et consequentiam.*
> *In alio codice dicitur*: *Illae sunt, in quarum utraque antecedentia et conse-*
> *quentia sunt.*
> **A**: *Superficies latera habentes alternata sunt, quarum latera proportionalia*
> *secundum antecessionem et consequentiam.*
> *In aliis tamen scripturis reperitur, quod alternate sunt ⟨quarum⟩ in una-*
> *quaque est antecedens et consequens.*
> Gerhard: *Aliter: superficies alternorum laterum sunt, quarum latera sunt*
> *proportionalia secundum antecedens et consequentiam.*
> *Figure rectilinee sunt alternorum laterum, cum in unaquaque earum fuerit*
> *antecedens in proportione et consequens.*
> Adelard: *Superficies mutekefie sunt inter latera quarum* incontinua *propor-*
> *tionalitas retransitive reperitur.*

Man erkennt, daß die Definitionen von **L**, **A** und Gerhard miteinander
übereinstimmen, während Adelard eine völlig andere Definition gibt, die
auf V.Def.5 zurückgeht.

Wie schon Murdoch [1963, 241] festgestellt hat, unterscheidet sich auch die Definition V.Def.5 bei Adelard von der Fassung in L, A und bei Gerhard:

Adelard: *Quantitates que dicuntur* continuam *proportionalitatem habere, sunt quarum eque multiplicia aut equa sunt aut eque sibi sine interruptione addunt vel minuunt.*

Gerhard: *Quantitates quarum quedam ad alias proportionales esse dicuntur, sunt quarum quasdam cum multiplicantur super alias addere possibile est.*

L [Besthorn und Heiberg 1893ff., III.2, 13]: *Magnitudines, quarum ratio alterius ad alteram exstare dicitur, eae sunt, quae cum multiplicantur, fieri potest, ut altera alteram superet.*

A [Curtze 1899, 161]: *Quantitates, inter quas dicitur esse proportio, sunt quarum possibile, cum multiplicantur, alias addere.*

Schließlich wollen wir noch bemerken, daß L, A und Gerhard mehr als eine Fassung zu Rate gezogen haben (siehe VI.Def.2).

Auch die beiden letzten Definitionen des Buchs V, die bei Adelard fehlen, stimmen in L und bei Gerhard überein:

Gerhard:

[xix] *Ordinata proportio est cum fuerit antecedens ad consequens sicut antecedens ad consequens et consequens ad rem aliam sicut consequens ad rem aliam.*

[xx] *Inordinata proportio est cum fuerit antecedens ad consequens sicut antecedens ad consequens et consequens ad rem aliam sicut res alia ad antecedens.*

L (V.Def.19, [Besthorn und Heiberg 1893ff., III.2, 27]): *Euclides dixit: Proportio ordinata oritur, si, ut antecedens ad sequentem, ita antecedens ad sequentem, et ut sequens ad aliud, ita sequens ad aliud.*

L (V.Def.20, [Besthorn und Heiberg 1893ff., III.2, 27]): *Euclides dixit: Proportio perturbata est, ubi antecedens ad sequentem est ut antecedens ad sequentem, et sequens ad aliud ut aliud ad sequentem* (muß heißen: *antecedentem.* Siehe A, [Curtze 1899, 169]).

In Buch III gibt L zwei Reihen von Definitionen. Die zweite Reihe, die in A fehlt, wurde vielleicht später hinzugefügt; sie ähnelt stark den Definitionen bei Adelard, obwohl es auch Unterschiede gibt. So ist L_2, Def.6: *segmentum circuli comprehendunt linea recta, quae chorda vocatur, et pars ambitus circuli, quae arcus vocatur,* identisch mit Gerhard, Def.vi: *Portio circuli est figura que a linea recta, que dicitur corda, et portione linee circuli circumflexe, que dicitur arcus, continetur.* Die erste Reihe hat große Ähnlichkeit mit den Definitionen bei Gerhard, aber auch hier gibt es bisweilen Unterschiede.

Die Postulate und Axiome von Buch I stimmen in **L** mit denen bei Gerhard überein. Allerdings fehlt unter den Postulaten bei Gerhard dasjenige, das in **L** (und **A**) das letzte ist: "Zwei Strecken umfassen keinen Flächenraum". Dieses Postulat wird sowohl bei Gerhard als auch in **L** als letztes Axiom gegeben, d.h. **L** und **A** geben es sowohl als Postulat als auch als Axiom. Das Axiom 5 (Wenn von Ungleichem Gleiches weggenommen wird, sind die Ganzen ungleich) fehlt bei Adelard, Robert von Chester und Hermann von Kärnten. Das folgende Axiom 6 (Die Doppelten von demselben sind einander gleich) wurde bei Adelard ersetzt durch: *Si fuerint due res alicui uni equales, unaqueque earum equalis erit alteri.* Da dieses Axiom sowohl bei Robert als auch bei Hermann mit denselben Worten formuliert wird, ist die Lesart der Adelard-Handschriften korrekt. In **A** fehlt das Axiom 7 (Die Halben von demselben sind einander gleich). Dieses Axiom findet man in **L** ebenso wie bei Gerhard und Adelard.

Zusammenfassend können wir sagen, daß die Definitionen in **L** eine größere Übereinstimmung mit Gerhard als mit Adelard zeigen. Eine ähnliche Beobachtung hat kürzlich De Young hinsichtlich der Enunziationen von Buch II gemacht [De Young 1991, 649]: "A comparison of the terminology used in these al-Ḥajjāj enunciations with the terminology used in Leiden ms. Or. 399/1 and the terminology of the Isḥāq-Thābit translation indicates that the al-Nayrīzī/al-Ḥajjāj enunciations are exactly like those found in the Isḥāq-Thābit versions." Aber auch hier muß man sich fragen, ob die Terminologie in **L** wohl von al-Ḥaǧǧāǧ herrührt.

IV

Es gibt auch Stellen, an denen Gerhard und **L** gegenüber Adelard übereinstimmen. Vergleichen wir z.B. V.11:

L:

> *Magnitudines, quae ad alias magnitudines eandem rationem habent, proportionales sunt.*
>
> *Exemplificatio: Sit A ad B ut G ad D et E ad Z ut G ad D. Dico igitur A esse ad B ut E ad Z.*
>
> *Demonstratio. Sumantur magnitudinum A, G, E quaelibet aequa multiplicia, et sint magnitudines H, T, K. Sumantur etiam magnitudinum B, D, Z, quaelibet aeque multiplicia, et sint magnitudines L, M, N. Tum, quoniam A est ad B ut G ad D, et magnitudinum A et G aeque multiplicia sumpta sunt, magnitudines H et T, atque etiam magnitudinum B et D aeque multiplicia, magnitudines L et M, perspicuum est ex propositione 4 libri quinti duas magnitudines H et T*

aut in excessu aut minores aut aequales esse duabus magnitudinibus L *et* M, *suo ordine sumptas. Rursus, quoniam* E *est ad* Z *ut* G *ad* D, *et duarum magnitudinum* E *et* G *aeque multiplicia sumpta sunt,* K *et* T, *atque etiam duarum magnitudinum* Z *et* D *aeque multiplicia sumpta sunt,* M *et* N, *ea de causa duae magnitudines* K *et* T *aut in excessu aut minores aut aequales sunt duabus magnitudinibus* N *et* M. *Itaque duae magnitudines* H *et* K *aut in excessu aut minores aut aequales sunt [magnitudinibus]* L *et* N. *Ergo est* A *ad* B *ut* E *ad* Z. *Quod erat demonstrandum.*

Gerhard:

Proportiones quantitatum que uni proportioni existunt equales, sunt equales.

Verbi gratia: Sit proportio a *ad* b *sicut proportio* g *ad* d, *et proportio* g *ad* d *sicut proportio* e *ad* z. *Dico igitur, quod proportio* a *ad* b *sicut proportio* e *ad* z.

Probatio huius: Quoniam sumam quantitatibus a *et* g *et* e *eque multiplicia, que sint* h *et* t *et* k, *et sumam quantitatibus* b *et* d *et* z *eque multiplicia, que sint* l *et* m *et* n. *Et quia proportio* a *ad* b *sicut proportio* g *ad* d, *et iam accepimus quantitatibus* a *et* g *eque multiplicia, que sunt* h *et* t, *et quantitatibus* b *et* d *eque multiplicia, que sunt* l *et* m, *ergo quantitates* h *et* t *aut simul addunt super quantitates* l *et* m, *aut simul equantur eis, aut simul minuunt ab eis. Et etiam quoniam proportio* g *ad* d *est sicut proportio* e *ad* z, *et iam sumpsimus quantitatibus* g *et* e *eque multiplicia, que sunt* t *et* k, *et quantitatibus* d *et* z *eque multiplicia, que sunt* m *et* n, *ergo quantitates* t *et* k *aut simul addunt super quantitates* m *et* n, *aut simul equantur eis, aut simul minuunt ab eis. Quantitates ergo* h *et* k *aut simul addunt super quantitates* l *et* n, *aut simul equantur eis, aut simul minuunt ab eis. Sed* h *et* k *sunt eque multiplicia quantitatum* a *et* e, *et* l *et* n *sunt eque multiplicia quantitatum* b *et* z. *Ergo proportio* a *ad* b *est sicut proportio* e *ad* z. *Et illud est quod voluimus demonstrare.*

Adelard:

Quantitatum proportiones que alicui uni equales, ipsas etiam proportiones sibi invicem equales esse.

Exempli gratia: Sit proportio a *ad* b *sicut proportio* g *ad* d, *proportio vero* g *ad* d *sicut proportio* h *ad* z. *Dico quia proportio* a *ad* b *sicut proportio* h *ad* z.

Rationis causa: Sumantur enim ad a *et* g *et* h *multiplicitates equales et sint* H *et* t *et* k. *Sumantur item ad* b *et* d *et* z *multiplicitates equales sintque* l *et* m *et* n. *Quanta itaque* a *ad* b *tanta* g *ad* d. *Sumpte autem erant ad* a *et* g *multiplicitates equales, scilicet,* H *et* t. *Atque ad* b *et* d *multiplicitates equales, suntque* l *et* m. *Quanta itaque* H *ad* l *tanta* t *ad* m. *Itemque quanta* g *ad* d *tanta* h *ad* z. *Sumpte vero ad* g *et* h *multiplicitates equales* t *et* k *atque ad* d *et* z *multiplicitates equales, suntque* m *et* n. *Quanta ergo* t *ad* m *tanta* k *ad* n. *Quanta itaque* H *ad* l *tanta* k *ad* n. *Sunt autem* H *et* k *multiplicitates equales ad* a *et* h, *et* l *et* n *sunt multiplicitates equales ad* b *et* z. *Quanta igitur* a *ad* b *tanta* h *ad* z. *Et hoc est quod demonstrare proposuimus.*

V

Im nächsten Beispiel stimmen die Beweise in L und bei Adelard überein,
weichen aber von Gerhard ab:

L (V.5):

> *Exemplificatio: Sit magnitudo* AB *multiplex magnitudinis* GD, *et abscidantur
> ab iis duae magnitudines* AE *et* GZ, *ita ut* AE [*magnitudinis*] GZ *toties
> multiplex sit, quoties* AB [*magnitudinis*] GD. *Dico igitur residuum* EB *residui*
> ZD *toties multiplex esse, quoties* AB *multiplex sit* [*magnitudinis*] GD.
>
> *Demonstratio. Coniungatur magnitudo* GH *magnitudini* GZ, *et fiat* EB
> [*magnitudinis*] GH *toties multiplex, quoties* EA [*magnitudinis*] GZ. *Tum
> perspicuum est ex propositione* 1 *libri quinti* AE [*magnitudinis*] GZ *toties
> multiplicem esse, quoties totum* AB *multiplex sit* [*magnitudinis*] HZ. *Atqui
> posuimus* AE [*magnitudinis*] GZ *toties multiplicem esse, quoties* AB *multiplex
> esset* [*magnitudinis*] GD. *Ergo magnitudo* AB *duarum magnitudinum* HZ *et*
> GD *aeque multiplex est. Ergo magnitudo* HZ *aequalis est magnitudini* GD.
> *Subtrahatur autem* GZ, *quae utrique communis est, et relinquitur* ZD *aequalis
> magnitudini* HG. *Ergo* EB [*magnitudinis*] ZD *toties multiplex est, quoties* AE
> *multiplex est* [*magnitudinis*] GZ, *et perspicuum est ex propositione* 1 *libri quinti
> residuum* EB *residui* ZD *toties multiplex esse, quoties totum* AB *multiplex sit*
> [*magnitudinis*] GD. *Quod erat demonstrandum.*

Adelard (V.5):

> *Exempli gratia: Sit* g d *pars* a b, *dueque earum minorationes* a h *et* g z. *Dico
> quia quod in* h b *reliqua ex comparatione* z d *relique sicut quod in* a b *ex
> comparatione* g d.
>
> *Rationis causa: Ponatur etenim quod in* a h *ex comparatione* g z *sicut quod
> in* h b *ex comparatione* g H. *Eritque multiplicitas* a h *ad* g z *sicut multiplicitas*
> a b *ad* H z. *Atqui quanta multiplicatio* a h *ad* g z *tanta multiplicatio* a b *ad* g
> d. *Quare* H z *sicut* d g. *Reiciaturque* g z *communis. Relinquitur itaque* H g
> *sicut* z d. *Quanta autem multiplicitas* a h *ad* g z *tanta multiplicitas* h b *ad* g H.
> *Atque* g H *sicut* z d. *Quanta itaque multiplicitas* a h *ad* g z *tanta multiplicitas* h
> b *ad* z d. *At vero quanta multiplicatio* a h *ad* g z *tanta multiplicatio* a b *ad* g d.
> *Quod igitur in reliqua* h b *ex comparatione* z d *sicut quod in* a b *ex
> comparatione* g d. *Et hoc est quod demonstrare intendimus.*

Gerhard (V.5):

> *Exempli causa: Sit quantitas* a b *quantitatis* g d *multiplex, et minuantur ab
> eis* a e; g z *et sit quod est in* a e *ex multiplicibus* g z *equale ei quod est in* a b *ex
> multiplicibus* g d. *Dico igitur, quod illud quod est in reliqua* e b *ex multiplicibus
> relique* z d *equale existit ei quod est in* a b *ex multiplicibus* g d.
>
> *Probatio eius: Quoniam ponam quod est in* a t *ex multiplicibus* z d *equale ei
> quod est in* a e *ex multiplicibus* g z, *ergo quod est in tota* e t *ex multiplicibus* g d
> *equale est ei quod est in* a e *ex multiplicibus* g z. *Sed iam fuit illud quod est in* a
> b *ex multiplicibus* g d *equale ei quod est in* a e *ex multiplicibus* g z. *Ergo quod*

est in e t *ex multiplicibus* g d *equale est ei quod est in* a b *ex multiplicibus* g d. *Ergo* e t *equalis existit* a b. *Sublato igitur communi* a e, *remanet* a t *equalis* e b. *Sed iam fuit quod est in* a t *ex multiplicibus* z d *equale ei quod est in* e a *ex multiplicibus* g z. *Ergo quod est in* e b *ex multiplicibus* z d *equale est ei quod est in* e a *ex multiplicibus* g z. *Sed iam fuit quod est in tota* a b *ex multiplicibus* g d *equale ei quod est in* a e *ex multiplicibus* g z. *Ergo quod est in* e b *ex multiplicibus* z d *ei existit equale quod est in tota* a b *ex multiplicibus* g d. *Et illud est quod ostendere voluimus.*

Bei Gerhard gibt es außer diesem Beweis noch einen zweiten, der mit dem Beweis in **L** und bei Adelard übereinstimmt. Dieser zweite Beweis wird bei Gerhard mit folgenden Worten eingeleitet: *Hoc preterea theorema aliter invenitur hoc scilicet modo.*

VI

Schließlich möchte ich noch einen Beweis geben, bei dem **L** gegenüber Adelard und Gerhard abweicht.

L (VI.21):

Figurae, quae eidem figurae similes sunt, etiam inter se similes sunt.

Exemplificatio. Sint duae figurae A *et* B *figurae* G *similes. Dico igitur alteram alteri similem esse.*

Demonstratio. Si duae figurae A *et* B *figurae* G *similes sunt, anguli duarum figurarum* A *et* B *angulis figurae* G *aequales sunt, suo quisque angulo correspondenti; i.e. anguli figurae* A *angulis figurae* G, *et anguli figurae* B *angulis figurae* G *similes sunt, suo quisque angulo correspondenti. Atqui ea, quorum quidque eidem aequale est, inter se aequalia sunt. Itaque anguli figurae* A *angulis figurae* B *aequales sunt.*

Rursus latera figurae A *lateribus figurae* G *proportionalia sunt, latera scilicet angulos aequales comprehendentia; et latera figurae* B *lateribus figurae* G *proportionalia sunt, latera scilicet angulos aequales comprehendentia. Tum si tollimus medium terminum, ut demonstratum est in propositione 11 libri quinti, latera figurae* A *lateribus figurae* B *proportionalia sunt, latera scilicet angulos aequales comprehendentia. Itaque anguli figurae* A *aequales sunt angulis figurae* B, *et latera angulos aequales comprehendentia proportionalia sunt. Ergo figura* A *similis est figurae* B. *Quod erat demonstrandum.* (Dies ist der Beweis im griechischen Euklid.)

Adelard (VI.21):

Cum fuerint alique superficies alicui uni similes, erunt etiam sibi invicem similes.

Exempli gratia: Sint due superficies a g *et* H k *superficiei* d z *similes. Dico quia due superficies* a g *et* H k *similes inter se.*

Rationis causa: Superficie etenim a g *simili superficiei* d h z *erit angulus* b
sicut angulus h *quantaque* a b *ad* d h *tanta* b g *ad* h z. *Item* H k *simili* d z *erit*
angulus t *sicut angulus* h. *Quantaque* d h *ad* H t *tanta* h z *ad* t k. *Angulusque*
h *sicut angulus* b. *Erat autem quanta* a b *ad* d h *tanta* b g *ad* h z. *Angulusque*
b *sicut angulus* t. *Quanta itaque* a b *ad* H t *tanta* b g *ad* t k. *Due igitur*
superficies a g *et* H k *inter se similes. Et hoc est quod demonstrare intendimus.*

Gerhard (VI.21):

> *Superficies, que uni superficiei existunt similes, vicissim sunt similes.*
> *Exempli causa: Sint superficies* a g *et* h k *superficiei* d z *similes. Dico ergo,*
> *quod superficies* a g *similis existit superficiei* h k.
> *Probatio eius: Quoniam superficies* a g *est similis superficiei* d z, *ergo angulus*
> b *est equalis angulo* e *et proportio* a b *ad* d e *sicut proportio* b g *ad* e z. *Et*
> *etiam quia* h k *est similis* d z, *ergo angulus* t *est equalis angulo* e *et proportio* d
> e *ad* h t *sicut proportio* e z *ad* t k. *Sed iam fuit proportio* a b *ad* d e *sicut*
> *proportio* b g *ad* e z. *Et angulus* b *equalis angulo* e *et angulus* t *equalis angulo*
> e, *ergo angulus* b *est equalis angulo* t. *Sed proportio* a b *ad* h t *sicut proportio*
> b g *ad* t k, *ergo superficies* a g *similis existit superficiei* h k. *Et illud est quod*
> *demonstrare voluimus.*

Wir sehen also, daß **L** an einigen Stellen mit Adelard gegenüber
Gerhard übereinstimmt, an anderen mit Gerhard gegenüber Adelard und
an noch anderen Stellen weder mit Adelard noch mit Gerhard. Hieraus
möchte ich schließen, daß **L** eine Bearbeitung ist, die aus verschiedenen
Elementen zusammengesetzt ist.

VII

Im zweiten Teil meines Aufsatzes möchte ich die Frage untersuchen, von
wem die alternativen Beweise stammen, die in der Übersetzung von
Gerhard von Cremona vorhanden sind. Zunächst möchte ich auf den
Aufsatz von G. De Young [1991] hinweisen. In ihm geht der Verfasser auf
einige Fragmente der al-Ḥağğāğ-Tradition ein, die er mit Hilfe der arabischen Handschriften Escorial, Derenborg 907, sowie Rabāṭ, Ḥasanīya 53
und 1101 ermittelt hat. Der Verfasser beginnt mit der Bemerkung, daß in
den Enuntiationen der ersten 9 Sätze von Buch II Isḥāq-Ṯābit von
"Rechteck" (*al-saṭḥ al-qāʾima az-zawāyā*) und "Quadrat" (*al-murabbaʿ*)
spricht, während in den al-Ḥağğāğ-Fragmenten vom "Multiplizieren einer
Strecke mit einer anderen" bzw. "mit sich selbst" gesprochen wird. Vor
kurzem hat Brentjes [1993] zwei von diesen drei Handschriften (Escorial =
E und Rabāṭ 53 = **R**) sehr gründlich untersucht und mit MS Paris, BN,
Persan 169H (= **P**) verglichen. Sie gelangte zum folgenden Ergebnis [7]:
"Im vorliegenden Artikel konnte gezeigt werden, daß **E/R** ungeachtet der

charakteristischen Unterschiede zu **P** grundlegende Gemeinsamkeiten mit **P** aufweist und folglich letztendlich auf eine gemeinsame Urquelle zurückgeht, die nach der Überlieferung sowie nach inneren Merkmalen eine der Versionen von al-Ḥaǧǧāǧ b. Yūsuf b. Maṭar war. Es konnte des weiteren nachgewiesen werden, daß **E/R** wie **P** eine Bearbeitung dieser Urquelle zu verkörpern scheint. In einem zentralen Punkt (II,3) unterscheiden sich jedoch beide Versionen, wobei auf der Grundlage der bei der Analyse von **P** aufgestellten Hypothesen und Argumente **E/R** in dieser Hinsicht als die ursprünglichere Version identifiziert wurde. Daraus scheint zu folgen, daß nicht beide Versionen von al-Ḥaǧǧāǧ selbst stammen können, da aus sprachlichen Gründen nach meiner Ansicht keine von ihnen direkt auf Version I, d.h. die Übersetzung von al-Ḥaǧǧāǧ zurückzuführen ist.

Aufgrund des archaischeren Ausdrucks bei der Beschreibung von Quadraten und Rechtecken in **P**, der erheblich größeren Verbreitung der entsprechenden Ausdrucksweise in **E/R** und weiteren, über die Bearbeitungsstufe von **P** hinausgehenden Veränderungen in **E/R** wurde geschlußfolgert, daß **P** diejenige der beiden vorliegenden Versionen ist, die letztendlich direkt von al-Ḥaǧǧāǧ herrührt. Das bedeutet, daß **E/R** von einem anderen Bearbeiter stammen dürfte. Schließlich konnten Argumente dafür erbracht werden, daß zwischen **P** und **E/R** eine andere, noch näher mit **P** verbundene *ḍarb*-Version existierte und daß es eine komplette ḍarb-Version der *Elemente* gegeben hat."

De Young fährt in seinem Aufsatz fort mit dem Material, das in Buch III.24, 32, 34–36 und IV.5 enthalten ist, und sagt [De Young 1991, 650]: "In each case, we find statements to the effect that Isḥāq did not include any other cases, but that al-Ḥajjāj included additional cases, which then follow." Über III.36 haben wir schon gesprochen, so daß wir diese Stelle hier übergehen können. Satz III.24 besagt: Wenn ein Kreisabschnitt gegeben ist, an ihn den Kreis zu zeichnen, dessen Abschnitt er ist. In der Isḥāq-Konstruktion wird nur ein Fall besprochen, nämlich derjenige, bei dem der Kreisabschnitt kleiner als der Halbkreis ist, d.h. nur der erste Fall der griechischen Konstruktion wird gegeben. Gerhard fügt nach III.30 einen Zusatz hinzu, der beginnt mit: *In vicessimo quarto theoremate repperi in quodam libro unam solam figuram, cum in hoc libro tres contineantur, que est quasi sufficiens.* In diesem Zusatz wird die Größe des Abschnittes außer Betracht gelassen. Es mag sein, daß diese Konstruktion auf die von Heron zurückgeht, denn in **L** [Besthorn und Heiberg 1893ff., II.2, 97] wird am Ende von III.24 gesagt: *Hanc propositionem postposuit Hero eamque propositionem XXXI fecit, quia ei propositum erat eam una figura demonstrare*

(*quia... demonstrare* fehlt in **A**) . Aber in dem in **L** nach III.31 gegebenen
Satz von Heron ([Besthorn und Heiberg 1893ff., II.2, 111, 113]; ebenso in **A**
[Curtze 1899, 135f.]) wird der gegebene Bogen halbiert (**L**, Satz III.29) und
nicht die Strecke, und auch die Verfahren bei Heron und Gerhard sind
verschieden. Da Heron den Satz III.29 verwendet hat, mag das der Grund
gewesen sein, warum Gerhard die Konstruktion erst nach III.30 hinzufügte
(siehe auch **L**: *Commentator dixit* [Besthorn und Heiberg 1893ff., II.2,
113]). Nachdem er die Konstruktion bei Isḥāq beschrieben hat, sagt De
Young [1991, 651]: "al-Ḥajjāj added a brief discussion of the case when the
given segment is equal to a semi-circle. In this case, we need only bisect
the line connecting the two endpoints in order to find the center and
radius of the desired circle. He follows this with a full proof of the case
when the given segment is greater than a semi-circle," d.h. er gibt die
griechische Konstruktion: zuerst der Abschnitt, der kleiner ist als der
Halbkreis; dann der Halbkreis; schließlich derjenige, der größer ist,
während die Reihenfolge bei Adelard, Gerhard und in **L** ist: zuerst der
Halbkreis; dann der Abschnitt, der größer ist als der Halbkreis; schließlich
derjenige, der kleiner ist.

Satz III.32 besagt: Über einer gegebenen Strecke einen Kreisabschnitt
zu zeichnen, der einen einem gegebenen geradlinigen Winkel gleichen
Winkel faßt. Isḥāq behandelt nur den Fall des spitzen Winkels, d.h. den
ersten Fall der griechischen Konstruktion, obwohl die Verfahren verschieden sind. Die Reihenfolge im griechischen Euklid ist: der Winkel ist
entweder spitz oder ein rechter oder stumpf. Ohne zu erwähnen, ob der
Winkel ein rechter, spitz oder stumpf ist, bringt Gerhard einen alternativen Beweis, den er einleitet mit: *Hoc preterea theorema in alio libro hoc
modo invenitur.* Es ist möglich, daß dieser Beweis von T̲ābit herrührt, denn
in dem Manuskript Teheran, Mālik 3586, wird gesagt "that Thābit found
three figures in the al-Ḥajjāj text and it was he who combined them into
one" [De Young 1991, 652]. **L** bietet ebenso wie Adelard und Gerhard (im
ersten Beweis) drei Fälle in derselben Reihenfolge wie in den Handschriften **E**/**R**: der Winkel ist entweder ein rechter oder stumpf oder spitz.
Jedoch enthält der Beweis des ersten Falles in **L** einen Zusatz, der
wahrscheinlich vom Verfasser hinzugefügt wurde [Besthorn und Heiberg
1893ff., II.2, 115]: *Et diametrus circulum in duas partes aequales dividit, ut
Simplicius in definitione libri primi demonstravit.*

Satz III.34 besagt: Schneiden im Kreis zwei Sehnen einander, so ist das
Rechteck aus den Abschnitten der einen dem Rechteck aus den Abschnitten der anderen gleich. Isḥāq gibt nach griechischem Muster nur den
zweiten Fall, daß die zwei Sehnen nicht durch den Mittelpunkt gehen und

einander nicht halbieren (den ersten Fall im griechischen Euklid: die zwei
Sehnen gehen durch den Mittelpunkt, hat er ausgelassen). Der alternative
Beweis von Gerhard enthält nur diesen Fall, wobei zu bemerken ist, daß
Gerhard von *superficies rectorum angulorum* spricht, was darauf hindeutet,
daß der Beweis auf Isḥāq/Ṯābit zurückgeht. Dagegen spricht Gerhard in
der Enuntiation von III.34 von *multiplicatio unius in aliam*, was auf eine
ḍarb-Version hinweist. Bei Adelard und im ersten Beweis von Gerhard
werden sechs Fälle unterschieden. In Fall 3 wird nachdrücklich gesagt, daß
die Strecke e g, und in Fall 4, daß die Strecke e b die längere ist, was man
auch in **L** findet. In Fall 5 wird nachdrücklich gesagt, daß die Strecken b e
und g e die längeren sind, was in **L** fehlt, und in Fall 6 wird der allgemeine
Beweis gegeben, der mit dem von Isḥāq und **L** (Fall 5) übereinstimmt. Zu
bemerken ist, daß dieser Beweis in **L** viel weitläufiger ist als bei den
anderen, da er auch die zweite Gleichheit herleitet. In Fall 6 in **L** wird der
Satz bewiesen für den Fall, daß die Sehnen einander senkrecht schneiden
und weder durch den Mittelpunkt gehen noch einander halbieren. Dieser
Fall fehlt bei allen anderen.

Satz III.35 besagt: Wählt man außerhalb eines Kreises einen Punkt und
zieht von ihm aus zum Kreis zwei Strecken, von denen die eine den Kreis
schneidet, die andere ihn berührt, so muß das Rechteck aus der ganzen
schneidenden Strecke und dem außen zwischen dem Punkt und dem
erhabenen Bogen angrenzenden Stück dem Quadrat über der Tangente
gleich sein. Isḥāq gibt wiederum nur einen Beweis, der mit dem alternativen Beweis von Gerhard und mit dem zweiten Beweis im griechischen
Euklid übereinstimmt (den ersten Beweis: die Strecke, die den Kreis
schneidet, geht durch den Mittelpunkt, hat er ausgelassen). Auch hier
können wir bei Gerhard denselben terminologischen Unterschied feststellen: *superficies rectorum angulorum* und *quadratum* im alternativen
Beweis, *multiplicatio totius linee in eam partem* und *multiplicatio linee in se
ipsam* in seinem eigenen Beweis. **E/R** gibt wie auch Adelard, Gerhard und
L drei Fälle. **L** verwendet aber im Gegensatz zu Adelard und Gerhard die
Worte "Rechteck" und "Quadrat." Die Reihenfolge ist bei allen dieselbe.

Satz IV.5 besagt: Einem gegebenen Dreieck den Kreis umzubeschreiben.
Auch in diesem Fall ist der alternative Beweis, den Gerhard bringt, Isḥāq
entnommen. Adelard, Gerhard und **L** unterscheiden wie **E/R** drei Fälle in
derselben Reihenfolge: für das rechtwinklige, für das stumpfwinklige und
für das spitzwinklige Dreieck. Schließlich ist noch zu bemerken, daß
Gerhard in den Sätzen III.32, 34−36 und IV.5 zuerst den Beweis gibt, der
mit dem Namen al-Ḥaǧǧāǧ verbunden ist, und dann den nach Isḥāq. In
den nun folgenden Sätzen ist dies umgekehrt.

Satz VIII.20 (= 21 Gerhard, 22 Heiberg) besagt: Bilden drei Zahlen eine geometrische Reihe und ist die erste eine Quadratzahl, so muß auch die dritte eine Quadratzahl sein.

Satz VIII.21 (= 22 Gerhard, 23 Heiberg): Bilden vier Zahlen eine geometrische Reihe und ist die erste eine Kubikzahl, so muß auch die vierte eine Kubikzahl sein.

An beiden Stellen geben die Handschriften Escorial und Rabāṭ 1101 zuerst die Ishāq-Ṭābit-Fassung (Rabāṭ 53 enthält nur die ersten fünf Bücher) und dann die Beweise nach al-Ḥaǧǧāǧ. Dieselbe Reihenfolge hat auch Gerhard. Adelard dagegen bringt die al-Ḥaǧǧāǧ-Fassung:

VIII.21 al-Ḥaǧǧāǧ [8]:

> When the first of any four continuously proportional numbers is a cube, [the fourth is a cube].
> An example of that is that the four numbers, *A, B, G, D*, are continuously proportional, and the first, namely *A*, is a cube. I say that the fourth, namely *D*, is a cube.
> The proof is that the ratio of *A* to *B* is as the ratio of *B* to *G* and as the ratio of *G* to *D*. Thus there fall between the two numbers, *A* and *D*, two [other] numbers, namely *B, G*, and they are continuously proportional. Thus *A, B* [*B* should here read *D*] are similar solid numbers. But *A* is a cube. Therefore, *D* is a cube.

Adelard VIII.21:

> *Exempli gratia: Sint quattuor numeri proportionaliter se comitantes* a *et* b *et* g *et* d. *Sitque* a *cubicus. Dico itaque quia* d *cubicus.*
> *Rationis causa: Quantus* a *ad* b *tantus* b *ad* g *tantusque* g *ad* d. *Sunt ergo inter* a *et* d *duo numeri proportionales suntque* b *et* g. *Quare* a *et* d *solidi et similes. Atqui* a *cubicus. Erit igitur* d *cubicus. Et hoc est quod demonstrare intendimus.*

Dasselbe gilt auch für die Sätze X.68–70 Heiberg (= 62–64 Adelard; 65–67 Gerhard; 64–66 al-Ḥaǧǧāǧ), die sich auf die Sätze X.57–59 Heiberg (= 51–53 Adelard; 54–56 Gerhard und al-Ḥaǧǧāǧ) statt auf X.39–41 Heiberg (= 36–38 Gerhard und al-Ḥaǧǧāǧ; bei Adelard fehlen diese Sätze) stützen. Nur die Handschrift Escorial enthält diese al-Ḥaǧǧāǧ-Sätze. Auch hier gibt Adelard die al-Ḥaǧǧāǧ-Fassung [Busard 1983a, 267f.; De Young 1991, 660]. Wir haben schon bemerkt, daß Gerhard in den Sätzen III.24, 32, 34–36 und IV.5 zuerst den Beweis nach al-Ḥaǧǧāǧ gibt und in den Sätzen VIII.20, 21 und X.68–70 zuerst den Beweis nach Ishāq-Ṭābit. Wenn wir annehmen, daß Gerhard nichts an seiner Vorlage geändert hat, muß diese Vorlage die Sätze ebenfalls in dieser Reihenfolge enthalten haben. Dies bedeutet, daß diese Vorlage ebenfalls eine Mischung von al-Ḥaǧǧāǧ- und Isḥāq-Ṯābit-Sätzen enthalten hat. Andernfalls kann ich nicht erklären, warum in Buch III gesagt wird: *in alio libro invenitur* und anschließend der Beweis von Isḥāq gegeben wird, während in den Büchern I und II wahrscheinlich eine Isḥāq-Ṯābit-Fassung vorliegt.

Hiermit möchte ich meine Bemerkungen zum Aufsatz von De Young beenden. Zum Schluß möchte ich noch untersuchen, welche alternativen Beweise der Gerhard-Text zusätzlich noch enthält.

VIII

Die Zusätze von Gerhard zu I.35 (*Huius preterea theorematis probatio aliter invenitur*) und I.44 (*Huius quoque theorematis dispositio invenitur hoc modo*) habe ich nirgendwo sonst gefunden.

Der am Ende von II.4 hinzugefügte Beweis (*Thebit dixit: In alia repperi scriptura quod alio ostenditur modo*) stimmt mit dem alternativen Beweis von II.4 im griechischen Euklidtext überein (siehe [Busard 1987, 57f.]).—Ich habe schon bemerkt, daß Gerhard und Isḥāq-Ṯābit in II.14 ebenso wie im griechischen Euklidtext den Beweis für eine beliebige gegebene geradlinige Figur bringen, während al-Ḥaǧǧāǧ, Adelard und L den Beweis nur für den Fall des Dreiecks geben. Auf diesen Beweis verweist Gerhard am Ende von II.14.

Am Ende von III.9 heißt es bei Gerhard: *Thebit dixit: Inveni in alia scriptura greca huius figure aliam probationem*. Dieser Beweis stimmt mit dem alternativen Beweis im griechischen Text überein [Busard 1987, 74, Heath 1956, II, 22]. A sagt hierzu: *Dixit Yrinus, quod nona figura consistit secundum hoc, quod dixit Euclides*. Aber der Zusatz zu III.10, der mit folgenden Worten eingeleitet wird: *Dixit Thebit: Inveni in alia greca scriptura aliam probationem*, wird in A und L Heron zugeschrieben. Er stimmt mit dem alternativen griechischen Beweis überein ([Busard 1987, 75]; L [Besthorn und Heiberg 1893ff., II.2, 45]; A [Curtze 1899, 120f.]).—Der Beweis, den Gerhard in III.11 für den Fall bringt, daß, wenn zwei Kreise einander außen berühren, die Verbindungsstrecke ihrer Mittelpunkte durch den Berührungspunkt geht, stimmt mit dem griechischen Beweis von III.12 überein [Busard 1987, 76, Heath 1956, II, 27f.]. In L [Besthorn und Heiberg 1893ff., II.2, 47] und A [Curtze 1899, 121f.] wird dieser Beweis Heron zugeschrieben.—Der Beweis in III.30 (= III.31 Heiberg) stimmt bei Gerhard für den Fall, daß der Winkel im Halbkreis ein rechter ist, mit dem alternativen griechischen Beweis überein, während der griechische Beweis mit dem Zusatz bei Gerhard übereinstimmt, der eingeleitet wird mit: *Dixit Thebit: Inveni in alia scriptura greca aliam huius figure probationem* [Busard 1987, 87f., Heath 1956, II, 62, 64].—Über die Sätze III.32, 34–36 und IV.5 habe ich schon gesprochen.

Der Beweis in IV.8 ist bei Gerhard identisch mit dem bei Adelard. Der alternative Beweis bei Gerhard, der griechische und der in **L** sind untereinander und von dem bei Adelard/Gerhard verschieden.—In IV.15 gibt Gerhard zuerst einen Beweis, der mit dem bei Adelard übereinstimmt, dann folgt ein alternativer Beweis und danach die Bemerkung, daß sich auf ähnlichem Wege, wie beim Fünfeck ausgeführt, einem gegebenen Kreis das Sechseck umbeschreiben sowie einem gegebenen Sechseck der Kreis ein- und umbeschreiben läßt. Diese Bemerkung folgt bei Adelard unmittelbar nach dem Beweis. Am Ende steht bei Adelard und bei Gerhard dann das Korollar. Die Differenzen zwischen dem griechischen Beweis, **L** und Gerhard sind größer als die Unterschiede zwischen Gerhard und Adelard, aber alle Texte folgen demselben Beweisgang. Der wichtigste Unterschied zwischen den beiden Beweisen bei Gerhard besteht darin, daß im alternativen Beweis nicht gezeigt wird, daß die Mittelpunktswinkel alle 60° betragen.

Bei Gerhard stimmt V.Def.18 überein mit Def.16 Adelard, Def.18 **L** und der alternativen Definition im griechischen Euklid [Busard 1987, 110]. Die alternative Definition bei Gerhard, die eingeleitet wird mit: *Dixit Thebit: in alia scriptura inveni*, stimmt mit der im griechischen Euklid überein. Nicht gefunden habe ich die zweite bei Gerhard hinzugefügte Definition: *Quattuor quantitates ordinatim sunt proportionales ut sumatur proportio antecedentis ad consequens et alterius antecedentis ad consequens.*—Den Satz V.5 habe ich schon erwähnt.—In V.18 stimmen Gerhard, Adelard und der griechische Euklid miteinander überein. **L** stimmt nur im ersten Teil seines Beweises mit Gerhard überein. Im zweiten Teil bringt **L** einen vollständigen Beweis, während bei Gerhard nur der folgende Satz steht: *Et ita ostenditur etiam quod non est proportio* a g *ad* g b *sicut proportio* d z *ad quantitatem que sit maior* z e. Nirgendwo gefunden habe ich Gerhards alternativen Beweis: *Huius preterea theorematis alia invenitur probatio que est huiusmodi* (siehe aber [Heath 1956, II, 172f.]).

Über VI.Def.2 und 3 habe ich schon gesprochen.—Von VI.Def.4 gibt Gerhard noch eine alternative Formulierung: *Dixit Thebit: Inveni in alia scriptura, quod altitudo est maior perpendicularis cadens a quolibet punctorum super basim aut super lineam que in eius existit rectitudine.* Diese Definition habe ich nirgendwo gefunden.—Von dem Zusatz zu VI.8 gibt Gerhard zwei Lesarten. Die erste stimmt mit der von Adelard überein. Von der alternativen Lesart stimmt der erste Teil mit dem Zusatz überein, der im griechischen Euklidtext steht; der zweite Teil (Außerdem ist zwischen der Grundlinie und einem beliebigen ihrer Abschnitte die dem Abschnitt anliegende Seite Mittlere Proportionale), der eingeleitet wird mit: *Quod tamen in greco non invenitur*, fehlt in den besten Theon-Handschriften [9]. In **L** fehlt der ganze Zusatz, aber in einer Randbemerkung wird gesagt: *Ṯābit ibn Qurrah dicit: Invenimus hic in quibusdam codicibus haec, quae in Graecis codicibus non invenimus*, und dann folgt der ganze Zusatz (**L**, Buch VI, [Besthorn und Heiberg 1893ff., III.2, 125]). - VI.12 Gerhard (= VI.9 Heiberg, VI.11 Adelard): *Huius preterea theorematis dispositio aliter invenitur hoc scilicet modo*, stimmt mit Adelard, **L** und dem griechischen Euklid überein. Der Beweisgang ist zwar in allen Texten derselbe, aber Adelard und der griechische Euklid enthalten eine kürzere, Gerhard und **L** eine längere Fassung. Der kürzeste und schlechteste Beweis von allen ist der alternative Beweis bei Gerhard: *Inveni in aliis scripturis aliam ab hac quam modo premisi probationem, que est huiusmodi.*—Der Beweis von VI.32 Gerhard (= VI.31 Adelard und Heiberg, VI.32 **L**) stimmt mit dem alternativen griechischen Beweis überein [Busard 1987, 151, Heath 1956, II, 270], während der alternative Beweis bei Gerhard mit Adelard und **L** übereinstimmt.

Bei Gerhard findet man im Anschluß an VII.Def.ix einen Zusatz, der nach De Young [1981, 289] von Ṯābit herrührt: "Thābit ibn Qurrah says we found in the Arabic text after the statement of the odd times even [that] if its half is even it is called even times even times odd and we did not find that in any of the Greek [texts]." Gerhard erwähnt Ṯābits Namen nicht: *Post hoc quod dicitur de impariter pari* repperi *in alia arabica scriptura hoc: Quando fuerit medietas impar, nominatur par impariter; et quando fuerit medietas par, nominatur par pariter et impariter. Neque* repperi *illud in greco.* —Nach VII.28 sagt Gerhard: *Thebit inquit: Hanc figuram et illam que sequitur post hanc repperi in greco post duas figuras que post eas sequuntur.* Diese Aussage betrifft die Sätze VII.29 und 30 bei Adelard und Gerhard, die bei Heiberg nach den Sätzen VII.31 und 32 stehen.

Nach dem Beweis von VIII.6, der endet mit: *Et illud est quod demonstrare voluimus*, sagt Gerhard noch: *Et secundum similitudinem eius quod antecessit in hac propositione ostenditur quod si primus fuerit numerans secundum, tunc ipse numerabit etiam postremum.* Mir ist nicht bekannt, wo diese Behauptung bewiesen wird; auch De Young [1981, 123−126, 407f., 626−628] gibt den Satz nicht. Auf dieser Behauptung beruht Gerhards Beweis von VIII.7. Nach diesem Beweis sagt Gerhard noch: *In alio libro erat: Sed si a numerat d, tunc ipse etiam numerat b.* Diesen Beweis findet man bei Adelard und im griechischen Euklid [Busard 1987, 181]. Er steht aber *nicht* in den Versionen **A** und **B** des Isḥāq-Ṭābit-Textes. Dies ist ein weiterer Hinweis darauf, daß Gerhard keine reine Isḥāq-Ṭābit-Fassung benutzte [De Young 1981, 127, 409, 628f.].—In Ṭūsīs *Taḥrīr* lesen wir nach VIII.15 folgendes [De Young 1984, 153]: "And concerning the order of some of these propositions, it is different from what was presented to us according to the ordering of Thābit. As for al-Ḥajjāj, he presented to us what was given in proposition 11 and 12 [by Thābit] in proposition 11 alone, and he presented to us as proposition 13 what was presented as proposition 2 [by Thābit]; and there are presented in [al-Ḥajjāj] as proposition 13 and 14 the proofs presented [by Thābit] in propositions 14 and 15; and proposition 15 [of al-Ḥajjāj] is lacking [from Thābit's version]. After that, the two of them are in agreement." Gerhard sagt am Ende des Buchs VIII folgendes: *Quoniam in alio libro repperi quod dicitur in XI et XII theorematibus in uno tantum theoremate contineri ne aliquid nobis deesset curam hic ipsum theorema ponere.* Da der dann folgende Beweis mit dem bei Adelard und demjenigen übereinstimmt, der in der arabischen Handschrift in Leningrad enthalten ist (siehe [De Young 1981, 634]), geht er wahrscheinlich auf al-Ḥaǧǧāǧ zurück. Übrigens enthält auch der griechische Euklid zwei Sätze. Der zweite davon wird nur in der Leningrader Handschrift (als Satz 13) gegeben, nicht aber bei Adelard und Gerhard. Der Beweis von VIII.13 bei Adelard (= VIII.14 Gerhard und Isḥāq-Ṭābit) stimmt mit dem bei Gerhard und in der Gruppe **B** des Isḥāq-Ṭābit-Textes überein; allerdings fehlt der Zusatz, den Gerhard und die Gruppe **B** aufweisen, bei Adelard und in der Leningrader Handschrift (siehe [De Young 1981, 637]: "but Leningrad lacks the porism which is found in most Isḥāq-Thābit texts"). Dasselbe gilt für den Beweis von VIII.14 bei Adelard (= VIII.15 Gerhard und Isḥāq-Ṭābit) und für den Zusatz nach dem Beweis [De Young 1981, 638], der in allen Isḥāq-Ṭābit-Fassungen vorhanden ist. Beide Zusätze fehlen auch im griechischen Euklid.—Der Satz VIII.15 bei Adelard stimmt mit VIII.16 Gerhard und VIII.15 der Leningrader Handschrift überein [De Young 1981, 640]. In den Gruppen **A** und **B** des Isḥāq-Ṭābit-Textes ist dieser Satz in den Zusätzen zu VIII.14 und 15 enthalten und fehlt deshalb hier. Da der Satz in einer Gerhard-Handschrift (Vat. Ross. 579) am Rande steht, ist es möglich, daß er später dem Gerhard-Text zugefügt worden ist. Wahrscheinlicher aber ist für mich die Annahme, daß diese Stelle ein weiterer Hinweis darauf ist, daß für den Gerhard-Text verschiedene Quellen benutzt wurden. Die beiden Zusätze sind im griechischen Euklid als Sätze VIII.16 und 17 aufgenommen worden. Der letzte Satz bei Ṭūsī stimmt nicht, denn in einer Randbemerkung wird noch gesagt [De Young 1984, 153]: "What he [Thābit] discussed in the case of propositions 17 and 18 is discussed in the reverse order [by the shaykh]; and he [Thābit] introduced the converse of propositions 24 and 25 in two propositions like them. Thus the [number of the] propositions became 27. As for what the shaykh discussed, it is consistent with the text of al-Ḥajjāj."—Die Reihenfolge der Sätze VIII.17 und 18 in den Gruppen **A** und **B** und bei Gerhard (dort = VIII.18 und 19) ist dieselbe, während sie bei Adelard (dort = VIII.17 und 18) und in der Leningrader Handschrift (dort = VIII.17 und 18) [De Young 1981, 641] umgekehrt ist und deshalb mit der von al-Ḥaǧǧāǧ übereinstimmt.—Weiterhin fehlen bei Adelard und in der Leningrader Handschrift die Sätze VIII.24 und 25 (= VIII.25 und 26 Gerhard), während die Gruppen **A** und **B** diese Sätze, die Anaritius Heron zuschreibt (**A**, [Curtze 1899, 195]; siehe auch [De Young 1981, 647]), enthalten.—Über die Sätze VIII.21 und 22 Gerhard (= VIII.20 und 21 Adelard) habe ich schon gesprochen.

Gerhard bringt in IX.4 einen Teil des Zusatzes, der sich in IX.5 befindet. Einige Isḥāq-Ṯābit-Handschriften enthalten den Zusatz sowohl in IX.4 als auch in IX.5 [De Young 1981, 653]. Im griechischen Euklid fehlt dieser Zusatz, aber bei Hermann von Kärnten und Robert von Chester, die zur Adelard-Tradition gehören (die erhaltenen Adelard-Handschriften haben eine Lücke von IX.1 bis X.35), ist dieser Zusatz in IX.5 vorhanden.—Ṭūsī sagt in IX.12: "in the text of al-Ḥajjāj this proposition precedes that which is before it" [De Young 1981, 660], und in IX.14: "this is proposition 20 in the text of al-Ḥajjāj" [De Young 1981, 662]. Gerhard folgt hier wiederum Isḥāq-Ṯābit, während die Leningrader Handschrift, Hermann und Robert dieselbe Reihenfolge wie al-Ḥaǧǧāǧ aufweisen. Der griechische Euklid stimmt mit Gerhard überein in der Reihenfolge der Sätze 11 und 12, mit der Leningrader Handschrift aber in Bezug auf Satz 20.—In einer Randbemerkung der Leningrader Handschrift zu IX.24 wird gesagt: "and by this it may be shown that if an odd number is subtracted from an odd number, the remainder is even" [De Young 1981, 667]. Diese Aussage bildet bei Hermann und Robert den Satz IX.25, in der Isḥāq-Ṯābit-Tradition und bei Gerhard IX.27, während der Satz in der Leningrader Handschrift fehlt.—Die Reihenfolge der Sätze, die in der Isḥāq-Ṯābit-Tradition und bei Gerhard IX.25 und 26 sind, ist bei Hermann, Robert und in der Leningrader Handschrift vertauscht (Hermann, Robert: IX.27 und 26; Leningrad: IX.26 und 25; siehe [De Young 1981, 668]).—Über die Sätze IX.30 und 31, die im griechischen Euklid fehlen, aber in allen arabischen Handschriften vorhanden sind, wird in den arabischen Handschriften Escorial, Oxford Thurston und Uppsala folgendes berichtet: "Thābit said: we did not find proposition thirty and thirty one in the Greek texts which we have, but we found them in the Arabic" [De Young 1981, 263, 539]. Bei Gerhard steht eine ähnliche Bemerkung. Auch Hermann und Robert bringen diese Sätze, die wahrscheinlich auf al-Ḥaǧǧāǧ zurückgehen.

Der griechische Beweis von X.6 stimmt mit dem bei Gerhard überein, während der alternative griechische Beweis [Busard 1987, 216] sich sowohl bei Gerhard (als alternativer Beweis) als auch bei Anaritius findet (A, [Curtze 1899, 217]).—Der griechische Beweis von X.14 [Busard 1987, 223] stimmt mit dem alternativen Beweis, den Gerhard in X.12 gibt, überein.—Nach dem Beweis von X.24 Gerhard (= X.29 Heiberg, X.17a Robert von Chester) sagt Gerhard: *Huius theorematis invenire alia probatio et est diversitas in hoc quod cum hic dicatur quod longior linea addit super breviorem equale quadrato linee communicantis sibi in longitudine, ibi dicitur incommunicantis sibi in longitudine.* Der Satz, auf den hier hingewiesen wird, ist X.30 Heiberg (= X.17b Robert, X.17 Hermann). Den Beweis dieses Satzes findet man auch bei Gerhard (in X.25); er sagt am Ende: *Probatio que premittitur huic theoremati et dicitur esse precedentis est huius sed secundum alium librum.* Da die Buchstaben für die verschiedenen Punkte im alternativen Beweis von Gerhard mit denjenigen bei Robert und Hermann übereinstimmen, ist der Beweis von Robert und Hermann wahrscheinlich der *secundum alium librum* [10].—Die Sätze X.27 und 28 Heiberg finden sich bei Pseudo-Ṭūsī in einem Anhang zu X.23: "Hier ist folgendes bereits geklärt: Wir sollen zwei quadriert kommensurable mediale Linien finden, die eine rationale Fläche umfassen (X.27), und sollen zwei mediale Linien finden, die eine mediale Fläche umfassen (X.28), was Ṭābit ibn Qurra in seiner Ausgabe als Propositionen 21 und 22 bringt. Ḥaǧǧāǧ läßt beide fort, da sie sich in den alten Ausgaben nicht finden" [Thaer 1936, 119]. Bei Gerhard sind diese Sätze vorhanden (X.21 und 22), während sie bei Hermann und Robert fehlen.—Nach dem Beweis von X.30 Gerhard (= X.33 Heiberg) sagt Gerhard: *Huius autem theorematis probatio secundum alium librum brevior invenitur. In propositione vero et exemplo et dispositione nichil mutatur nisi quod non est ibi de duplo superficiei unius in aliam aliquid et in figura littere secundum loca mutantur, quia* e *est in loco* d *et* d *in loco* z *et* z *in loco* e, *cetera non mutantur.* Tatsächlich sind in X.25 Robert (= X.25 Hermann) die Buchstaben in der Weise geschrieben, wie sie hier angegeben sind, und es gibt auch eine gewisse Übereinstimmung zwischen den Beweisen (siehe auch [Curtze 1899, 313f.], und [Kunitzsch 1985, 126]).—Nach dem alternativen Beweis sagt Gerhard: *Refert Thebit, qui transtulit hunc librum de greco in arabicam linguam, se invenisse quod additur ante figuram tricesimam primam huius partis in quibusdam scriptis grecis cuiusdam Iohanicii Babiloniensis, quod tamen non est de libro.* Der dann folgende Hilfssatz stimmt überein mit dem vor X.33 gegebenen griechischen Lemma [Busard 1987, 238].—Nach dem Beweis von X.40 Gerhard (= X.43 Heiberg) sagt Gerhard: *Hic additur quedam figura que non est huius libri neque invenitur in translatione Thebit, sed quia est necessaria theoremati in quo dicitur quod linea binomia non dividitur in suas lineas et in sua nomina nisi in uno puncto tantum et aliis que ipsum sequuntur, ideo adiuncta est et dicitur esse cuiusdam qui vocabatur iezidi.* Nach Kunitzsch [1985, 127] könnte dieser Satz übereinstimmen mit einem Satz bei Anaritius (**A**, [Curtze 1899, 250–252]). Dies trifft aber nicht zu, denn es handelt sich um zwei verschiedene Beweise. Eher ist dieser Satz eine Bearbeitung des griechischen Hilfssatzes, der nach X.41 Heiberg (= X.38 Gerhard) eingefügt worden ist [Busard 1987, 243]. Gerard hat ihn wahrscheinlich nach X.40 Gerhard gesetzt, weil er im nächsten Satz (X.41 Gerhard = X.44 Heiberg) angewandt wird. Auch die Aussage von Gerhard: *quia est necessaria theoremati in quo dicitur quod linea binomia non dividitur in suas lineas et in sua nomina nisi in uno puncto tantum et aliis que ipsum sequuntur,* deutet darauf hin, daß der Satz nach X.38 gehört, denn in X.39 Gerhard wird bewiesen: Eine Binomiale zerfällt nur in einem Punkt in Glieder, in X.40 Gerhard: Eine Erste Bimediale zerfällt nur in einem Punkt, usw. (X.41–44 Gerhard).—Über die Sätze X.65–67 Gerhard habe ich schon gesprochen.—In X.89 Gerhard (= X.91 Heiberg, X.86 Adelard) wird im Anschluß an den Beweis gesagt: *Ultima autem probatio huius theorematis invenitur aliter.* Es ist nicht klar, ob hiermit der ganze Beweis gemeint ist. Falls nur der Teil bei Gerhard mit dem alternativen Beweis gemeint ist (Anfang: *Sed linea z d* [Busard 1983b, 312, Z. 28] ... bis Ende), stimmt dieser Teil mit Adelard überein (Anfang: *Atqui a d longitudine* [Busard 1983a, 282, Z. 978] ... bis Ende).—Die Enuntiation von X.109 Gerhard ist identisch mit der bei Isḥāq-Tābit. Der erste Teil dieses Satzes stimmt mit dem Zusatz nach X.111 Heiberg [Busard 1987, 301f.] überein. Dann wird gesagt: *Aliter secundum alium librum hoc idem.* Dieser Teil stimmt mit Adelard X.106a überein. Dann folgt: *Linea que est residuum non est ea que est binomium.* Der Beweis dieses Satzes stimmt mit dem griechischen überein [Busard 1987, 301]. Dann folgt eine Alternative für einen Teil dieses Beweises (*Cum ergo diviserimus ... sub eius diffinitione*), die mit einem Teil von X.106b Adelard übereinstimmt (Anfang: *Cumque separaverimus* [Busard 1983a, 296, Z. 1359] ... bis Ende). Dann folgt ein Teil, der eingeleitet wird mit: *Hec autem probatio in alio invenitur libro.* Dieser Teil stimmt mit X.107a Adelard überein; allerdings ist das Ende bei Adelard gekürzt. Dann folgt der erste alternative griechische Beweis

[Busard 1987, 306], und es wird noch gesagt: *Hoc item aliter disponitur et probatur hoc modo.* Dieser Teil stimmt mit X.107b Adelard überein. Gerhard beendet den Satz mit: *Propositum huius probationis tale est: Ex lineis medialibus plures sunt et infinite quarum nulla est in termino alterius que est ante eam neque in eius ordine.* Bei Adelard heißt es: *Estque possibile esse linearum mediarum surdas multas et innumerabiles nec est ex eis linea in termino eius quod erat ante se nec in suo ordine.*

Am Ende der Definitionen von Buch XI sagt Gerhard: *Duo ex his elementis in alio libro sic inveniuntur.* Die erste Definition ist eine Alternative für Definition 3 und stimmt mit Heibergs Definition 4 überein; die zweite ist eine Alternative für Definition 6 und stimmt mit Adelard, Definition 6, überein.—Nach dem Beweis von XI.30 sagt Gerhard: *In alio libro inveni quod corpus* b h *completur sic.* Diese alternative Version stimmt mit XI.30 Adelard überein, und auch die beigefügten Figuren sind identisch.

In Buch XIII.1–3 werden für jeden Satz jeweils zwei verschiedene Beweise gegeben. Sehr wahrscheinlich summiert sich dadurch die Zahl der Sätze in Buch XIII sowohl bei al-Ḥaǧǧāǧ als auch bei Isḥāq-Ṯābit auf 21 statt auf 18.

IX

Hiermit möchte ich meine Untersuchung der arabisch-lateinischen Übersetzung von Gerhard von Cremona beenden. Was hat nun diese Untersuchung ergeben? Das wichtigste Ergebnis für mich ist die Erkenntnis, daß es arabische Handschriften gibt, die nicht nur eine, sondern zwei verschiedene Versionen enthalten, und daß die Einteilung des Isḥāq-Ṯābit-Textes in die beiden Gruppen **A** und **B**, die für die Bücher VII–IX gültig ist, sich nicht ohne weiteres auf die anderen Bücher übertragen läßt. Es würde mich z.B. nicht wundern, wenn es sich herausstellen würde, daß ein Teil von Buch III in dem Text, den Gerhard benutzte, auf eine al-Ḥaǧǧāǧ-Version und Buch III in der Handschrift Leningrad auf eine Isḥāq- oder Isḥāq-Ṯābit-Version zurückginge, Buch VII aber in der Vorlage Gerhards auf eine Isḥāq- oder Isḥāq-Ṯābit-Version und in der Leningrader Handschrift auf eine al-Ḥaǧǧāǧ-Version. Ich möchte ein ähnlich gelagertes Beispiel aus meiner Erfahrung im lateinischen Bereich geben: Die Handschrift London, British Library, Burney 275, enthält auf ff.293r–302r die Sätze II.1–VII.2 der sogenannten Adelard II-Version, auf ff.302r–308r die Sätze VII.3–VIII.25 der sogenannten Adelard I-Version und auf ff.308r–335r die Sätze IX.1–XV.2 der sogenannten Adelard III-Version. Warum dies so ist, weiß ich nicht; vielleicht enthielt die Vorlage, von der der Schreiber kopierte, nicht mehr Sätze der Version Adelard I, und der Abschreiber ergänzte den Text mit Hilfe anderer Fassungen. Etwas Derartiges erwarte ich auch im arabischen Bereich. Ehe man anfängt, gewisse Handschriften, wie z.B. Leiden 399/1, zu edieren, halte ich es daher für weit besser, nach dem Vorbild von M. Clagett, der etwas Ähnliches für den lateinischen Bereich mit seinem bahnbrechenden Aufsatz [Clagett 1953] geleistet hat, zuerst festzustellen, welche Teile in welchen arabischen Handschriften miteinander übereinstimmen. Erst danach wird sich eine Edition lohnen.

ANMERKUNGEN

[1] Sie wird von Sonja Brentjes untersucht. Nach Angaben von Frau Brentjes ähnelt der Text sehr demjenigen, der in L überliefert wird.

[2] In der zweiten arabischen Handschrift, die diesen Text überliefert (siehe Anm. 1), sind diese Passagen vorhanden.

[3] Neuerdings wurde Buch I nach diesen drei Handschriften neu ediert von [Tummers 1984, II, 121–190].

[4] Die letzte Definition fehlt in **A**; siehe [Curtze 1899, 138].

[5] In diesem hinzugefügten Satz wird auf I.12 verwiesen. In Wirklichkeit handelt es sich um VI.12.

[6] Dieses Ergebnis wird auch durch die Untersuchung von Engroff bestätigt. Siehe [De Young 1984, 149].

[7] [Brentjes 1993, 66]. Siehe auch ihre Bemerkungen zu den lateinischen Übersetzungen von Hermann von Kärnten und Adelard [Brentjes 1993, 59, 62, 64f.].

[8] [De Young 1981, 180]. Zur englischen Übersetzung von VIII.20 siehe [De Young 1984, 154f.].

[9] Siehe [Heath 1956, II, 211]. [Busard 1987, 132] enthält den ganzen Zusatz.

[10] Siehe auch [Curtze 1899, 300–302], wo derselbe Beweis zusammen mit einem Zahlenbeispiel gegeben und dann gesagt wird: *Communicantes vero in longitudine in elementis ostendimus.*

LITERATURVERZEICHNIS

Besthorn, R. O., and Heiberg, J. L. 1893ff. *Codex Leidensis* **399**, 1. *Euclidis Elementa ex interpretatione Al-Hadschdschadschii cum commentariis al-Narizii. Arabice et latine ediderunt.* I, 1893; II.1, 1900; II.2, 1905; III.1, 1910; III.2, 1932. Kopenhagen: Gyldendal.

Brentjes, S. 1993. Varianten einer Ḥaǧǧaǧ-Version von Buch II der *Elemente*. In *Vestigia Mathematica, Festschrift für H. L. L.Busard*, M. Folkerts and J. P. Hogendijk, Eds., pp. 47–67. Amsterdam: Rodopi.

Busard, H. L. L. 1967. The translation of the *Elements* of Euclid from the Arabic into Latin by Hermann of Carinthia (?): Books I–VI, *Janus* **54**, 1967, 1–140, und separat veröffentlicht (Leiden: E. J. Brill, 1968); Books VII–IX: *Janus* **59**, 1972, 125–187; Books VII–XII (Amsterdam: Mathematisch Centrum, 1977).

— 1983a. *The First Latin Translation of Euclid's Elements Commonly Ascribed to Adelard of Bath*. Toronto: Pontifical Institute of Mediaeval Studies.

— 1983b. *The Latin Translation of the Arabic Version of Euclid's Elements Commonly Ascribed to Gerard of Cremona*. Leiden: New Rhine Publishers.

— 1987. *The Mediaeval Latin Translation of Euclid's Elements made Directly from the Greek*. Stuttgart: Franz Steiner.

—, and Folkerts, M. 1992. *Robert of Chester's (?) Redaction of Euclid's Elements, the so-called Adelard II Version*. Berlin: Birkhäuser.

Clagett, M. 1953. The medieval latin translations from the Arabic of the elements of Euclid, with special emphasis on the versions of Adelard of Bath. *Isis* **44**, 16–42.

Curtze, M. 1899. *Anaritii in decem libros priores Elementorum Euclidis commentarii*. In *Euclidis Opera Omnia*, I. L. Heiberg, and H. Menge, Eds., Suppl. Leipzig: Teubner.

De Young, G. 1981. *The arithmetic books of Euclid's Elements in the Arabic tradition*. Ph.D. Dissertation (unpublished), Harvard University, Cambridge, MA.

— 1984. The Arabic textual traditions of Euclid's *Elements*. *Historia Mathematica* **11**, 147–160.

— 1991. New traces of the lost al-Ḥajjaj Arabic translations of Euclid's *Elements*. *Physis* **28**, 647–666.

Heath, T. L. 1956. *The Thirteen Books of Euclid's Elements*, 3 vols. New York: Dover.

Klamroth, M. 1881. Ueber den arabischen Euklid. *Zeitschrift der Deutschen Morgenländischen Gesellschaft* **35**, 270–326.

Kunitzsch, P. 1985. Findings in some texts of Euclid's Elements. In *Mathemata. Festschrift für Helmuth Gericke*. M. Folkerts and U. Lindgren, Eds., pp. 115–128. Stuttgart: Franz Steiner.

Murdoch, J. E. 1963. The medieval language of proportions: Elements of the interaction with Greek foundations and the development of new mathematical techniques. In *Scientific Change. Symposium on the History of Science, University of Oxford. 9–15 July 1961*, A. C. Crombie, Ed., pp. 237–271. Heinemann, London.

— 1971. Euclid: Transmission of the *Elements*. In *Dictionary of Scientific Biography*, C. Gillispie, Ed., 4, pp. 437–459. New York: Scribner's.

Sezgin, F. 1974. *Geschichte des arabischen Schrifttums*, Vol. 5. Leiden: E. J. Brill.

Thaer, C. 1936. Die Euklid-Überlieferung durch Aṭ-Ṭûsî. *Quellen und Studien zur Geschichte der Mathematik, Astronomie und Physik, Abt. B: Studien* **3**, (2), 116–121.

— 1962. *Euklid: Die Elemente*. Darmstadt: Wissenschaftliche Buchgemeinschaft.

Tummers, P. M. J. E. 1984. *Albertus (Magnus)' commentaar op Euclides' Elementen der Geometrie*, 2 vols. Nijmegen.

The *Book of Assumptions,*
by Thābit ibn Qurra (836–901)

Yvonne Dold-Samplonius

Mathematisches Institut, University of Heidelberg,
D-69120 Heidelberg, Germany

The *Book of Assumptions* (Kitāb al-Mafrūḍāt) by Thābit ibn Qurra [1940] exists in two versions. We know the original version by Thābit ibn Qurra (836–901). The only extant manuscript dates from the 11th century. The other version is the *Redaction* of the same by Naṣīr al-Dīn al-Ṭūsī (1201–1274), which exists in numerous manuscripts, of which the oldest date from the 14th century.

What is the relation between Thābit's treatise and al-Ṭūsī's redaction? The purpose of this paper is to compare the two versions and to discover how much of the original treatise is left in al-Ṭūsī's redaction. In addition a survey of the mathematical content of the treatise is given. An edition of this treatise is in preparation.

1. ON THE TITLE

Sezgin [1974, 491] lists two treatises with the title *Kitāb al-Mafrūḍāt*: (1) by Ps. Archimedes and (2) by Thābit ibn Qurra. The first treatise, which has been discussed before [Dold-Samplonius 1977, 1978], exists in two copies of unequal length. The shorter version, consisting of 19 propositions and entitled *Elements of Geometry* (K. fi l-Uṣūl al-handasīya), is said to be by Archimedes. An addition to the title reads: "Thābit ibn Qurra translated the treatise from Greek into Arabic for the astronomer/astrologer Abu l-Ḥasan 'Alī ibn Yaḥyā, companion to the Caliph." Thābit ibn Qurra [Dictionary of Scientific Biography (DSB) XIII, 1976, 288–295], who lived from 221 to 288 H (836–901 A.D.), was an outstanding scholar and translated many scientific treatises from Greek and Syriac into Arabic. His mathematical writings, the most studied of his works, comprise important contributions to number theory and spherical trigonometry. Thābit's attempt to prove Euclid's fifth postulate is based on the concept of motion and influenced Ibn al-Haytham (965–ca. 1040). Al-Khayyāmī (1048–1131), and, later, al-Tūsī objected against this procedure as being foreign to geometry. These attempts eventually led to the creation of non-Euclidean geometry. In astronomy Thābit was one of the first reformers of the Ptolemaic system, and in mechanics he was a founder of statics. Abu l-Hasan 'Alī ibn Yaḥyā (d. 888), the son of the astronomer Yaḥyā ibn Abi Manṣūr [DSB 1976 XIV, 537–538], was the companion to the Caliph al-Mutawakkil (847–861). He was eminent in Baghdad and had a large library in which Abu Ma'shar studied. The translation must have been made before 888, perhaps even before 861.

In the other, longer version no translator is mentioned. It bears the title *Kitāb al-Mafrūdāt* with Aqāṭun as the author. A marginal note at the end of the treatise states that "the amount of propositions by Aqāṭun is forty-three, counting improvements and additions." This longer version appears to be the more authentic one [Dold-Samplonius 1977, 1–8, cf. also Jones 1986, 603–604]. Which Greek mathematician is hidden behind the name Aqāṭun is not clear. Treatise 2, by Thābit ibn Qurra, is the subject of this paper.

What is the best rendition of the title "Kitāb al-Mafrūdāt?" *Mafrūd*, with the plural *mafrūdāt*, is the passive participle of the verb *farada*. This verb has various meanings, which we can express by the following: to decide, determine, decree, appoint, assign, assume, presume, and suppose. Hence its passive participle *mafrūd*, might mean "something that has been determined, appointed, etc.," as well as "something having been assumed, supposed, etc." We shall see that different meanings occur in the same treatise.

In Ibrāhīm ibn Sinān's treatise *On Analysis and Synthesis* (M. fī Ṭarīq at-Taḥlīl wa-t-Tarkīb fī l-masā'il al-handasīya) [1948, Saidan 1983, 73–143], a classification of mathematical propositions is worked out [Bellosta 1991]. Ibrāhīm ibn Sinān [DSB VII, 1973, 2–3], the grandson of Thābit ibn Qurra, divided the problems into two large classes, each of which is in turn divided into subclasses: "problems whose hypotheses and conditions are sufficient" and "problems whose hypotheses require a modification." For hypotheses the Arabic text reads *mafrūdāt*. Saidan [1983, 138] explains *mafrūdāt* in this context as "This is what nowadays is called *mu'ṭayāt*, i.e. 'data'." However, "data" is too strong. Bellosta's translation "hypotheses," or rather "assumptions," seems more accurate here.

Apart from this general level, the word *mafrūd* is used in several places in this treatise. In these examples the meaning of *mafrūd* is almost, or actually, interchangeable with the meaning of *ma'lūm*, known. This is remarkable! Examples where the word *mafrūḍ* occurs are

> A line is drawn through two points till point E, which is "determined". ... [Ibn Sinān 1948, 8, l. 8/9, Saidan 1983, 81, l. 3/4]
> A line is "known" in position and size [Ibn Sinān 1948, 8, l. 12/13, Saidan 1983, 81, l. 6] as well as the first example of the treatise: To divide an assumed (*mafrūḍ*) line in a known (*ma'lūm*) ratio. [Ibn Sinān 1948, 6, l. 9/10, Saidan 1983, 79, l. 16/17]

In the last case we can just as well translate "to divide a known line in a assumed ratio." We can substitute one term with the other without changing the meaning of the sentence in the least. This was also true for Arabic mathematicians, as is shown in the following example [Rashed 1993, 165, l. 5]: "Angle BAG and triangle D are known (*ma'lūmān*) and the ratio E to Z is given (*mafrūḍāt*)...," found in proposition 8 from the *Book on the Synthesis of Problems* (K. Tarkīb al-masā'il), analyzed by Abū Sa'd al-'Alā' ibn Sahl [Sezgin 1974, 341/342, Rashed 1993, 159–188]. Here we can also interchange and write, instead, "angle BAG and triangle D are 'assumed' and the ratio E to Z is 'known.'" The use of the two termini technici has been interchanged in comparison with the treatise by Ibn Sinān above. Hence already in that time the meaning of *mafrūḍ* and its verb varied. In other places of Ibn Sahl's treatise we have the same picture, *mafrūḍ* is used for "assumed" or "supposed," as well as for "known" or "given," etc. Examples include the following:

> ... such that the product of the known (*ma'lūm*) line-segment together with the extension and the extension be equal to a given (*mafrūḍ*) surface [Rashed 1993, 163, l. 8/9]
> Si on "suppose donnés" (*furiḍa*) un angle à cotés rectilignes et un point à l'intérieur de cet angle, de sorte que la droite joignant ce point et son extrémité, le partage en deux moitiés et une droite; nous entendons faire passer une droite par le point pour que la droite interceptée par l'angle soit égale à la droite "donnée" (*mafrūḍ*). [Translation by Rashed: Rashed 1993, 175, l. 10–12]

Hence we have various possibilities to translate the title. In Thābit's case the title is often translated as "Data." This seems appropriate, because the contents of Thābit's treatise are in the spirit of Euclid's *Data*. However, Euclid's *Data* is always rendered in Arabic by K. *al-Mu'ṭayāt*, which is the literal translation of the Greek title δεδόμενα. Dodge [1970, 636] translated it, in Archimedes' case, as "things determined," whereas Heath [1963, 286] wrote, here, *Book of Data*. Jones [1986, 603] interpreted Aqāṭun's K. *al-Mafrūḍāt* as "Lemmata," for which, however, the standard

Arabic translation is *K. al-Ma'khūdhāt*. Considering all this, I have translated the title as "Book of Assumptions."

2. TRANSMISSION OF THE TEXT

The text of Thābit's *Book of Assumptions* has come down to us in the original version, by Thābit himself, and in the *Redaction* (taḥrīr), by Naṣīr al-Dīn al-Ṭūsī. Thābit's version is extant in an unique manuscript dating from the 5th century H (~ 11th century A.D.), which is contained in Aya Sofya manuscript 4832. It was first mentioned by Max Krause [1936, 453], who also remarked that one folio, containing propositions 3 to 9, is missing. This missing page was located by Jan Hogendijk, while looking through the microfilm of the entire manuscript, as folio 47. Hogendijk explained to me that "The pages have been numbered twice, first in the extreme upper left, and again, after the ms. had been rebound, more downward, nearer to the text. In the second numeration the original folio 37 has now become folio 36, folio 38 has become folio 37 etc., and the original folio 36 has been wrongly rebound between the original folii 47 and 48. The original folio 36 was apparently the beginning of a new quire (Latin: *quaternio*), quire number 4 (d), the Arabic letter d is still visible under the number 36. Such a page can easily fall off. Each quire probably contained 12 folii, the numbers a, b, and g (in Arabic letters) probably started on the frontpage, f. 12, and f. 24". Apparently the first page of the fourth quire, being loose, was rebound as the last page of the same quire. The signature of Thābit's treatise should accordingly be Aya Sofya 4832, fols. 35v, 47r–v, 36r–39r.

Thābit's treatise is also readily available in the redaction by Naṣīr al-Dīn al-Ṭūsī (1201–1274) [DSB XIII, 1976, 508–514], one of the best known and most influential figures in Islamic intellectual history. He was scientific advisor to the Mongol ruler Hūlāgū (1217–1265), whom he convinced to have a major observatory constructed at Marāgha. Al-Ṭūsī's writings concern nearly every branch of the Islamic sciences, from astronomy to philosophy and from the occult sciences to theology. Many of these became authoritative works in the Islamic world. His innovative ideas in astronomy probably reached Copernicus (1473–1543) through Byzantine intermediaries. Sezgin [1974, 271] lists several manuscripts of al-Ṭūsī's redaction, the oldest dating from the 8th century H (14th century A.D.), namely, Topkapi Saray 3456/14, 60r–61r (apparently incomplete), Köprülü 930/13, 207v–214r, and 931/13, 125r–128v (complete?).

The treatise was often included in the *Intermediate Books* (K. Mutawassiṭāt), so called (by the Arabic authors) because of their intermediate place in instruction between Euclid's *Elements* and Ptolemy's *Almagest* [Berggren, to appear]. This collection contains mostly treatises by Greek authors dealing with spherical geometry and its application to astronomy, as well as auxiliary mathematics, such as Thābit's *Book of Assumptions* and Euclid's *Data*. According to Steinschneider [1865, 457], Thābit revised most of the intermediate books and possibly translated some. It is quite possible that the corpus of the *Intermediate Books* was first established by Thābit. Libri [1838, 1, 298] mentions a note by Thābit in a Latin translation, *Liber quem edidit Thebit, filius Chore, de his quae indigent expositione antequam legatur Almaghestus* (Paris, BN Suppl. lat. 49, f. 24). The collection probably became gradually enlarged and is currently known in the standard edition, by Naṣīr al-Dīn al-Ṭūsī, who revised and redacted most of the treatises and added one by himself. It is found in practically all collections of Oriental manuscripts, often even in several copies (so, for example, three copies are found in Oxford). Therefore many copies of Thābit's treatise in the redaction by Naṣīr al-Dīn al-Ṭūsī are avaiable to us. An uncritical printed version was published in Hyderabad/Deccan in 1940.

3. COMPARISON OF THE TWO VERSIONS

In the short introduction to Thābit's treatise, al-Ṭūsī wrote, "This treatise has thirty-six propositions and in some copies thirty-four, in the latter case the propositions four and twenty-three are not included." This conveys the impression that al-Ṭūsī was a careful worker. Now the question arises: What is the relation between Thābit's treatise and al-Ṭūsī's redaction of the same? Putting it differently and more generally: Can we use al-Ṭūsī's redaction of a treatise by author X to make conclusions about the mathematics of author X?

The *first difference* between Thābit's treatise and al-Ṭūsī's redaction is easily seen: the sequences of the propositions are completely different in the two versions. The correlation between the numerations of the propositions is given in Table 1. The reverse correlation can be found in Section 4, in which the propositions are numbered according to al-Ṭūsī, with Thābit's numbers added in parentheses.

Al-Ṭūsī rearranged the propositions of Thābit's treatise thematically: The first group contains P. 1–12 and P. 14. P. 1–12 are all the construction

Table 1.

Correlation of the Propositions in the Two Versions

Thābit	al-Ṭūsī	Thābit	al-Ṭūsī	Thābit	al-Ṭūsī
1	15	13	24	25	33
2	16	14	27	26	20
3	17	15	28	27	34
4	18	16	25	28	35
5	1	17	7	29	9
6	2	18	29	30	10
7	19	19	8	31	21
8	3	20	30	32	36
9	5	21	31	33	4
10	6	22	32	34	11
11	13	23	26	35	12
12	22	24	23	36	14

problems of the treatise. P. 14 is closely connected with P. 12 but is not a construction problem itself. Some of these construction problems seem very simple, e.g.,

> **P. 1.** *We want to trisect the right angle ABG. Let us construct on BG the equilateral triangle DBG and bisect angle DBG by line BE. Then each of the angles ABD, DBE, and EBG is one third right. QEF.* (Fig. 1).

This proposition is already found in Pappus' *Collection*, Book 4, P. 39 [Hogendijk 1979, 5–6]. There it is imbedded between the proposition for trisecting an acute and that for trisecting an obtuse angle, and applied in the latter problem, whereas the case here is simple. Angle trisection was of eminent importance for Muslim mathematicians. Hogendijk [1981, 432] related that the geometer Abū Jaʿfar al-Khāzin and his younger contemporary al-Sijzī "did not hear that any one of the ancients made any contribution to its solution in the geometrical way, with which the solution can be

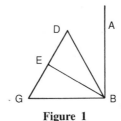

Figure 1

constructed." Both asserted that Thābit ibn Qurra was the first to trisect the angle. Thābit also composed a treatise, *The Trisection of the Rectilinear Angle* [Knorr 1989, 277–281 (translation)/361–366 (Arabic text)]. The trisections in this text are exactly the same as the trisection from the *Collection* of Pappus [Hogendijk 1979, 10]. It is probable [Hogendijk 1981, 432] that Thābit's treatise is heavily dependent on the *Trisection of the Rectilineal Angle* by Aḥmad ibn Shākir, one of the three Banū Mūsā (mid ninth century). Knorr [1983] thinks that all these angle trisections originate in Greek mathematics. In particular he claims that Thābit's source was the geometric compilation in three books by Menelaus (second century A.D.). In *On the Measurement of Plane and Curved Figures*, a geometric miscellany, most of which is devoted to proving Archimedes' results on the measurement of the circle and the sphere [Knorr 1989, 535–543], the Banū Mūsā trisected an acute angle by a *neusis* method, as well as an obtuse angle [Clagett 1964, 345]. The latter case is solved by bisecting the angle, thus obtaining two acute angles. Trisection of the right angle is not mentioned. Perhaps this is the reason why Thābit treated only the trisection of the right angle here. Others propositions of this group are more complex, e.g.,

> **P. 7.** *Side BG of triangle ABG is extended till an arbitrary point D. We want to draw a line DZ with Z on AB, such that the area of triangle DZB is equal to the area of triangle ABG.* (Fig. 2).
> **Construction.** Draw the parallelogram BDEZ, with its surface twice the surface of triangle ABG (E I.44). Line DZ is the required line.
>   Because the area parallelogram BDZE = 2 area triangle ABG = 2 area triangle ZBD.

The construction is a direct application of Euclid I.44. The method of "application of areas," introduced by Euclid in this theorem, was one of the most powerful methods underlying Greek geometry [cf. Heath 1956, 343–345]. It originates with the Pythagoreans and found its most elegant use in Apollonius' *Conics*.

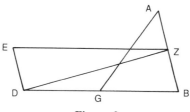

Figure 2

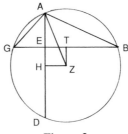

Figure 3

The next two groups are formed by P. 15 to 19 (P. 15 lemma) and by P. 22 to 25 and 27/28. In these two groups some elements of a triangle are given, from which the remaining elements have to be derived. The solutions use only lines and angles in the first group, but also areas in the second group. A simple example is

> **P. 16.** *Triangle ABG is rectangular and isosceles. When the base BG is known, both sides are known. And conversely.* Commentary in the treatise: This is obvious.

More demanding is P. 27, which can also be reckoned to the next group. Also, P. 28 could be classified with this fourth group, which then consists of P. 27 to 36 (P. 31 lemma), all concerning circles.

> **P. 27.** *When the sides of triangle ABG are known, then the diameter of the circumscribed circle is known.* (Fig. 3).
> **Solution.** We draw the perpendicular AE and extend it until D on the circumference.
>   AB, AG, and BG are known, whence (by Euclid *Data* 76) AE is known, hence EG and BE are known.
>   BE · EG = AE · ED, hence ED is known.
>   Let Z be the center of the circle, so ZH halves AD and ZT halves BG, hence AH and BT are known. GE is known, so ET is known. Angle AHZ is 90°, hence AZ and thus the diameter are known.

The main step in this proof is the use of Euclid's Data 76 [Thaer 1962, 52, 71] in the beginning: "When we draw in a triangle known in shape the perpendicular from the top to the base, then the ratio of the perpendicular to the base is known." The application of this proposition is neither shown nor mentioned. Thābit explained, "Thus the perpendicular of triangle ABG is known, because its sides are known." This reference has been omitted by al-Ṭūsī, who is, as usual, very concise (cf. third difference). The rest of the proof is simple. The *Data* was probably translated into Arabic around the mid ninth century by Isḥāq b. Ḥunain. This translation was corrected by Thābit and later redacted by al-Ṭūsī. The treatise was included in the *Intermediate Books*. Both scientists must have been familiar with it.

There remain P. 13, 20, 21, and 26, all dealing with "geometrical algebra." A corollary is added to P. 13, which is not found in the only extant version of Thābit's version. Because al-Ṭūsī does not mention in his introduction that this corollary is sometimes not included, it is probably his own addition. P. 20 is explained in detail below.

The *second difference* lies in the formulation of the propositions. Thābit enunciated some propositions (Thābit 1, 2, 3, 4, 7, 11, 12, 21, 23, and 26) first in a general way, followed by an example with proof, whereas al-Ṭūsī gave only the example with proof in these cases. For example,

P. 20. Thābit formulated in P. 26 (Ṭūsī 20), [*For*] *Every line-segment, which is known and extended by a line-segment, when the product of the extended line-segment and its extension is known, then the extended line-segment is known,* after which the example follows.

In al-Ṭūsī's redaction this proposition reads, *Line-segment AB is known, and is extended by BG. The product of AG and GB is known. Then both AG and GB are known.* (Fig. 4).

The solution is practically the same in both versions: Halve AB in point D; because BD^2 and $AG \times GB$ are known, it follows (by Euclid II.6) that DG^2, whence also DG is known. DB, and therefore BG is known. AB is also known, whence AG is known as well.   Q.E.D.

This solution is very concise and does not mention the underlying theorem, namely, Euclid II.6. The reader is apparently supposed to be familiar with Thābit's theory, which he developed in another treatise, *Qawl fī taṣḥīḥ masā'il al-jabr bi-l-barāhīn al-handasīya* (*On the Verification of Problems in Algebra by Means of Geometrical Proofs*) [Luckey, 1941]. Here he gave a general solution for al-Khwārizmī's cases IV–VI of the equation $cx^2 + bx = a$, using Euclid II.6 and for case V, Euclid II.5, without adding numerical examples [Dold-Samplonius 1987, 73–75].

For case IV Thābit formulated: "Square plus roots equal numbers: $x^2 + bx = a$." After an elaborate geometrical proof, in which Euclid II.6 is applied and explicitly mentioned, Thābit continued: "This procedure is analogous to the procedure of the algebraists (*aṣḥāb al-jabr*) in the solution of this problem" [NB, Here as well as in the title the term *al-jabr*

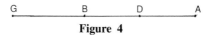

Figure 4

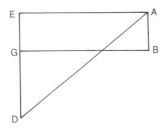

Figure 5

is used without its counterpart, *al-muqābala*). He explained this by confronting the successive corresponding steps in the geometrical and the algebraical solution [Dold-Samplonius 1987, 74].

The *third difference* is that al-Ṭūsī is more concise. This is illustrated by a comparison between Thābit P. 1 and al-Ṭūsī P. 15.

> **P. 15.** al-Ṭūsī. *The lines AB and GD are two perpendiculars drawn from the extremities of line-segment BG in two directions, their sum is known. The points A and D are joined, then AD is also known.* (Fig. 5).
> **Proof.** Let us draw AE parallel to BG and extend GD until E. Hence EG, that is AB, is known and their sum ED is known. AE, that is BG, is known. Angle E is right, whence AD is known. QED.

In Thābit's version this proposition reads, "For every known straight line-segment from whose extremities we draw two straight and known line-segments in two different directions at a right angle, when we thereupon join the extremities of the two drawn segments by a straight line, then this is known. Example: Line-segment GB is known and straight ... ." This is a more elaborate and explicit enunciation. In this particular case it might be clearer to the reader, because the conditions are given one by one, starting with the line one would draw first. Also in other cases al-Ṭūsī exaggerated and left out too much, as in the proof of P. 27 and in the enuciation of some propositions.

P. 15 is applied as a lemma in the proofs of P. 18 and P. 19.

Agreement

The figures in Thābit's version (5th century H) and in the printed edition of al-Ṭūsī's redaction contain the same kind of mistakes. This edition was printed in 1940 in Hyderabad/Deccan. Whether this agreement is fundamental I do not know. I have not yet seen the three oldest manuscripts containing al-Ṭūsī's redaction, which date from the 8th century H (~ 14th century A.D.). Of the manuscripts I consulted, Oxford Arch. Seld. 45 and Paris BN 2467 (16th century A.D.), have different figures, Oxford Marsh 709 has rather similar figures, and Oxford Arch. Seld. 45 contains empty spaces for the figures, but no figures.

4. CONTENTS OF THE PROPOSITIONS

In the following, all the propositions of the Book of Assumptions are reproduced in the order of al-Ṭūsī's redaction. The correlating number of Thābit's treatise is added in parentheses. Conditions, which should have been given, are added in square brackets.

P. 1 (Thābit 5). *We want to trisect the right angle ABG.*

P. 2 (Thābit 6). *We want to divide line AB in three parts, such that the squares of the two extreme sections together equal the square of the middle section.*

P. 3 (Thābit 8). *We want to draw a line, from angle A of triangle BAG, dividing BG in two parts [at point T], such that the ratio of AT to TG equals the [given] ratio of d to e.*

P. 4 (Thābit 33). *Another method for the preceding proposition.*

P. 5 (Thābit 9). *Let, in triangle ABG, base BG be longer than side AG. We want to draw line AD from A until BG, such that AD and DG together are equal to BD.*

P. 6 (Thābit 10). *We want to draw, in triangle ABG, the line AD from angle A until BG, such that AD and DG together are equal to DB and BA together.*

P. 7 (Thābit 17). *Side BG of triangle ABG is extended until an arbitrary point D. We want to draw a line, DZ, with Z on AB, such that the area of triangle DZB is equal to the area of triangle ABG.*

P. 8 (Thābit 19). *We want to draw, from point A of triangle ABG, the line AD with D on the extension of BG, such that the line segments AD plus DG equal the line segments AG plus GB.*

P. 9 (Thābit 29). *We want to draw, in triangle ABG, two lines. One halves the other, and the other cuts off one third from the first.*

P. 10 (Thābit 30). *Assumed triangle ABG, with AD drawn arbitrarily inside it. We want to draw, in the interior, a line, YTK, such that the ratio of YT to TK is equal to the [given] ratio of e to w.*

P. 11 (Thābit 34). *Let us draw, in circle ABG, an arbitrary chord, AG. We want to draw the lines AD and DG in arc ADG in the ratio of the [given] lines EZ to HT.*

P. 12 (Thābit 35). *Assumed circle ABG, with diameter AB and point G on the circumference. We want to draw, from point G, a chord, which the diameter intersects in the [given] ratio of d to e.*

P. 13 (Thābit 11).   *Line segment AB has been divided in G. From the longer segment, AG, a segment, GD, equal to the shorter segment BG, is cut off, so AD is cut off. We assert that the product of AB and AD equals the square of AD plus twice the product of AD and DG, i.e.,*

$$(a + b)(a - b) = (a - b)^2 + 2b(a - b), \quad a > b.$$

Corollary (not in Thābit).   *When a line segment has been divided like line segment AB, the difference between the squares of the two segments equals the product of the whole line segment and the difference of the two segments. When two of these three are known, the third one is also known, i.e.,*

$$a^2 - b^2 = (a + b)(a - b), \quad a > b.$$

P. 14 (Thābit 36).   *Assumed circle ABG, with diameter AB and chord GD. Tangent AE is drawn at point A. Lines BG and BD are extended until the points E and Z [on AE]. We assert that triangles BGD and BZE are similar to each other.*

P. 15 (Thābit 1).   *The lines AB and GD are two perpendiculars drawn from the extremities of line segment BG in two directions; their sum is known. The points A and D are joined. Then AD is also known.*

P. 16 (Thābit 2).   *Triangle ABG is rectangular and isosceles. When the base BG is known, both sides are known. And conversely.*

P. 17 (Thābit 3).   *Angle A of triangle ABG is right and angle G is 30°. When one of the sides is known, the two remaining sides are known.*

P. 18 (Thābit 4).   *BA is drawn from one of the extremities of line BG at 45°. GH is drawn perpendicularly from the other extremity. The three are known. AH is joined. Then AH is known.*

Addendum.   *When BA is drawn at 30°, or 60°, then AH is likewise known.*

P. 19 (Thābit 7).   *The sides of the quadrilateral ABGD and the diagonal, which joins A and G, are known. Then the other diagonal is known.*

P. 20 (Thābit 26).   *Line segment AB is known and is extended by BG. The product of AG and GB is known. Then both AG and GB are known. I.e., if $(b + x)x = a$ and if a and b are positive and known, then x is known.*

P. 21 (Thābit 31).   *AB and ED are perpendiculars on the line BD. All three are known. If [G lies on BD and] AG and GE are equal to each other, then they are also known. I.e.,*

if $z^2 = a^2 + x^2 = b^2 + y^2$, $x + y = c$, for known a, b, and c, then z is known.

P. 22 (Thābit 12).   *The surface of the isosceles triangle ABG is known. The [equal] side[s], i.e., AB and AG, is known. Then the base is known.*

P. 23 (Thābit 24).   *Sides AB and AG of triangle ABG are equal to each other. Angle A is 30° and the surface is known. Then the sides are known.*

P. 24 (Thābit 13).   *Triangle ADG is rectangular [with D as the right angle]. Its sides are known. At A on line AG, angle GAB has been constructed like angle AGD. DG is extended until it meets AB in B. Then DB and AB are both known.*

P. 25 (Thābit 16).   *The sides of triangle ABD are known. On DA, angle DAG has been constructed like angle DAB. BD is extended until it meets AG in G. Then GA and GD are both known.*

P. 26 (Thābit 23). *Line segment AB is divided in G. The product of AG and GB and the ratio of AG to GB are known. Then the segments are known and the line segment is known. I.e.,*

If $xy = a$ and $x : y = b$, then x, y, and $x + y$ are known.

P. 27 (Thābit 14). *When the sides of triangle ABG are known, the diameter of the circumscribed circle is known.*

P. 28 (Thābit 15). *In circle ABGD the chords AB and GD are parallel to each other, but not known. From their extremities [the diagonals] AG and BD are joined. When AG divides BD in two known segments, i.e, BE and ED, which form the two triangles [ABE and GDE] known in size, then the two chords and the diameter are known.*

P. 29 (Thābit 18). *The diameter BG of circle BDG and the tangent BA are known. When the division of BG in H is known and when AH is drawn [intersecting the circumference in T], AH, AT, and TH are all known.*

P. 30 (Thābit 20). *Given two points, D and E, on diameter AB of circle ABG, DE is known. When we draw the perpendiculars DZ and EH in the two points, they are known. We assert that the diameter is known.*

P. 31 (Thābit 21). *Triangle ABG is rectangular with B as the right angle. Side BG is known, and the sum of the sides AB and AG is known. We assert that AB and AG are both known.*

P. 32 (Thābit 22). *The perpendicular line DG has been erected on diameter AB of circle ABG [at point D]. The sum of AD and DG is known and likewise the sum of BD and DG is known. We assert that the diameter is known.*

P. 33 (Thābit 25). *The diameter of circle ABG is known. The chords AB and GD intersect at T in a right angle. AB is known and the ratio of GT to TD is known. We assert that GD is known.*

P. 34 (Thābit 27). *The perpendicular line EG has been erected on diameter AB of circle ABG [at point E]. AE and the difference of BE and GE are known. We assert that the diameter is known.*

P. 35 (Thābit 28). *The diameter and chord AB of circle ABGD are known. At A, angle GAB is constructed as 60°. BG is drawn. Then BG and GA are both known.*

P. 36 (Thābit 32). *The chord BD in circle ABGD is known. The diameter AG intersects it at E in a right angle. The difference of AE and EG is known. We assert that the diameter and the two segments are known.*

5. CONCLUSION

The contents of the two versions are similar. However, the mathematics is not identical, although there is a substantial agreement. There must have been an idea behind the sequence of the propositions in Thābit's treatise. My suggestion is that Thābit's treatise consists of selected notes resulting from his work as a translator and redactor of classical Greek works, in particular of the *Intermediate Books*. In al-Ṭūsī's redaction not much is left of the original style, or, to put it another way, the spirit of the treatise is lost. As long as one is aware of this, al-Ṭūsī's redaction may be regarded as a faithful interpretation of Thābit's original.

ACKNOWLEDGMENTS

An earlier version of this chapter was to be published in the "Proceedings of the Second International Symposium at Aleppo (Syria)," which never appeared. Doctor J. P. Hogendijk (Utrecht) suggested a new look at this material and provided me with the missing page of Thābit's treatise. I am most grateful for this as well as for his comments on the chapter. Other helpful suggestions of a technical and linguistic nature were made by Professors D. A. King (Frankfurt/Main) and E. R. Fadell (Madison, WI), respectively. I alone am responsible for any remaining mistakes and errors.

REFERENCES

Manuscripts

Thābit ibn Qurra, '*Kitāb al-Mafrūḍāt*'. Aya Sofya 4832, ff. 35v, 47r-v, 36r-39r. (11th century A.D.).
Naṣīr al-Dīn al-Ṭūsī, Redaction of the '*Kitāb al-Mafrūḍāt*'. Paris BN 2467, ff. 68v-72v (16th century A.D.);
—, Oxford Bodleian Arch. Seld. A45, ff. 159v-164v;
—, Oxford Bodleian Arch. Seld. A46, ff. 11v-14r;
—, Oxford Bodleian Marsh 709, ff 64v-70r.

Published Sources

Bellosta, H. 1991. Ibrāhīm ibn Sinān: On analysis and synthesis. *Arabic Sciences and Philosophy* **1**, 211–232.
Berggren, J. L. to appear. Spheres of influence: The transmission of the science of spherics from the Greek to the Islamic world. *Actes du Colloque* (1989) *pour l'Histoire des Mathématiques Arabes*. Paris: Institut du Monde Arabe.
Clagett, M. 1964. *Archimedes in the Middle Ages*. Vol. I: *The Arabo-Latin Tradition*. Madison: University of Wisconsin Press.
Dictionary of Scientific Biography (DSB). 1970–1980. 16 vols. New York: Schribner's.
Dodge, B. 1970. *The Fihrist of al-Nadīm* (Bayard Dodge, Trans.) 2 vols. New York and London: Columbia University Press.

Dold-Samplonius, Y. 1977. *Book of Assumptions by Aqāṭun*. Thesis. Amsterdam: Mun. University of Amsterdam.

— 1978. Some remarks on the 'Book of Assumptions by Aqāṭun.' *Journal for the History of Arabic Science* **2**, 255–263.

— 1987. Developments in the solution to the equation $cx^2 + bx = a$ from al-Khwārizmī to Fibonacci. In *From Deferent to Equant: A Volume of Studies in the History of Science in the Ancient and Medieval Near East in Honor of E. S. Kennedy*, D. A. King and G. Saliba, Eds., vol. 500, pp. 71–87. New York: Annals of the New York Academy of Sciences.

Heath, Sir T. L. 1908. The thirteen books of '*Euclid's Elements.*' Translated with introduction and commentary. Cambridge, UK: Cambridge University Press. Reprinting of the second edition of 1925, New York: Dover, 1956. Cited from this reprinted edition.

— 1931. *A Manual of Greek Mathematics*. Oxford: Oxford University Press. Cited from the reprinted edition, New York: Dover, 1963.

Hogendijk, J. P. 1979. *On the Trisection of an Angle and the Construction of a Regular Nonagon by means of Conic Sections in Medieval Islamic Geometry*, Preprint No. 113. Utrecht: University of Utrecht, Department of Mathematics.

— 1981. How trisections of the angle were transmitted from Greek to Islamic geometry. *Historia Mathematica* **8**, 417–438.

Ibrāhīm b. Sinān b. Thābit b. Qurra al-Harrānī 1948. A treatise on the method of analysis and synthesis, and the other geometrical operations (Arabic). In *Rasā'ilu Ibn-i-Sinān*, Based on the Unique Compendium of Mathematical and Astronomical Treatises in the Oriental Public Library, Bankipore (Arabic Ms. No. 2468/V, III, XXVI, IV, XXVII, II), Treatise **2**. Hyderabad-Deccan: Osmania Oriental Publications Bureau.

Ibn Isḥāq al-Nadīm. 1970. *Kitāb al-Fihrist*. CF Dodge [1970].

Jones, A. 1986. *Pappus of Alexandria*, Book 7 of the '*Collection,*' 2 vols. New York: Springer-Verlag.

Knorr W. 1983. On the transmission of geometry from Greek into Arabic. *Historia Mathematica* **10**, 71–78.

— 1989. *Textual Studies in Ancient and Medieval Geometry*. Basel and Boston: Birkhäuser.

Krause, M. 1936. Stambuler Handschriften islamischer Mathematiker. *Quellen und Studien zur Geschichte der Mathematik, Astronomie und Physik*, Abt. B **3**, 437–532.

Libri, G. 1838. *Histoire des sciences mathématiques*, 2 vols. Paris: Renouard.

Luckey, P. 1941. Thābit ibn Qurra über den geometrischen Richtigkeitsnachweis der Auflösung der quadratischen Gleichungen. *Sitzungsberichte der Sächsischen Gesellschaft der Wissenschaften, Math.-Phys. Klasse* **93**, 93–114.

Rashed, R. 1993. *Géométrie et Dioptrique au Xe Siècle: Ibn Sahl, al-Qūhī et Ibn al-Haytham*. Paris: Les Belles Lettres.

Saidan, A. S. 1983. *The Works of Ibrāhīm ibn Sinān (Arabic)*. Kuwait: Al-Majilis al-Waṭanīli-l-thiqāfa wa-l-funūn wa-l-adab.

Sezgin, F. 1974. *Geschichte des arabischen Schrifttums*. Vol. V: *Mathematik. Bis ca. 430 H*. Leiden: E. J. Brill.

Steinschneider, M. 1865. Die mittleren Bücher der Araber und ihre Bearbeiter. *Zeitschrift für Mathematik und Physik* **10**, 456–498.

Thābit ibn Qurra 1940. Kitāb al-Mafrūdāt. In *Naṣir al-Dīn al-Ṭūsī, Majmū al-Rasā'il*, Vol. II, Treatise **2**. Hyderabad-Deccan: Osmania Oriental Publications Bureau. Russian translation by B. A. Rosenfeld in Thābit ibn Qurra, *Mathematical Treatises*, pp. 45–54. Moscow: Akademia Nauk, 1984. (An English edition is in preparation by the author.)

Thaer, C. 1962. *Die Data von Euklid*. Berlin and Heidelberg: Springer-Verlag.

Partielle Differentiation im Briefwechsel Eulers mit Niklaus I Bernoulli—eine Miszelle [*]

Emil A. Fellmann

Euler-Archiv, Arnold Böcklinstrasse 37, CH-4051 Basel, Switzerland

The article gives a short biography of Niklaus I Bernoulli (1687–1759) and discusses two letters of the correspondence between Bernoulli and Euler. On November 10, 1742, Euler wrote to Bernoulli saying that a theorem of Bernoulli's on partial differentiation was also claimed by Bouguer and Fontaine. In his answer of April 6, 1743, Bernoulli rejects this statement and adds an analytical proof for the construction of orthogonal trajectories.

I

Fast jeder Mathematiker, sicher aber jeder Mathematikhistoriker kennt die seit 1911 in drei Serien erschienenen rund 70 Bände der "großen Euler-Ausgabe", die heute bis auf drei noch ausstehende Bände [1] komplett vorliegt. Da diese drei Serien jedoch lediglich die bereits früher einmal gedruckten Bücher und in mehreren Periodika erschienenen Abhandlungen Eulers enthalten, faßte die Euler-Kommission der Schweizerischen Akademie der Naturwissenschaften 1967 den bedeutsamen Entschluss, diesen Serien I–III mit einer vierten die Krone aufzusetzen. Diese *Series quarta* zerfällt in zwei Teile A und B. Der Teil IV A soll den auf neun Bände veranschlagten Briefwechsel [2] Eulers enthalten, der Teil IV B seine wissenschaftlichen Notiz- und Tagebücher in ca. sieben Bänden.

Aktuell befindet sich der Doppelband O.IV A,2 in der letzten Vorbereitungsphase. Er wird bzw. wurde von E. A. Fellmann und G. K. Mikhajlov bearbeitet und enthält die Korrespondenz Eulers mit seinem großen Lehrer Johann I Bernoulli [3] (Teil 1) und mit dessen Neffen Niklaus I Bernoulli [4] (Teil 2).

Hier soll uns bloß ein Detail aus dem Briefwechsel Eulers mit dem um zwanzig Jahre älteren Niklaus I Bernoulli beschäftigen. Die ganze Korre-

spondenz umfaßt elf—größtenteils sehr umfangreiche und inhaltsschwere
—Briefe, die sich auf Eulers Initiative vom Januar 1742 bis Mitte 1745
alternierend folgen [5]. Die uns hier interessierenden Korrespondenzstücke
sind durch die Nummern R 238 und R 239 gekennzeichnet, doch bevor wir
uns dem eigentlichen Gegenstand, der partiellen Differentiation, zuwen-
den, sollen einige Akzente aus dem Leben und Wirken des von der
Mathematikhistorie bisher arg vernachlässigten, sehr tüchtigen und bedeutenden Mathematikers Niklaus I Bernoulli beleuchtet werden.

II

Niklaus I Bernoulli wurde am 20. Oktober 1687 in Basel als Sohn des
Malers und Ratsherrn gleichen Namens geboren und starb daselbst am 29.
November 1759. Er war der Neffe der beiden Leibniz-Schüler Jakob und
Johann Bernoulli und Vetter der Söhne des letzteren, Niklaus II, Daniel
und Johann II. Mit diesen fünf Persönlichkeiten gehörte Niklaus I
Bernoulli zum "mathematischen Kern" dieser Familie, welche die Entwicklung der Mathematik, der Mechanik und der Physik im 18. Jahrhundert grundlegend beinflusste. Der größte Mathematiker dieses Jahrhunderts, Leonhard Euler, gehörte zwar nicht dieser Familie an, doch war er
in jungen Jahren direkter Schüler von Johann I Bernoulli, ebenso wie
dessen drei genannten Söhne, deren Freundschaft er frühzeitig teilhaftig
wurde, und wie Niklaus I, der zuvor Jakob Bernoullis Schüler an der
Basler Universität gewesen war.

Das wissenschaftliche Leben von Niklaus I Bernoulli verlief weniger
glücklich als dasjenige seiner Onkeln und Vettern. Eine ausführliche und
gut dokumentierte Biographie dieses bedeutenden Gelehrten und hochbe-
gabten Mathematikers steht bis heute noch aus, und sein Werk ist erst
unvollständig analysiert; ja bis zur Stunde ist sogar noch nicht einmal eine
vollständige Bibliographie seiner Publikationen im Druck zugänglich, und
seine im Handschriftenarchiv der Basler Universitätsbibliothek deponierte
Korrespondenz ist trotz ihrer großen Bedeutung für die Wissenschaftsgeschichte größtenteils noch unediert. Man findet sogar in einschlägigen
Enzyklopädien und biographischen Sammelwerken Fehler und große
Lücken (so beispielsweise auch im *Dictionary of Scientific Biography* [6]),
und im brandneuen Schweizer Lexikon [7] wurde er beschämenderweise
einfach übergangen.—So sollen hier wenigstens einige Hauptstationen
und -fakten seines Lebens aufgelistet werden [8].

Ursprünglich sollte Niklaus I Bernoulli Maler werden, doch studierte er Jurisprudenz und Mathematik, letzteres unter der Leitung seines Onkels Jakob, der seit 1687 bis zu seinem frühzeitigen Tod 1705 den mathematischen Lehrstuhl an der Basler Universität innehatte. Jakob gestattete Niklaus als seinem offensichtlich bevorzugten Schüler, seine noch unveröffentlichten Manuskripte zu studieren, im speziellen das Manuskript seines 1713 in Basel postum erschienenen Hauptwerkes *Ars conjectandi* [9]. Mit der Verteidigung von Jakob Bernoullis fünfter *Reihendissertation* [10] wurde Niklaus 1704 zum *magister artium* promoviert. Später wurden die unendlichen Reihen für Niklaus Bernoulli zum Objekt tiefschürfender Untersuchungen, insbesondere in seiner Korrespondenz, zuerst mit Leibniz, dann—viel später—auch mit Euler.

Nach dem Tod von Jakob Bernoulli wurde dessen Lehrstuhl von seinem jüngeren Bruder Johann—einst ebenfalls sein Schüler und später bekanntlich sein streitbarer Widersacher—besetzt, wodurch Niklaus einen neuen Lehrmeister für seine wissenschaftlichen Forschungen erhielt, denn Johann Bernoullis Interessengebiet war die Infinitesimalmathematik samt ihren geometrischen und mechanischen Anwendungen. Dennoch legt Niklaus im Jahre 1709—in Verfolgung seines Zieles, den Grad eines *doctor utriusque juris* zu erwerben—eine Dissertation über Anwendungen der "Kunst des Vermutens" auf juristische Probleme vor [11], als deren geistiger Vater zweifelsfrei Jakob Bernoulli angesehen werden darf und bei welcher es sich in Wirklichkeit um eine Art Fortsetzung von Jakobs Hauptwerk handelt. Diese *Dissertatio* brachte Niklaus das Lizentiat der Jurisprudenz ein.

Anläßlich eines längeren Aufenthaltes in Paris (1710) machte Niklaus Bernoulli die Bekanntschaft von P. Varignon [12], und während dieser Zeit bot sich ihm—durch die Initiative Johann Bernoullis—die Chance, einen Irrtum in Newtons *Principia* (1687) richtigzustellen [13]. Niklaus reiste im Sommer 1712 über Holland nach London und traf dort im Oktober mit A. de Moivre zusammen, der ihn bei Newton einführte. Der junge Bernoulli traf Newton in der Folge dreimal und wurde von diesem—welch ungewöhnliche Ehre für den jugendlichen Basler—sogar zweimal zum Diner eingeladen [14] und zwei Jahre später (gleich nach Johann I Bernoulli) erst noch als Fellow der Royal Society gewählt. Auch dem Royal Astronomer Edmond Halley wurde Niklaus Bernoulli vorgestellt, und er konnte unter anderem auch die persönliche Bekanntschaft von Christian Goldbach [15] machen, der nicht allein eine schillernde Persönlichkeit, sondern auch ein vielseitig gebildeter und mathematisch interessierter Gelehrter war.

Ende 1712 begab sich Niklaus Bernoulli von London aus nach Paris (resp. nach Epernay), wo er sich für ein Vierteljahr als Gast auf dem Schloß von Pierre Rémond de Montmort aufhielt und diesem seinem Freund bei der Schlußredaktion der zweiten Ausgabe des *Essay* [16] kräftig zur Hand ging, nachdem die beiden bereits eine ansehnliche Korrespondenz unterhalten hatten, von welcher einige Kernstücke in dieses wichtige wahrscheinlichkeits- und spieltheoretische Werk eingingen [17].

Nach etlichen Komplikationen wurde Niklaus I Bernoulli als Nachfolger von Jakob Hermann auf den mathematischen Lehrstuhl nach Padua berufen, wo er im Dezember 1716 seine Vorlesungen aufnahm und beim Mathematiker und Astronomen G. Poleni logierte. Nicht sehr glücklich in Italien, nahm Niklaus Bernoulli den Tod seines Vaters am Weihnachtstag 1716 zum Anlaß einer Reise nach Basel, blieb ein paar Monate hier und promovierte daselbst im Mai 1717 zum Dr. jur., wonach er sich nach Padua zurück begab; nicht für sehr lange Zeit, denn er reichte 1719 sein Entlassungsgesuch ein und reiste im Oktober desselben Jahres in seine Heimat zurück. Nach einer vergeblichen Bewerbung (1720) um einen juristischen Lehrstuhl an der Basler Universität—er scheiterte am Losverfahren [18] —erhielt er 1722 die Professur für Logik und schließlich durch Tausch (1731) die Lehrkanzel für Lehnrecht. Nach dem Tod seines Onkels Johann I Bernoulli wurde Niklaus die mathematische Professur angeboten, doch verzichtete er darauf zugunsten seines Vetters Johann II ("Jean"), des Sohnes von Johann I Bernoulli.

Niklaus I Bernoulli starb am 29. November 1759 an einem Schlaganfall. Er war Mitglied der Berliner Akademie (1713), der Londoner Royal Society (1714) und der Akademie zu Bologna (1724). Viermal versah er das Amt des Rektors der Basler Universität. Sein Epitaph steht in der Kirche zu St. Peter in Basel neben denjenigen von Johann I, Daniel und Johann II Bernoulli, während dasjenige von Jakob den Hauptpfeiler im Kreuzgang des geschichtsträchtigen Münsters zu Basel ziert—"*habent sua fata fratelli*"!

III

Der Brief an Niklaus I Bernoulli vom 10. November 1742 (R 238) ist der fünfte der Kollektion. Er handelt u.a. von der Zerlegung der Sinusfunktion in eine unendliche Reihe und in ein unendliches Produkt, dann von der Zerlegung von Polynomen in reelle Faktoren und von der Interpretation rationaler Funktionen. Nach einigen Aufgaben zur *partitio numerorum* bespricht Euler die Integration linearer homogener Differentialgleichungen mit konstanten Koeffizienten sowie gewisse Extrema eines Variationsintegrals und Methodisches zur (sogenannten) Eulerschen Differentialgleichung. Im ersten Drittel des Briefes äußert sich nun Euler zu dem uns hier interessierenden Thema wie folgt [19]:

> Die französischen Mathematiker haben vor einigen Jahren ein analytisches Theorem hochgerühmt, welches nach dem Erfinder, bald nach Bouguer, bald nach Fontaine benannt wurde. Darin wurde nämlich eine besondere Eigenschaft der Differentialausdrücke, die zwei Variable enthalten, dargelegt, nämlich derjenigen, die durch Differentiation einer endlichen Grösse entstanden sind. Doch dieses Theorem kann gewiss auf verschiedene Arten vorgestellt werden; ich werde es aber wie folgt aussprechen: Geht aus der Differentiation einer endlichen Grösse, d.h. einer Funktion von x und y, $P\,dx + Q\,dy$ hervor, so ist das Differential von $P\,dx$, wenn nur y als variabel angenommen worden ist, immer gleich dem Differential von $Q\,dy$, wenn man x allein als variabel gesetzt hat. Da sie also über die Entdeckung dieses sehr nützlichen Theorems miteinander stritten und mir dieses mitteilten, antwortete ich ihnen freilich unverzüglich, dieser Satz sei mir schon längst bekannt gewesen, da ich ihn ja unter anderem schon in den Band VII der *Commentarii* eingerückt hätte, dennoch gebühre der Ruhm der Entdeckung nicht mir, sondern Ihnen ... . Ich erinnere mich nämlich, dass Sie vor langer Zeit, als man über die Orthogonaltrajektorien diskutierte, die wahre Grundlage dieses Satzes aufgezeigt hatten. Da es sich nämlich um das Differential von $\int P\,dx$ handelte, wenn ausser x auch a (als Parameter) variabel angenommen wird, zeigten Sie, daß das gesuchte Differential $P\,dx + da\int R\,dx$ wird, wobei $R\,da$ das Differential von P für konstantes x ist. Dies ist, nur in andern Worten ausgedrückt, schon das eigentliche Theorem, um dessen Entdeckung sich die Herren Bouguer und La Fontaine stritten. Sei nämlich $\int R\,dx = Q$, sodass das gesuchte Differential $P\,dx + Q\,da$ ist, so wird das Differential von $P\,dx$ (nur a sei variabel) $= R\,da\,dx$, und das Differential von $Q\,da$ wird auch $= R\,da\,dx$ wegen $Q = \int R\,dx$. Es folgt aber aus diesem Lehrsatz, der Ihrem Konto gutzuschreiben ist, eine ganze Menge vorzüglicher Hilfsmittel für die Integralrechnung, die Sie entweder schon kennen oder die Sie leicht selbst sehen werden.

Euler streift hier einige Sätze der Theorie der Funktionen zweier unabhängiger Variablen $f(x,y)$: die exakte Bedingung des totalen Differentials, Ableitung des Integrals bezüglich eines Parameters, die Gleichheit

$$\frac{\partial^2 f}{\partial x\,\partial y} = \frac{\partial^2 f}{\partial y\,\partial x}$$

sowie verwandte Sätze zu diesen Themata. Im Jahre 1739 wurde Euler mehrfach darüber informiert, daß Pariser Mathematiker neue Forschungen in dieser Richtung unternähmen, worüber er von Daniel Bernoulli,

und dieser vor allem durch Clairaut unterrichtet worden sei. Beispielsweise nennt Daniel Bernoulli in seinem Brief an Euler vom 14. November
1739 (R 130) die Ableitungsregel eines Integrals nach einem Parameter
Theorema Bouguerianum. Clairaut bemerkte in seinem ersten Brief an
Euler vom 17. August 1740 (R 386: O.IV A,5, p. 68), daß dieselbe Methode
schon von Fontaine ausgesprochen worden sei. Am (29.) 18. Dezember
1739 (R 131) erklärte Euler seinem Freund Daniel Bernoulli: "Das
überschriebene *Theorema Bouguerianum* ist mir nicht nur längstens bekannt
gewesen und von mir viel weiter extendiret worden [,] sondern dasselbe ist
schon vor vielen Jahren als das *Problema trajectoriarum orthogonalium*
ventilirt worden...". Euler bezieht sich hier auf den zweiten Band seiner
Mechanica von 1736, Prop. 14, sowie auf seine Abhandlung *De infinitis
curvis ejusdem generis*..., die der Petersburger Akademie am (21.) 10. Mai
1734 vorgelegt, mit einem *Additamentum ad dissertationem de infinitis
curvis ejusdem generis* ergänzt und 1740 in den *Commentarii* (für 1734–1735)
gedruckt worden waren. In gleichem Sinne antwortete Euler Clairaut am
30. Oktober 1740 (R 387: O.IV A,5, p. 71–78).

Betreffs der Differentialeigenschaften von Funktionen zweier Variablen
gibt nun Niklaus I Bernoulli in seinem Antwortbrief an Euler vom 6. April
1743 (R 239) mit bewundernswerter Bescheidenheit zu, daß er sich ander-
weitig dieser Eigenschaft bedient habe, jedoch weist er den Ruhm, ihr
Entdecker zu sein, zurück. Für ihn sei dies nämlich kein zu beweisendes
Theorem, sondern er habe es als Axiom behandelt, das evident zum
Differentialbegriff gehöre. Dazu zitiert er zwei Satzanfänge aus dem
zweiten Teil seiner Arbeit [20] von 1721 und gibt zwei geometrische
Figuren, wo diese Eigenschaft unmittelbar in die Augen springt. Während
seine Konstruktion der Orthogonaltrajektorien, wie sie in der Abhandlung
[21] von 1719 dargestellt ist, auf der besagten Eigenschaft der Differentiale
basiert, gibt er nun einen *analytischen* Beweis, der zuvor nicht publiziert
worden ist. Dieser—recht ausführliche—Beweis muß als eine neue
Ergänzung oder gar als Fortsetzung von Niklaus Bernoullis Arbeit von
1719 betrachtet werden, und allein schon dieser Umstand rechtfertigt
zweifellos die (deutsche) Wiedergabe dieser Briefstelle (cf. die Faksimile-
Abbildung, Abb. 1, mit den beiden Figuren):

Schon seit langer Zeit habe ich die *Memoiren der königlichen Akademie* zu
Paris nicht mehr gesehen. Daher war mir der zwischen den Herren Bouguer
und de la Fontaine [22] ausgetragene Streit über die Entdeckung des, wie Sie
sagen, von mir überkommenen Lehrsatzes unbekannt. Ich schreibe mir
freilich in dieser Sache keine Erfinderehre zu, habe ich doch jene Eigenschaft der Differentiale, die zu dieser Kontroverse Anlass gab, nicht als

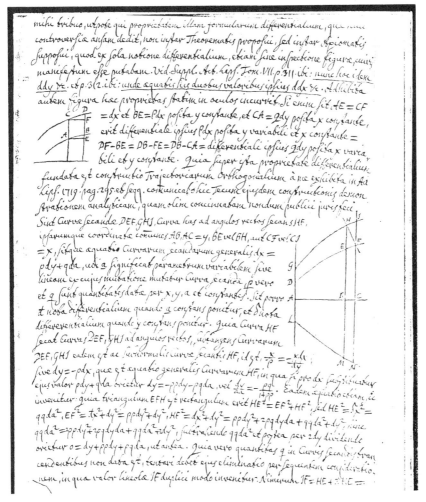

Abbildung 1. Dritte Seite des Briefes von Niklaus I Bernoulli an Leonhard Euler vom 6. April 1743 (R 239).—Archiv Petersburg, f. 136, op. 2, Nr. 7, Bl.79–80r. Die Reproduktionsvorlage entstammt dem Euler-Archiv Basel.

Theorem dargestellt, sondern als Axiom vorausgesetzt, welches ich für jedermann allein aus dem Begriff der Differentiale auch ohne Betrachtung einer Figur als einsehbar hielt. Siehe *Leipziger Akten*, Suppl. Bd. VII, dort p. 311: *nunc hoc idem ddy etc.* und p. 312: *unde aequatis his duobus valoribus ipsius ddx etc.* Aber bei Verwendung einer Figur springt diese Eigenschaft sofort in die Augen. Ist nämlich $AE = CF = dx$ und $BE = P\,dx$ bei konstantem y und $CA = Q\,dy$ bei konstantem x, so wird das Differential von $P\,dx$ bei variablem y und konstant gesetztem x gleich $DF - BE = DB - FE = DB$

$- CA =$ dem Differential von $Q\,dy$ für variables x und konstant gesetztem y. Da die von mir in den *Leipziger Akten* 1719, p. 295ff. vorgelegte Konstruktion der orthogonalen Trajektorien auf diese Eigenschaft der Differentiale gegründet ist, werde ich Ihnen hier einen analytischen Beweis dieser Konstruktion mitteilen, den ich einst verfasst, aber noch nicht publiziert habe. Die zu schneidenden Kurven seien *DEF*, *GHI*, die Kurve, welche diese im rechten Winkel trifft, *HF*, ihre gemeinsamen Koordinaten *AB*, $AC = y$ und *BE* oder *BH*, bezw. *CF* oder $CI = x$. Es sei die allgemeine Gleichung der zu schneidenden Kurven $dx = p\,dy + q\,da$, wobei a den variablen Parameter, d.h. die Strecke bedeutet, bei deren Veränderung die zu schneidende Kurve sich ändert. Aber p und q sind Grössen, die durch x, y, a sowie Konstanten gegeben sind. Ferner sei $đ$ das Differentialsymbol, wenn a konstant gesetzt, und δ das entsprechende, wenn y festgehalten wird. Da die Kurve *HF* die Kurven *DEF*, *GHI* im rechten Winkel schneidet, ist der Subtangens der Kurven *DEF*, *GHI* derselbe wie die Subnormale der schneidenden Kurve *HF*, d.h.

$$\frac{x}{p} = \frac{-x\,dx}{dy} \quad \text{oder} \quad dy = -p\,dx,$$

und dies ist die allgemeine Gleichung der Kurven *HF*. Setzt man darin für dx seinen Wert $p\,dy + q\,da$ ein, so entsteht

$$dy = -pp\,dy - pq\,da \quad \text{oder} \quad \frac{dy}{da} = -\frac{pq}{1 + pp}.$$

Dieselbe Gleichung findet man auch so: Da das Dreieck *EFH* rechtwinkling ist, wird $HE^2 = EF^2 + HF^2$, aber $HE^2 = \delta x^2 = qq\,da^2$, $EF^2 = đx^2 + dy^2 = pp\,dy^2 + dy^2$, $HF^2 = dx^2 + dy^2 = pp\,dy^2 + 2pq\,dy\,da + qq\,da^2 + dy^2$ und daher $qq\,da^2 = 2pp\,dy^2 + 2pq\,dy\,da + qq\,da^2 + 2dy^2$. Bei Subtraktion von $qq\,da^2$ und nachträglicher Division durch $2dy$ kommt wie vorher $0 = dy + pp\,dy + pq\,da$ heraus. Da aber die Grösse q bei transzendenten Kurven, die zu schneiden sind, nicht gegeben ist, muss ihre Elimination durch folgende Betrachtung versucht werden, wobei man den Wert des Linienelements *IF* auf zweifache Weise findet. In der Tat ist $IF = HE + đHE = \delta x + đ\delta x$. Aber es ist auch $IF = \delta CF = \delta BE + \delta đBE = \delta x + \delta đx$. Nach beidseitiger Subtraktion von δx wird $đ\delta x = \delta đx$, d.h. (da $\delta x = q\,da$ und $đx = p\,dy$) $đq\,da = \delta p\,dy$ und daher $đq = \delta p\,dy/da$, dessen Integral man wenigstens durch Quadraturen erhalten kann, falls x nicht in die Grösse δp eingeht. Es muss aber in der Integration eine aus a und anderen Konstanten zusammengesetzte Grösse derart addiert werden, dass im Falle $AB = y = 0$, *HE*, d.h. $q\,da$ gleich $GD = \delta AD$ wird. Die Strecke *AD* aber ist wegen der gegebenen Lage der zu schneidenden Kurven durch a und Konstanten bestimmt. Setzt man diese gleich E, so wird $q = dE/da$ im Falle $y = 0$. Wenn die eben gefundene Differentialgleichung $đq = \delta p\,dy/da$ mit der oben gefundenen Gleichung der Kurve *HF*

$$\frac{dy}{da} = \frac{-pq}{1 + pp}$$

verglichen wird, findet man

$$\frac{-\partial q}{q} = \frac{p\,\delta p}{1 + pp}.$$

Diese Gleichung dient dazu, die Kurve *LMN* für jede beliebige zu schneidende Kurve *GHI* zu finden, sodass die weitergeführte Ordinate *MB* die Kurve *GHI* in irgendeinem Punkt *H* der gesuchten Trajektorie *HF* schneidet, indem sie eine Fläche von gegebener Grösse *ALMB* abtrennt. Es sei die Ordinate der zu konstruierenden Kurve *BM* = *z* und die dazugehörige Abszisse *AB* = *y*. Man nenne die Fläche *ALMB* = *S*, und es sei allgemein *dS* = *z dy* + *u da*, so gilt hier $\partial\delta S = \delta\partial S$ wie oben $\partial\delta x = \delta\partial x$, d.h. $\partial u\,da$ = $\delta z\,dy$. Da aber $\delta S = u\,da$ ist und im Falle *y* = 0 alle Flächen *ALMB* verschwinden, wird auch δS verschwinden, und im Falle *y* = 0 wird eben *u* = 0. Es werde *z* = (1 + *pp*)/*pn* gesetzt und die abzutrennende Fläche *ALMB* = *C* − *A*, wobei *C* eine konstante Grösse und *A* eine aus *a* und Konstanten zusammengesetzte Grösse bedeutet, die noch zu bestimmen ist. Es sei *dA* = *b da*, und es wird *dS* = *z dy* + *u da* = *dC* − *dA* = −*b da* oder

$$\frac{dy}{da} = \frac{u + b}{-z} = \frac{-pq}{1 + pp},$$

daher *z* = (*u* + *b*)(1 + *pp*)/*pq* = (1 + *pp*)/*pn* und *u* + *b* = *q*/*n*, also ∂u = $(n\,\partial q - q\,\partial n)/nn$. Da im Falle *y* = 0 auch *u* = 0 ist, wird in diesem Fall *b* = *q*/*n* = (wenn man *m* = *n* setzt im Falle *y* = 0) *dE*/*m da* und *b da* = *dA* = *dE*/*m*. Aber vorher haben wir die Gleichung $\partial u\,da = \delta z\,dy$ oder

$$\frac{dy}{da} = \frac{\partial u}{\delta z} = \frac{-pq}{1 + pp} = \frac{-q}{nz}$$

gefunden. Daher ist

$$\frac{\delta z}{z} = \frac{-n\,\partial u}{q} = \frac{-\partial q}{q} + \frac{\partial n}{n} = \left(\text{da ja } \frac{-\partial q}{q} = \frac{-p\,\delta p}{1 + pp}\right)$$

$$[=]\frac{p\,\delta p}{1 + pp} + \frac{\partial n}{n} = \left(\text{wegen } z = \frac{1 + pp}{pn}\right)[=]\frac{2p\,\delta p}{1 + pp} - \frac{\delta p}{p} - \frac{\delta n}{n},$$

d.h. (weil $dn = \bar{d}n + \delta n$)

$$\frac{dn}{n} = \frac{p \, \delta p}{1 + pp} - \frac{\delta p}{p} = \delta \, \text{Log.} \, \frac{\sqrt{1 + pp}}{p},$$

und dies ist genau das, was die in den *Leipziger Akten* am erwähnten Ort angegebene Konstruktion vorschreibt ... .

IV

In seinem ersten Brief an Euler vom 13. Juli 1742 (R 235) schrieb Niklaus Bernoulli unter anderem:

> Es schmerzt mich aufs höchste, mich—entgegen meiner inneren Neigung—schon seit langem nicht mehr mit mathematischen Dingen beschäftigen zu können, da ich daran durch verschiedene Geschäfte, abgesehen von den akademischen, gehindert werde. Sie gehen fehl in der Annahme, ich pflegte Ihre eleganten Entdeckungen und Schriften—die durchzulesen, wie es sich geziemte, ich noch keine Zeit hatte—meiner Prüfung zu unterziehen. Wenn ich mich schon einmal daran mache, etwas Mathematisches zu lesen oder zu prüfen, so tue ich das nur auf Anfrage hin oder bei erlangter Freizeit.

Doch zeigt die ganze Korrespondenz mit Euler deutlich, daß Niklaus Bernoullis mathematische Eingebung—auch unter diesen erschwerten Umständen und auch nach langer Unterbrechung seiner mathematischen Forschungen—keineswegs verblaßt war, selbst im Hinblick auf sein vorgeschrittenes Alter.

Der letzte Brief Eulers an N. Bernoulli vom 11. Juli 1745 (R 244) blieb unbeantwortet. Offensichtlich wurde die Fortsetzung dieser anstrengenden Korrespondenz für Niklaus Bernoulli unmöglich—trotz allen Interesses, das sie ihm bot und das sie noch immer besitzt für die Leser des 20. Jahrhunderts.

ANMERKUNGEN

[∗] Die vorliegende Arbeit ist ein Teilprodukt der Editionsarbeit an der *Series quarta* im Rahmen der Gesamtausgabe von Leonhard Eulers *Opera omnia*, besorgt von der Euler-Kommission der Schweizerischen Akademie der Naturwissenschaften. Die Herausgabe der Series quarta wird wesentlich mitfinanziert vom Schweizerischen Nationalfonds zur Förderung der wissenschaftlichen Forschung.

[1] Dies betrifft die Bände O.II,26 und 27 sowie O.III,10 (cf. den Verlagsprospekt Birkhäuser 1991: *Leonhard Euler, Opera Omnia*).—Eine kurze Geschichte der EULER-Ausgabe mit chronologischen Editionstabellen findet sich in [Biermann 1983] für 1783–1907 und in [Burckhardt 1983] für 1907–1983. Zum neuesten Stand s. [Fellmann und Im Hof 1993].

[2] Der 1975 erschienene Band O.IV A,1 (Birkhäuser, Basel) gibt eine Übersicht sowie Résumés aller ca. 3000 erhalten gebliebener Briefe von Eulers Korrespondenz. Alle in der vorliegenden Abhandlung herangezogenen Briefe werden gemäß IV A,1 mit ihren Résumé-Nummern mit vorangestelltem R gekennzeichnet. Die ersten bis heute erschienenen Korrespondenzbände sind O.IV A,5 und 6. Sie enthalten Eulers Briefwechsel mit Clairaut, d'Alembert und Lagrange (1980) resp. mit Maupertuis und König Friedrich II (1986).

[3] Ein knapp gehaltener Längsschnitt durch die in dieser Korrespondenz abgehandelten *mathematischen Gegenstände* liegt vor in [Fellmann 1984], in überarbeiteter Fassung in [Fellmann 1985] und durch nicht-mathematische Subjekts in [Fellmann 1992].

[4] Im Interesse der Transparenz der genealogischen Verhältnisse sei hier ein Stammbaum der *Mathematiker* Bernoulli wiedergegeben. Darin möge auch Leonhard Euler als *geistiger Sohn* Johann Bernoullis Platz finden. Die Namen der beiden in der vorliegenden Abhandlung auftretenden Hauptpersonen sind in diesem Schema doppelt eingerahmt):

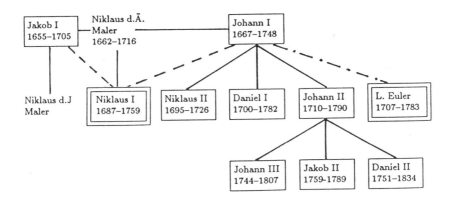

[5] Sie tragen gemäß O.IV A,1 die Nummern R 234–244. Die Briefe sind durchwegs lateinisch geschrieben und im letzten Jahrhundert auch so veröffentlicht worden, und zwar R 234, 236, 238, 240, 242, 243, 244 in [Euler 1862], R 235, 237, 239, 241 in [Fuss 1843].—In der neuen, textkritischen Ausgabe in O.IV A,2 werden die lateinischen Brieftexte von einer deutschen Übersetzung begleitet und auch deutsch kommentiert.

[6] Unter den rühmlichen Ausnahmen sind zu nennen [Bernoulli–Sutter 1972], [Engelsman 1984], [Gottwald, Ilgauds, und Schlote 1990].

[7] *Schweizer Lexikon 91*, Bd. 1, Luzern 1991.

[8] Eine ausführlichere Werkbiographie zu Niklaus I Bernoulli gedenken wir in der Einleitung zum Briefwechsel-Band O.IV A,2 zu veröffentlichen.

[9] Zur ziemlich verworrenen Editionsgeschichte cf. [Kohli 1975]. In diesem Zusammenhang sei wiederholt, daß Niklaus I Bernoulli keineswegs—wie in vielen Darstellungen zu lesen ist—der "Herausgeber" dieses Buches war; seine Mitarbeit reduziert sich auf die Abfassung eines (sehr knappen) Vorwortes und einer Errata-Liste, die fast gänzlich während seiner Abwesenheit angefertigt wurde und die er nicht einmal initiiert hatte.

[10] *Positionum de seriebus infinitis*... *pars quinta*, Basileae 1704, später abgedruckt in Jakob Bernoullis zweibändigen (durchpaginierten) *Opera*, Genf 1744, Nr. 101, p. 955–975. An der Publikation dieser *Opera* war Niklaus I Bernoulli wesentlich beteiligt.

[11] *Dissertatio de usu artis conjectandi in jure*, Basileae 1709. Diese Arbeit ist abgedruckt in [Bernoulli, 1975, 287–326] und von K. Kohli reichlich kommentiert [Kohli 1975, 541–556].

[12] Eine vorzügliche Kurzbiographie zu Varignon gab P. Costabel in [Bernoulli, 1988, 5–14]. Es handelt sich hier um den ersten der auf drei Bände veranschlagten Korrespondenz J. Bernoullis mit Varignon; der zweite ist 1992 erschienen und der dritte wird momentan von Jeanne Peiffer (Paris) vorbereitet.

[13] Cf. [Whiteside 1981, 49–51, 312–314].

[14] Cf. [Wollenschläger 1931–1932, 270f.]. Zu Leben und Werk von A. de Moivre greife man zu der noch immer maßgeblichen Studie (Dissertation) von [Schneider 1968–1969].

[15] Goldbach, nach welchem die heute jedem Mathematiker bekannte zahlentheoretische "Vermutung" benannt ist, wurde später eines der ersten Mitglieder der Petersburger Akademie, wie Johann I, Niklaus II und Daniel Bernoulli und—noch später—Leonhard Euler.—Cf. die neulich von A. und W. Purkert aus dem Russischen (1982) ins Deutsche übersetzte Goldbach-Biographie [Juškevič und Kopelevič 1994].

[16] P. R. de Montmort, *Essay d'analyse sur les jeux de hasard*, Paris 1708, 2. A. 1713.

[17] Montmorts Werk ist anhand des Briefwechsels mit Niklaus und Johann Bernoulli gründlich analysiert worden von Julian Henny in [1975].

[18] Über die—auf den ersten Blick etwas befremdliche—Wahlart der Professoren cf. [Burckhardt 1916].

[19] Den Text bringen wir hier erstmalig in deutscher Übersetzung, desgleichen (*infra*) die Replik Niklaus Bernoullis, die für sich eine kleine Abhandlung darstellt.

[20] *Acta eruditorum* 1721, Supplementband 7, pp. 311f.

[21] *Acta eruditorum* VI, 1719, pp. 295–304.

[22] Der Zugang zu Fontaines Arbeiten zu diesem Thema wird erleichtert, wenn nicht gar erst ermöglicht durch Greenbergs gründliche Studien [1981, 1982, 1984].

LITERATURVERZEICHNIS

Bernoulli, J. 1975. *Die Werke*, Bd. 3. *Naturforschende Gesellschaft in Basel*. Basel: Birkhäuser.

Bernoulli, J. I. 1988. *Der Briefwechsel*, Bd. 2. *Naturforschende Gesellschaft in Basel*. Basel: Birkhäuser.

Bernoulli-Sutter, R. 1972. *Die Familie Bernoulli*. Basel: in Kommission bei Helbing und Lichtenhahn.

Biermann, K.-R. 1983. Aus der Vorgeschichte der Euler-Ausgabe 1783–1907. In *Leonhard Euler 1707–1783*, Beiträge zu Leben und Werk, Gedenkband des Kantons Basel-Stadt, S. 489–500. Basel: Birkhäuser.

Burckhardt, A. 1916. Über die Wahlart der Basler Professoren, besonders im 18. Jahrhundert. *Basler Zeitschrift für Geschichte und Altertumskunde* **15**, 28–46.

Burckhardt, J. J. 1983. Die Eulerkommission der Schweizerischen Naturforschenden Gesellschaft—ein Beitrag zur Editionsgeschichte. In *Leonhard Euler 1707–1783*, Beiträge zu Leben und Werk, Gedenkband des Kantons Basel-Stadt, S. 501–510. Basel: Birkhäuser.

Engelsman, S. B. 1984. *Families of Curves and the Origins of Partial Differentiation*. Amsterdam: North-Holland.

Euler, L. 1862. *Opera postuma*, P. H. Fuss und N. Fuss, Hrsgg., Bd. 1. St. Petersburg: Eggers & Socius.

Fellmann, E. A. 1984. Über einige mathematische Subjekts im Briefwechsel Leonhard Eulers mit Johann Bernoulli. In *Zum Werk Leonhard Eulers—Vorträge des Euler-Kolloquiums im Mai 1983 in Berlin*, E. Knobloch u.a., Hrsgg., S. 39–66. Basel: Birkhäuser.

— 1985. Über einige mathematische Subjekts im Briefwechsel Leonhard Eulers mit Johann Bernoulli. *Verhandlungen der Naturforschenden Gesellschaft in Basel* **95**, 139–160.

— 1992. *Non-Mathematica* im Briefwechsel Leonhard Eulers mit Johann Bernoulli. In *Amphora – Festschrift für Hans Wussing zu seinem 65. Geburtstag*, S. S. Demidov, M. Folkerts, u.a., Hrsgg., 189–228. Basel: Birkhäuser.

—, und Im Hof, H. C. 1993. Die Euler-Ausgabe—Ein Bericht zu ihrer Geschichte und ihrem aktuellen Stand. *Jahrbuch Überblicke Mathematik 1993*, 185–198.

Fuss, P. H., Hrsg. 1843. *Correspondance mathématique et physique de quelques célèbres géomètres du XVIIIème siècle*, 2. St. Pétersbourg: Académie Impériale des Sciences.

Gottwald, S., Ilgauds, H.-J., Schlote, K.-H. Hrsgg. 1990. *Lexikon bedeutender Mathematiker*. Thun, Frankfurt/M: Verlag Harri Deutsch.

Greenberg, J. L. 1981. Alexis Fontaine's "Fluxio-differential method" and the origins of the calculus of several variables. *Annals of Science* **38**, 251–290.

— 1982. Alexis Fontaine's integration of ordinary differential equations and the origins of the calculus of several variables. *Annals of Science* **39**, 1–36.

— 1984. Alexis Fontaine's route to the calculus of several variables. *Historia Mathematica* **11**, 22–38.

Henny, J. 1975. Niklaus und Johann Bernoullis Forschungen auf dem Gebiet der Wahrscheinlichkeitsrechnung in ihrem Briefwechsel mit Pierre Rémond de Montmort. In [Bernoulli, 1975, 457–507].

Juškevič, A. P., und Kopelevič, Ju. Kh. 1994. *Christian Goldbach 1690–1764*. Basel: Birkhäuser. (*Vita Mathematica* Bd. 8.)

Kohli, K. 1975. Zur Publikationsgeschichte der Ars Conjectandi. In [Bernoulli, 1975, 391–401].

Schneider, I. 1968–1969. Der Mathematiker Abraham de Moivre (1667-1754). *Archive for History of Exact Sciences* **5**, 177–317.

Whiteside, D. T. Ed. 1981. *The Mathematical Papers of Isaac Newton*, Vol. 8. Cambridge, UK: Cambridge University Press.

Wollenschläger, K. 1931–1932. Der mathematische Briefwechsel zwischen Johann I Bernoulli und Abraham de Moivre. *Verhandlungen der Naturforschenden Gesellschaft in Basel* **43**, 151–317.

Zur Rezeption der arabischen Astronomie im 15. und 16. Jahrhundert

Eberhard Knobloch

Institut für Philosophie, Wissenschaftstheorie, Wissenschafts- und Technikgeschichte, Technische Universität Berlin, D-10587 Berlin, Germany

This article deals with the reception of Arabic astronomy from the time printed Latin translations of Arabic astronomical treatises were available at the end of the 15th century. Two aspects were of special interest: determining the positions of Venus and Mercury, and criticizing the Ptolemaic system of eccentricities and epicycles, which at least facilitated the construction of a new world system. The seven sections of the article discuss the views of Ptolemy, al-Biṭrūjī, Jābir, Regiomontanus, Copernicus, Clavius, and Ibn Rushd.

EINLEITUNG

Ausgangspunkt meiner folgenden Überlegungen sind drei historische Fakten:

1. Das copernicanische Werk "Über die Umwälzungen der himmlischen Sphären" entstand infolge der Unzufriedenheit des Copernicus mit der ptolemäischen Astronomie: Copernicus lehnte die Ausgleichspunkte ab [Copernicus 1990, 2–5], wie lange vor ihm der islamische Gelehrte Ibn al-Haiṯam (965–1041) [Sezgin 1978, 34].

2. Alle bedeutenden Wegbereiter des copernicanischen Weltsystems, Copernicus selbst, einflußreiche Gegner des neuen Systems haben sich mit den Theorien der islamischen Astronomen intensiv auseinandergesetzt: Peurbach mit eben diesem Ibn al-Haiṯam, mit al-Farġānī (gestorben nach 861), al-Battānī (vor 858–929), Ṯābit ibn Qurra (836–901) in seinen "Neuen Planetentheorien" [Aiton 1978], Peurbach und Regiomontan in der "Epitome" des Almagest von Ptolemaios darüber hinaus auch mit al-Biṭrūġī (fl. 1190) [Regiomontanus 1972, 55–274]. Regiomontan hielt in Padua Vorlesungen über al-Farġānī, deren Einführung erhalten ist [Regiomontanus

1972, 41–53], verfaßte eine Widerlegung der Theorie von al-Biṭrūǧī, die erst 1951 ediert wurde [Carmody 1951]. Rheticus studierte und kommentierte zu Beginn seiner wissenschaftlichen Laufbahn al-Farġānī [Rheticus 1982, 11]. Copernicus zitiert über die bereits genannten islamischen Autoren hinaus az-Zarqāllu (gestorben 1100) und Ibn Rušd (1126–1198). Clavius (1538–1612), der tonangebende Mathematiker und Astronom der Jesuiten seiner Zeit, zieht direkt oder indirekt vierzehn islamische und drei jüdische Autoren heran [Knobloch 1990]:

al-Ma'mūn (Regierungszeit 813–833)

al-Farġānī (gestorben nach 861)

Abū Ma'šar (787–880)

Ṯābit ibn Qurra (836–901)

al-Battānī (vor 858–929)

Ibn Sīnā (980-1037)

Muḥammad al-Baġdādī (vielleicht gestorben 1037)

Ibn al-Haiṯam (965–1041)

Pseudo-Ibn al-Haiṯam = Abū 'Abd Allāh Muḥammad ibn Mu'āḏ
(wahrsch. 11. Jh.)

$$\text{Hali} \begin{cases} \text{entweder 'Alī ibn Abī r-Riǧāl (11. Jhd.)} \\ \text{oder 'Alī ibn Riḍwān (998–1061 od. 1069)} \end{cases}$$

az-Zarqāllu (gestorben 1100)

Ǧābir ibn Aflaḥ al-Išbīlī (fl. 1. H. 12. Jh.)

Ibn Esra (ca. 1090–1167)

al-Biṭrūǧī (fl. ca. 1190)

Ibn Rušd (1126–1198)

Jacob ben Machir ibn Tibbon (ca. 1236–1305)

Abraham ben Samuel Zacuto (ca. 1450–ca. 1522)

Pseudo-Ṭūsī (um 1300)

3. Im 12. Jahrhundert entstand unter den islamischen Philosophen Spaniens eine Diskussion über das ptolemäische Weltbild. Die ptolemäische Theorie wurde in Einzelheiten kritisiert, so von Ǧābir

ibn Aflaḥ aus Sevilla, oder insgesamt zugunsten anderer Weltbilder abgelehnt. Ibn Ṭufail (gest. 1185) erklärte die Lehre von den Epizykeln und der Exzentrizität für absurd [Sezgin 1978, 36]. Seine Nachfolger al-Biṭrūǧī, in Sevilla wirkend, und Ibn Rušd aus Cordoba versuchten demgemäß, ein homozentrisches Weltsystem an die Stelle des ptolemäischen zu setzen.

Zwar waren viele Schriften dieser Autoren schon im 13. Jahrhundert ins Lateinische übersetzt worden, so daß ihr Werk im westlichen Europa durch Handschriftenkopien bekannt wurden. Die weit verbreitete, Anfang des 13. Jahrhunderts verfaßte "Sphaera" des Johannes von Sacrobosco (gest. um 1250) zog bereits al-Farġānīs Schrift in entscheidendem Maße heran. Aber eine wesentlich breitere Rezeption setzte erst ein, als seit Ende des 15. Jahrhunderts vorhandene oder neu angefertigte lateinische Übersetzungen gedruckt wurden. Dazu zählen die folgenden in chronologischer Reihenfolge ihres Erscheinens angeführten Ausgaben:

Abū l-Ḥasan ʿAlī ibn Abī r-Riǧāl (Übersetzer: Aegidius de Thebaldis und Petrus de Regio, die die kastilische Übersetzung von Jehuda ben Moses (1256) übersetzten): *Praeclarissimus liber completus in iudiciis astrorum, quem edidit Albohacen Haly filius abenragel.* Venedig 1485 (1501, 1525). Eine Überarbeitung dieser lateinischen Übersetzung wurde von Anton Stupa herausgegeben: *Albohazen Haly Filii Abenragel libri de iudiciis astrorum, summa cura et diligenti studio de extrema barbarie vindicati, ac latinitati donati, per Antonium Stupam Rhoetum Praegalliensem. Additus est huic authori Index capitum singularum octo partium, seu librorum, quo lector facilius inveniat quaestionem sibi oblatam.* Basel 1551.

Abu Maʿšār (Übersetzer: Hermann von Carinthia): *Introductorium in astronomiam octo continens libros partiales.* Venedig 1489.

al-Farġānī (Übersetzer: Johannes von Sevilla): *Rudimenta astronomica.* Ferrara 1493.

ʿAlī ibn Riḍwān (Übersetzer: Aegidius de Thebaldis): *Glossae in Ptolemaei Quadripartitum.* Venedig 1493 (1519) (Erwähnt als Handschrift bei [Houzeau und Lancaster 1887, I, 711]; erwähnt als Druck bei [Suter 1900, 109] und [Gabrieli 1924, 506]; eine mögliche Teilübersetzung, kein Druck, ist erwähnt bei [Sezgin 1979, 44]).

Ibn Rušd (Übersetzer: u.a. Jacob Mantino): *Paraphrases, expositiones, commentarii* zusammen mit der lateinischen Ausgabe der Werke von Aristoteles: Venedig 1495–1497, 1550–1552, 1562–1574 und sehr oft.

Ibn Sīnā (Übersetzer: Dominicus Gundisalvi): *Avicennae peripatetici philosophie ac medicorum facile primi opera in lucem redacta*. Venedig 1495 (1508, 1546).

Abraham ben Samuel Zacuto: *Almanach perpetuum*. Leiria 1496.

Ibn Esra (Übersetzer: Pietro d'Abano): *Abrahe Avenaris judei de re judiciali opera astrologica, a Petro de Abano in latinam traducta*. Venedig 1507.

al-Biṭrūǧī (Übersetzer: Calo Calonymos, der die hebräische Übersetzung von Moses ben Tibbon übersetzte): *Alpetragii arabi planetarum theorica physicis rationibus probata, nuperrime latinis litteris mandata a Calo Calonymos Hebraeo Neapolitano*. Venedig 1531.

Ǧābir ibn Aflaḥ (Übersetzer: Gerhard von Cremona): In: Peter Apian. *Instrumentum primi mobilis. Accedunt iis Gebri filii Affla Hispalensis Astronomi vetustissimi pariter et peritissimi, libri IX de astronomia, ante aliquot secula Arabice scripti, et per Giriardum Cremonensem latinitate donati, nunc vero omnium primum in lucem editi*. Nürnberg 1534.

az-Zarqāllu (Übersetzer: Gerhard von Cremona): *Canones Arzachelis in tabulas Toletanas a mag. Gerardo Cremonensi ordinati. Descriptio scaphaeae, quae etiam codicibus Arabice scriptis exstat, typis impressa est: Scapheae recentis res doctrinae patris Abrysakh Azarchelis* (!) *summi astronomi a Ioanne Schonero Carolo-Stadio*. Nürnberg 1534.

al-Farġānī (Übersetzer: Johannes von Sevilla), al-Battānī (Übersetzer: Platon von Tivoli): In: *Rudimenta astronomica Alfragrani. Item Albategnius astronomus peritissimus de motu stellarum, ex observationibus tum propriis, tum Ptolemaei, omnia cum demonstrationibus geometricis et additionibus Ioannis de Regiomonte. Item oratio introductoria in omnes scientias mathematicas Ioannis de Regiomonte, Patavii habita, cum Alfraganum publice praelegeret. Eiusdem utilissima introductio in Elementa Euclidis. Item Epistola Philippi Melanthonis nuncupatoria, ad Senatum Noribergensem. Omnia iam recens prelis publicata*. Nürnberg 1537.

Abū 'Abd Allāh Muḥammad ibn Mu'āḏ = Pseudo-Ibn al-Haiṭam (Übersetzer: Gerhard von Cremona): In: *Petri Nonii Salaciensis, de Crepusculis liber unus, nunc recens ornatus et editus. Item Allacen Arabis vetustissimi, de causis crepusculorum Liber unus a Gerardo Cremonensi iam olim Latinitate donatus, nunc vero omnium primum in lucem editus*. Lissabon 1542. Erneut herausgegeben in: *Opticae Thesaurus Alhazeni Arabis libri septem, nunc primum editi. Eiusdem liber de crespusculis et nubium ascensionibus. Item Vitellionis Thuringopoloni libri X.* ed. F. Risner. Basel 1572.

Meine Ansicht ist, daß diese Rezeption mit zum Wandel des Weltbildes beitrug, indem man sich mit den partiellen oder globalen Kritiken an Ptolemaios auseinandersetzte. Die Unizität der ptolemäischen Theorie wurde in Frage gestellt. Vor diesem Hintergrund ist auch die vorübergehende Renaissance der Theorie von den homozentrischen Sphären in Italien zu sehen. Man denke nur an Girolamo Fracastoro, Alessandro Achillini, Giovanni Battista della Torre, Giovanni Battista Amico, Torquato Tasso, Agostino Nifo u.a. im 16. Jahrhundert [Lattis 1991].

Auf diese Weise wurde der Boden für eine Geisteshaltung bereitet, die Rheticus bei der Veröffentlichung seines "Ersten Berichtes" über das neue Weltsystem des Copernicus mit den schönen Worten des Alkinoos beschrieb:

Δεῖ δ'ἐλευθέριον εἶναι τῇ γνώμῃ τὸν μέλλοντα φιλοσοφεῖν.

"Wer philosophieren will, muß in seinem Urteilvermögen frei sein."

Unter diesem Blickwinkel soll untersucht werden, wie eine Reihe von Autoren die Frage nach der Anordnung und der Bewegung der Himmelskörper behandelt hat. Das besondere Augenmerk gilt dabei der Argumentation von:

1. Ptolemaios: *Almagestum*
2. al-Biṭrūǧī: *Planetarum theorica physicis rationibus probata*
3. Ǧābir: *Libri IX de astronomia*
4. Regiomontan: *Epitoma in almagestum Ptolomei*
5. Copernicus: *De revolutionibus orbium caelestium*
6. Clavius: *Commentarius in Sphaeram Ioannis de Sacro Bosco*
7. Ibn Rušd: Kommentare zu Aristoteles, *De caelo, Metaphysica, Meteorologica.*

1. PTOLEMAIOS: *ALMAGESTUM*

Über die Anordnung der Himmelskörper herrschte bereits im Altertum keine einhellige Meinung. Ptolemaios erörtert in Buch IX, Kapitel 1 des "Almagest" die Reihenfolge der Sphären der Sonne, des Mondes und der fünf Planeten. Die Uneinigkeit betraf die Lokalisierung der Planeten

Venus und Merkur. Diese Lokalisierung machte den eigentlichen Spiel-
raum der vorcopernicanischen Astronomie aus [Blumenberg 1975, 276].
Was Wunder, daß Rheticus 1540 schreibt [Rheticus 1982, 60]: "Praeterea,
dii immortales, quae digladiatio, quanta lis fuit de orbium Veneris et
Mercurii situ, et quomodo sint ad solem collocandi," "Außerdem, un-
sterbliche Götter, welch erbitterten Kampf, welch großen Streit gab es
über die Lage der Bahnen von Venus und Merkur und darüber, wie sie in
Bezug auf die Sonne zu lokalisieren sind."
Nach Ptolemaios gab es damals die zwei Ansichten:

1. Die älteren Astronomen, bei ihm $\mu\alpha\theta\eta\mu\alpha\tau\iota\kappa o\acute{\iota}$ genannt, setzen
   Venus und Merkur *unter* die Sonnensphäre. (Bei Copernicus [1949,
   Buch I, Kapitel 10] sind dies Ptolemaios und viele Jüngere).
2. Einige der *späteren* setzen diese Planeten *über* die Sonnensphäre.
   Deren Grund sei: Nie habe ein Vorübergang dieser Planeten vor der
   Sonne stattgefunden.

Ptolemaios lehnt diesen zweiten Standpunkt aus zwei Gründen ab:

Erster Grund: Nichtbeobachtbarkeit. Die unteren Planeten können in
einer anderen Bahn liegen als derjenigen, die durch die Sonne und unser
Auge geht. Deshalb gibt es in diesem Fall keinen sichtbaren Vorübergang.
So treten auch meistens keine Finsternisse zur Zeit der Konjunktionen im
Falle des Mondes ein. (In der Tat müssen Sonne und Mond zugleich in der
Nähe der Knotenlinie bei Neumondstellung sein, um eine Sonnenfinster-
nis zu verursachen).

Zweiter Grund (in Form eines vollständigen Syllogismus): Unentscheid-
barkeit. Die Frage ist nicht entscheidbar. Entfernungen lassen sich allein
durch wahrnehmbare Parallaxen bestimmen. Keiner der beiden Planeten
zeigt aber eine solche Parallaxe. Also ist die Frage nicht entscheidbar.

Deshalb plädiert Ptolemaios für die älteren Astronomen. Allerdings
dürfe die Nähe der beiden Planeten zur Erde nicht so groß sein, daß die
Annäherung eine bemerkenswerte Parallaxe zur Folge habe.

Halten wir fest: Die Anordnungsfrage ist für Ptolemaios mit astronomischen Mitteln im strengen Sinn unentscheidbar [Blumenberg 1975, 276].
Deshalb hält er sich an den älteren Standpunkt, eine prinzipiell vernünftige
Entscheidung aus wissenschaftstheoretischer Sicht: das neue System muß
sich legitimieren, um das alte ablösen zu können.[1]

2. AL-BIṬRŪǦĪ: *PLANETARUM THEORICA PHYSICIS RATIONIBUS PROBATA*

Regiomontan wiederholt in der von Copernicus herangezogenen Epitome zu Buch IX, Satz 1 des "Almagest" die von Ptolemaios angeführten beiden Ansichten mit dem Finsternisargument für die zweite und fügt hinzu [Regiomontanus 1972, 192]:

> Alpetragius autem, qui motuum diversitates, atque eorum apparentes velocitates incurtatione quadam accidere putabat, sub Marte Venerem, sub qua solem, deinde Mercurium statuebat. Minus enim incurtat Venus a motu primo quam sol, ex parte quidem epicycli. Mercurius autem plus quam sol.

al-Biṭrūǧī aber, der glaubte, die Verschiedenheiten der Bewegungen und deren sichtbare Geschwindigkeiten kämen durch eine gewisse Verzögerung (engl. lag, arabisch naqṣ [Carmody 1951, 123], taqṣīr [Samsó 1978, 33]) zustande, versetzte die Venus unter den Mars, unter diese die Sonne, darauf den Merkur. Weniger nämlich bleibt die Venus gegenüber der ersten Bewegung als die Sonne zurück, jedenfalls was den Epizykel angeht, der Merkur aber mehr als die Sonne.

al-Biṭrūǧī vertrat danach das System: Mars, Venus, Sonne, Merkur. Sollen wir glauben, daß al-Biṭrūǧī seine Theorie mit Bezug auf die von ihm abgelehnten Epizykel begründete? Wir müssen schon jetzt vermuten, daß es sich um einen Zusatz von Regiomontan handelt. Um hier Klarheit zu erhalten, ist zunächst genauer auf al-Biṭrūǧīs Werk einzugehen.

al-Biṭrūǧī verfaßte sein "Kitāb fī l-hai'a" (Buch über den Kosmos oder Astronomie) im Zeitraum zwischen 1185 und 1217, dem Tod seines Lehrers Ibn Ṭufail und der Beendigung der lateinischen Übersetzung durch Michael Scotus (gest. 1235). Diese Übersetzung wurde erst 1952 erstmalig von Carmody herausgegeben [Al-Biṭrūǧī 1952], stand allerdings in handschriftlicher Kopie Regiomontan zur Verfügung, der sie stark mit Marginalien versah. Heute hat als Textgrundlage die vorzügliche, von Goldstein besorgte Ausgabe des arabischen Textes und der hebräischen Übersetzung von Moses ben Samuel ben Jehuda ben Tibbon aus dem Jahre 1259 zu dienen [Al-Biṭrūǧī 1971].

1531 erschien die 1528 vollendete lateinische Übersetzung von Calo Calonymos, die sich ihrerseits auf diese hebräische Übersetzung stützte. Im Titel heißt es dort: "Planetentheorie des Arabers al-Biṭrūǧī nach physikalischen Gründen geprüft." al-Biṭrūǧī ist der herausragende Astronom unter den spanischen Aristotelikern, Zeitgenosse von Ibn Rušd, der mit naturphilosophischen Gründen—daher der Titel—im Sinne des Aristoteles die ptolemäischen Epizykel und Exzenter ablehnte.

Wie er im siebten Buch sagt, nimmt er die Geschwindigkeit der Planetenbewegungen und ihre Nähe zum ersten Beweger, der neunten Sphäre, als Prinzip, Grund und wahre Ursache für die Anordnung der Himmelskörper: Im Sinne der täglichen Bewegung ist die oberste Sphäre am schnellsten. Denn was dem ersten Beweger näher ist, hat größere Kraft und ist schneller. Die Kraft des ersten Bewegers nimmt mit der Entfernung zu ihm ab, bis hin zur Erde, die völlig ruht.

Es verdient hervorgehoben zu werden, daß die Geschwindigkeitsgröße, also die Zeit, als Ordnungskriterium im "Commentariolus" von Copernicus explizite wieder auftritt [Blumenberg 1975, 278], nur eben auf die siderischen Umläufe bezogen (Zekl in [Copernicus 1990, 8–9]), so daß Saturn der langsamste, Merkur der schnellste Planet wird.

Er, al-Biṭrūǧī, habe gefunden, daß die Verzögerung der Venus größer als die der Marssphäre, geringer als die der Sonnensphäre ist. Deshalb liege die Venus zwischen den beiden Sternen. Die Verzögerung der Sonne (1° pro Tag) sei geringer als die des Merkur und des Mondes (33° pro Tag) [Carmody 1952, 565], aber größer als die der Venus. Demnach haben die vier Sphären Saturn, Jupiter, Mars, Venus eine Ordnung, die drei übrigen Sphären eine andere. Dies stimme seiner Ansicht nach mit ihrer natürlichen Ordnung überein, obwohl er sich damit bewußt gegen die Ansicht vieler antiker Gelehrter stelle.

Er vertritt also tatsächlich die aus Regiomontans Angaben erschlossene Reihenfolge: Mars, Venus, Sonne, Merkur. Von einem "ex parte epicycli" ist, wie zu vermuten war, nicht die Rede. Regiomontan interpretierte al-Biṭrūǧīs Theorie mit Hilfe der Begriffe der von ihm vertretenen ptolemäischen Theorie. Venus durchläuft danach ihren Epizykel fünfmal in acht Jahren von West nach Ost [Hallyn 1987, 95f.], das heißt, ihre synodische Periode beträgt ungefähr 584 Tage. Dagegen durchläuft die Sonne die Ekliptik von West nach Ost in ungefähr 365 Tagen. Interpretiert man diese Bewegungen als verzögerte Bewegungen von Ost nach West entsprechend der Bewegung der neunten Sphäre, so ist tatsächlich die Verzögerung der Venus geringer als die der Sonne, oder mit anderen Worten: die Venus bewegt sich schneller von Ost nach West als die Sonne.

Seine Verzögerungstheorie macht für al-Biṭrūǧī die Annahme unakzeptabler rückläufiger Bewegungen unnötig. Er lehnt damit das ptolemäische Weltbild samt dessen Argumenten ab. Ptolemaios habe gesagt, die rückwärtige Bewegung werde am besten erklärt [Al-Biṭrūjī 1971, 124], wenn man die Sonne in die Mitte setze, "was am natürlichsten sei." In Wahrheit hatte Ptolemaios nur zur Rechtfertigung der Mittelstellung gesagt, daß eine solche Anordnung "in natürlicher Weise," $\varphi \upsilon \sigma \iota \kappa \acute{\omega} \tau \varepsilon \rho o \nu$

(Adverb des Komparativs), die beiden Gruppen von Planeten voneinander trenne (Almagest IX, 1). Al-Biṭrūǧī kritisiert jedoch, daß Ptolemaios keine Gründe für dieses "am natürlichsten" angegeben habe. Dieser habe damit gezeigt, daß er kein Naturphilosoph war, auch wenn er ein Mathematiker gewesen sei.

Er weist das Argument des Ptolemaios zurück, es müsse bei dessen Reihenfolge—Sonne oberhalb von Venus und Merkur—nicht notwendig Venus- und Merkurdurchgänge geben. Ǧābir habe in seinem "Iṣlāḥ al-Maǧasṭī" (Verbesserung des Almagest) gezeigt, daß nach den eigenen Prinzipien, das heißt den verwendeten Beobachtungsdaten von Ptolemaios, die Planeten unsere Sehstrahlen zur Sonne kreuzen müssen. Freilich bedeutet dies nach Ǧābir noch keine Bedeckung der Sonne, wie wir sehen werden. Wir kommen auf Ǧābir, einen berühmten Kritiker von Ptolemaios, zurück.

Tatsächlich hat al-Biṭrūǧī darin Recht, daß die Tafeln des Ptolemaios Venus- und Merkurdurchgänge nach sich ziehen. Aber halten wir fest: al-Biṭrūǧī entkräftet ein Argument, das Ptolemaios zur Verteidigung seines Systems angeführt hatte, das er selbst, al-Biṭrūǧī, ablehnt. Al-Biṭrūǧī akzeptiert deshalb aber nicht etwa die Implikation

Die Lokalisierung impliziert Bedeckungen,

sondern er ersetzt den Vordersatz und verwendet

Die Beobachtungsdaten implizieren Bedeckungen.

Sodann lehnt er überhaupt das Verfinsterungsargument ab, da von den Bedeckungen unzulässigerweise auf deren Beobachtbarkeit geschlossen werde:

Die Beobachtungsdaten implizieren keine Beobachtung von Bedeckungen durch Verfinsterungen.

Das Verfinsterungsargument wäre nur richtig, wenn die zwei Planeten von einem anderen Körper erleuchtet würden, wie der Mond von der Sonne. Wenn die Planeten selbst leuchten, so verursachen sie keine Verfinsterung. Als Beweis dafür, daß sie nicht von der Sonne erleuchtet werden, sondern ihr Licht von irgendeiner anderen Quelle erhalten, nimmt er die Tatsache, daß die Helligkeit stets gleich groß ist, unabhängig von ihrer Entfernung von der Sonne. Erhielte der Merkur sein Licht von der Sonne, erschiene er stets halbmondförmig, da er nie weit von der Sonne entfernt sei. Gleiches gelte für die Venus.

Tatsächlich ist die Helligkeit der Venus fast konstant [Al-Biṭrūǧī 1971, 33 Anm. 12], und tatsächlich konnte man die vorhandenen Merkur- und

Venusphasen nur nicht beobachten. Aber als Galilei mit Hilfe seines Fernrohrs 1610 die Venusphasen entdeckt hatte, nahm er diese als Beweis für die Richtigkeit des copernicanischen Systems [2].

3. ĞĀBIR: *LIBRI IX DE ASTRONOMIA*

al-Biṭrūǧī hatte zur Widerlegung der ptolemäischen Verteidigung gegen das Verfinsterungsargument Ğābir aus Sevilla herangezogen. Aber eben nur insoweit: Ğābir vertrat nämlich in seinem Iṣlāḥ al-Maǧasṭi—seit 1534 in der lateinischen Übersetzung Gerhard von Cremonas zugänglich—das von Ptolemaios und auch von al-Biṭrūǧī abgelehnte System, das Ptolemaios den jüngeren Astronomen zuschrieb. Unter den 15 Irrtümern des Ptolemaios listet er als siebten zu Beginn seiner Schrift auf:

> Et erravit in principio tractatus noni sui libri: in hoc quod posuit ordinem duorum orbium Veneris et Mercurii sub orbe Solis. Nam illud quod dant radices suae, est, quod ambo sunt supra orbem Solis necessario. Et similiter erravit in sermone suo, quod ambo non vadunt per lineas, quae transeunt per visus nostros et per solem.

> Ptolemaios irrte zu Beginn der neunten Abhandlung seines Buches darin, daß er die Stellung der zwei Bahnen von Venus und Merkur unter die Bahn der Sonne versetzte. Denn das, was seine eigenen Prinzipien ergeben, ist, daß beide notwendigerweise über der Bahn der Sonne sind. Und entsprechend irrte er in seiner Rede darin, daß beide nicht die Geraden kreuzen, die durch unsere Augen und die Sonne gehen.

Genauer führt Ğābir diese Argumentation im siebten Kapitel "Über die fünf Planeten" aus. Aber halten wir fest: Ğābir behauptet nicht, daß es zu einer Bedeckung der Sonne durch die beiden Planeten kommt, sondern daß diese aufgrund der ptolemäischen Grundlagen *obere* Planeten sind, die die Sehstrahlen zur Sonne kreuzen.

4. REGIOMONTAN: *EPITOMA IN ALMAGESTUM PTOLOMEI*

Regiomontan läßt al-Biṭrūǧīs Theorie in der "Epitome" unkommentiert. Dagegen wiederholt er das ptolemäische Argument der Nichtbeobachtbarkeit und fügt dafür ein weiteres Argument, ein relatives Größenargument, hinzu, das er auf al-Battānīs Beobachtungsdaten stützt [Regiomontanus 1972, 192]. Danach beträgt der sichtbare Durchmesser der

Venus nur ein Zehntel des Sonnendurchmessers [Al-Battānī 1899–1907 I, 121 (Kap. 50)]. (In Wirklichkeit ist es ein Dreißigstel.) Die Bedeckung betrüge nur ein Hundertstel der Sonnenscheibe, wäre also unmerklich klein (insensibile):

Die Stellung Sonne—Venus—Auge hat keine beobachtbaren Konsequenzen.

Bisher hat also Regiomontan folgendermaßen argumentiert:

1. Die Lokalisierung der Planeten Venus, Merkur unter der Sonne führt nicht notwendigerweise zu einer Bedeckung:

   Die Lokalisierung impliziert keine Bedeckungen.

2. Eine Bedeckung ist aber auch nicht ausgeschlossen. Denn wenn es eine solche geben sollte, wäre sie wegen der relativen Größenverhältnisse nicht zu beobachten. Eine Bedeckung führt also nicht notwendigerweise zur Lokalisierung (durch Beobachtung) unter der Sonne:

   Die Bedeckung impliziert keine entsprechende Lokalisierung.

Damit ist nach Regiomontan gezeigt, daß das Finsternisargument für das Lokalisierungsproblem unbrauchbar ist. Deshalb führt er ein neues, *das Vakuumargument*, ein:

Die größte Mondentfernung vom Weltzentrum beträgt fast 64 Erdradien [Al-Battānī 1899–1907 I, 60f. (Kap. 30)], die kleinste Sonnenentfernung 1070 Erdradien oder mehr [3]. Die kleinste Entfernung zwischen beiden Sternen betrüge also mehr als 1006 Erdradien.

"Hoc autem spacium natura non sinit vacuum," "Diesen Raum aber läßt die Natur nicht leer", kommentiert Regiomontan diesen numerischen Befund. Deshalb müsse ihn notwendigerweise irgendein Himmelskörper einnehmen:

Frustra enim tanta moles in celo permitteretur.

Ohne Zweck würde nämlich eine solche Größe im Himmel zugelassen.

Deshalb wird mit "natürlicher Angemessenheit, Zweckmäßigkeit", "naturalis commoditas", dieser Raum für die Bahnen von Venus und Merkur beansprucht. Regiomontan zieht also ein teleologisches Argument heran, das freilich schon in den Ὑποθέσεις τῶν πλανωμένων, den "Hypothesen über die Planeten" des Ptolemaios, und in der Ὑποτύπωσις τῶν ἀστρονομικῶν ὑποθέσεων, dem "Entwurf der astronomischen Hy-

pothesen" des Proklos, anzutreffen ist [Copernicus 1990, 233 Anm. 115].
Seine volle Wirksamkeit hatte es im Nominalismus des 14. Jahrhunderts
entfaltet, auch wenn es genuin aristotelisch war [Blumenberg 1975, 177].

Zugleich gibt Regiomontan zu, daß die Frage offen bleiben muß, welcher
von den beiden Sternen höher bzw. tiefer steht, da sowohl die notwendigen
Instrumente wie die Genauigkeit für die zahlenmäßige Erfassung der
Merkurbewegungen fehlen. Dasselbe müsse man von der Venus glauben.

Es bleibt festzuhalten, daß er das "horror-vacui"-Argument heranzieht,
das al-Biṭrūǧī—gemäß aristotelischer Lehre—zur Ablehnung exzentri-
scher Sphären anführte [Samsó 1978, 33].

5. COPERNICUS: *DE REVOLUTIONIBUS ORBIUM CAELESTIUM*

Sowohl Copernicus [Rosen 1961] wie Clavius, was noch zu zeigen ist,
bezogen ihre Kenntnisse über al-Biṭrūǧī aus Regiomontans "Epitome,"
nicht aus der 1531 erschienenen lateinischen Übersetzung des Werkes von
al-Biṭrūǧī. Demgemäß erwähnt Copernicus wie Regiomontan kommentar-
los die Anordnung des arabischen Autors. Die "jüngeren" Astronomen
des Ptolemaios nennt Copernicus—anders als bisher—beim Namen: Es ist
Platon mit seinem "Timaios" (38d). Dies geschieht jedoch nicht beim
Referieren des Vakuumargumentes. Regiomontan bleibt ungenannt, dien-
te hier auch nicht als Vorlage, da für die größte Mond- und kleinste
Sonnenentfernung andere Werte ($64\frac{1}{6}$, also mehr als 64, und 1160 Erdradien) zugrunde gelegt werden. Dagegen stammt die Bezugnahme auf
al-Battānī um des Größenvergleichs willen zwischen Venus und Sonne
sicher wieder aus der "Epitome."

Von besonderem Interesse ist seine folgende Bemerkung:

> Quamvis et Averroes in Ptolemaica Parafrasi nigricans quiddam se vidisse
> meminuerit, quando Solis et Mercurii copulam numeris inveniebat exposi-
> tam.

> Obwohl sich sogar Averroes in seiner Paraphrase zu Ptolemaios erinnert,
> etwas Schwärzendes gesehen zu haben, als er eine durch Zahlen
> nachgewiesene Vereinigung von Sonne und Merkur fand.

Diese Anspielung konnte bis vor kurzem nicht verifiziert werden
[Copernicus 1990, 234f. Anm. 120]. Sie findet sich in der von Jakob Anatoli
1213 angefertigten hebräischen Übersetzung des "Compendium des Almagest" (Muḫtaṣar al-Maǧisṭī). [4] Der Titel der im arabischen Original

nicht erhaltenen Schrift stammt vom Sohn des Ibn Rušd [Sezgin 1978, 93; Steinschneider 1892, 54]. Wir müssen auf sie im Zusammenhang mit Clavius zurückkommen. Hier sei nur betont, daß damit Ibn Rušd die platonisch-aristotelische Anordnung von Merkur über der Sonne in Frage gestellt hätte, da eine solche "Verfinsterung" die Lokalisierung von Merkur unter der Sonne impliziert hätte. Um dieser überraschenden Konsequenz willen führt Copernicus das Zitat an.

6. CLAVIUS: *COMMENTARIUS IN SPHAERAM IOANNIS DE SACRO BOSCO*

In seinem 1570 erstmalig und 1611 letztmalig unter seiner Verantwortung erschienenen Kommentar zur Sphaera des Johannes von Sacrobosco verfaßte Christoph Clavius einen Abschnitt "De ordine sphaerarum coelestium," "Über die Anordnung der Himmelssphären" innerhalb des Kommentars zum 1. Kapitel des Johannes (Clavius 1611, 42−47). Seine Absicht war, die von Ptolemaios und Regiomontan vertretene Anordnung zu bestätigen, und zwar aufgrund der drei Kriterien, die die Astronomen anzuwenden pflegen:

1. *diversitas aspectus*, das heißt die "Verschiedenheit des Anblicks"; gemeint ist die Parallaxe, der Winkel zwischen wahrem und beobachtetem Ort eines Sterns, der entsteht, weil der Beobachter auf der Erdoberfläche steht, nicht im Erdzentrum;
2. *velocitas* und *tarditas motus*, das heißt Schnelligkeit und Langsamkeit der Bewegung;
3. *eclipses seu occultationes planetarum*, das heißt Finsternisse oder Bedeckungen der Planeten.

Die Kriterien 1 und 3 hatte Ptolemaios angeführt, das Kriterium 2 al-Biṭrūǧī. Dies hat in unserem Zusammenhang das größte Interesse. Clavius führt aus [1611, 44]:

> Quo magis coelum a natura, et conditione primi mobilis recedit, eo etiam in inferiori est loco ponendum.

> Je mehr ein Himmel von der Natur und der Beschaffenheit des primum mobile zurückweicht, auf einen desto tieferen Platz ist er auch zu setzen.

Da sich der Mond am schnellsten von West nach Ost bewegt und damit am meisten von der Bewegung und Beschaffenheit des *primum mobile* abzuweichen scheint, erhält er den untersten Platz. Denn das *primum mobile* [Clavius 1611, 29, 36] bewegt sich am schnellsten in umgekehrter Richtung: Über die Pole der Welt vollführt es durch den Äquinoktialkreis in 24 Stunden eine Umdrehung von Ost nach West.

Die Anzahl der Sphären und damit die Ordinalzahl des *primum mobile* hatte Clavius im Laufe der verschiedenen Auflagen seines Kommentars vergrößert. Noch in der Ausgabe von 1581 vertrat er unter Einbeziehung der aristotelischen Elementenlehre die Theorie, daß zwischen dem Caelum Empyreum und dem achten Himmel, das heißt dem Firmament oder der Fixsternsphäre, nur zwei Sphären anzunehmen seien, der neunte Himmel und das *primum mobile*.

1589 erschienen in Venedig die "Novae coelestium orbium theoricae congruentes cum observationibus N. Copernici" von Giovanni Antonio Magini (1555–1617) [Magini 1589], der auf ptolemäischer Grundlage die copernicanischen Angaben zu den Umlaufperioden der vier Bewegungen der achten Sphäre übernahm [Krayer 1991, 77]. Auf Maginis Präzessionstheorie stützt sich Clavius mit den Worten [1611, 36]:

> Nunc, ut quatuor in eodem coelo octavo motus observatos tueamur, opus est supra illud non solum duos orbes mobiles collocare, sed tres, ut iam non solum 10 coeli mobiles cum Alphonsinis, sed omnino undecim concedendi sint, si phaenomena coelestia certa ratione et probabiliter, ita ut nihil absurdi ex assumptis hypothesibus sequatur, servare velimus et tueri.

> Um nun die vier in demselben achten Himmel beobachteten Bewegungen zu bewahren, muß man über jenem nicht nur zwei bewegliche Kreisbahnen ansetzen, sondern drei, so daß nunmehr nicht nur zehn bewegliche Himmel gemäß den Alphonsinischen Tafeln zuzugestehen sind, sondern insgesamt elf, wenn wir die Himmelsphänomene nach einer sicheren Methode und auf wahrscheinliche Weise retten und bewahren wollen, so daß nichts Absurdes aus den angenommenen Hypothesen folgt.

Die Formulierung ist bezeichnenderweise vorsichtig. Eine solche bestimmte Methode, die nicht absolute Wahrheit, sondern Wahrscheinlichkeit beanspruchte, war gemäß der jesuitischen Studienordnung Sache der Mathematik, nicht der Naturphilosophie. Um so leichter konnte sich Clavius zu einer Modifikation seiner bisherigen Theorie bereit erklären.

Im Abschnitt "Über die Perioden der Himmelsbewegungen" [Clavius, 1611, 29f.] legte Clavius noch insgesamt zehn Himmel zugrunde. Die neunte, auf das primum mobile folgende Sphäre bewege sich sehr langsam um den Tierkreis, nach König Alfons in 49.000, nach Ptolemaios in 36.000, nach al-Battānī in 23.760 Jahren. Dieser "orbis" ziehe die unteren acht Sphären mit sich.

Der Fixsternhimmel erhalte den obersten Ort, "quoniam tardissime contra motum primi mobilis fertur" [Clavius 1611, 44], "weil er sich am langsamsten gegen die Bewegung des primum mobile bewegt," danach Saturn, Jupiter und die übrigen. Über die Reihenfolge von Sonne, Venus, Merkur könne auf diese Weise nichts Gewisses gesagt werden. Sie seien zwar über dem Mond anzusetzen, da sie sich langsamer von Westen nach Osten bewegen, und unter dem Firmament, Saturn, Jupiter, Mars, da sie sich schneller entgegen dem primum mobile bewegen. Aber sie selbst vollenden fast in gleicher Zeit ihre Bewegungen von West nach Ost, so daß nicht sicher gewußt werden kann, wer höher, wer tiefer zu setzen sei. "Immo," fährt Clavius unter Bezugnahme auf Regiomontans "Epitome" fort,

Alpetragius, ut testatur Ioan. Regiom. lib. 9 Epitomes propos. 1. ex hac ratione colligit, sub Marte positum esse caelum Veneris, et sub hac, caelum Solis, deinde Mercurii, ac postremo Lunam, propterea quod Venus ratione epicycli tardius peragat cursum suum quam Sol, et Sol tardius quam Mercurius, Luna denique citissime omnium periodum suam absolvat.

Allerdings schließt al-Biṭrūǧī, wie Joh. Regiomontan Buch 9 der Epitome, Satz 1 bezeugt, aus dieser Überlegung, unter dem Mars sei der Venushimmel gesetzt, und darunter der Sonnenhimmel, darauf der des Merkur, und schließlich der Mond, weil die Venus—zieht man ihren Epizykel zum Vergleich heran—ihren Lauf langsamer vollendet als die Sonne, und die Sonne langsamer als der Merkur, weil der Mond schließlich am schnellsten von allen seine Periode vollendet.

Clavius hat also Regiomontans Bemerkung bis hin zum Venusepizykel herangezogen und doch entscheidend anders formuliert—erinnern wir uns daran, daß die Monströsität des Venus-Epizykels eine treibende Kraft der copernicanischen Theorie war [Copernicus 1990, 236 Anm. 124]. Er ersetzte die "incurtatio", das Zurückbleiben, gegenüber dem ersten Beweger durch "langsamere Bewegung hinsichtlich des Epizykels", der sich in umgekehrter Richtung zur Bewegungsrichtung des primum mobile in synodischer Periode einmal um sich selbst dreht.

So entsteht zunächst ein scheinbarer Widerspruch zwischen al-Biṭrūǧī und Clavius. Nach al-Biṭrūǧī steht die Venus über der Sonne, weil sie *schneller* als diese ist, da ihr Zurückbleiben gegenüber der täglichen Rotation des ersten Bewegers geringer ist. Nach Clavius wäre dies der Fall —er selbst vertritt die ptolemäische Anordnung der Venus unter der Sonne—, weil die Venus *langsamer* ist, weil nämlich ihre Längenbewegung längs dem Tierkreis von West nach Ost geringer ausfällt, so daß der Mond,

von dem bei Regiomontan nicht die Rede war, zum schnellsten Planeten
avanciert.

Zugleich bleibt die Schwierigkeit, daß das Heranziehen des Venus-Epizykels eine Vermischung von synodischen und siderischen Umlaufzeiten
bedeutet, die bereits Regiomontan vorgenommen hatte. Clavius ließ den
Grundgedanken von al-Biṭrūǧī außer acht, wonach die Annahme zueinander entgegengesetzter Bewegungsrichtungen abzulehnen ist, ein Standpunkt, den al-Biṭrūǧī durch eine auf die tägliche Himmelsrotation bezo-
gene Verzögerungstheorie einnehmen konnte. Die West-Ost-Bewegungen
der Himmelskörper waren danach optische Täuschungen [Sezgin 1978,
36f.]. Die Übersetzung der fremden Theorie in die eigene Theorie führt
bei Clavius zu einer Verfälschung.

Um die der ptolemäischen Anordnung widersprechenden Systeme, also
auch die von al-Biṭrūǧī, abtun zu können, führt Clavius neben den schon
bekannten Argumenten weitere an. Er widerlegt also nicht al-Biṭrūǧīs
Verzögerungstheorie, die er aufgrund der von ihm benutzten Quelle nicht
versteht oder nicht zur Kenntnis nimmt, jedenfalls nicht thematisiert,
sondern er will die Richtigkeit des ptolemäischen Systems unmittelbar
aufzeigen. Diese impliziert die Falschheit des anderen, also al-Biṭrūǧīs
System.

Welche Argumente verwendet er? Es sind im ganzen fünf:

1. Das auf al-Battānīs Größenangaben basierende 'horror-vacui'-Argu-
   ment.
2. Das Trennungsargument: "motus solis est regula, et mensura motuum aliorum planetarum," "Die Bewegung der Sonne ist die
   Richtschnur und das Maß für die Bewegungen der anderen Planeten,"
   was ohne Kontext leicht im Sinne eines heliozentrischen Systems
   umgedeutet werden kann. Gemeint ist: Während Mars, Jupiter, Saturn "ratione epicycli", "durch ihren Epizykel" mit der Sonnenbewegung übereinstimmen, werden Mond, Merkur, Venus durch ihre
   Deferenten der Sonnenbewegung angeglichen (conformantur), wie in
   den "Theoricae planetarum" erklärt werde. Die Sonnentheorie dieser
   "Planetentheorien" wurde vor kurzem von Baldini erstmalig veröffentlicht, nachdem bis dahin an ihrer Existenz gezweifelt worden
   war [Baldini 1992, 469–564].
   Die Sonne trennt also die beiden verschiedenen Arten der uni-
   formitas der Bewegung mit der Sonne, sie trennt das, was auf
   verschiedene Weise mit ihr ähnlich ist.

3. Das commoditas-Argument: "Sol est rex, et quasi cor omnium planetarum," "Die Sonne ist der König und gleichsam das Herz aller
Planeten."
Dies scheint mir durchaus eine Reminiszenz an das copernicani-
sche Wort sein zu können [Copernicus 1949, 26; 1990, 136]: "In
medio vero omnium residet sol", "In der Mitte aber von allen
verweilt die Sonne".
Clavius kannte das copernicanische Werk sehr gut. Meine Hypothese stimmt mit der grundlegenden Bedeutung überein, die Blumenberg der interplanetaren Heliozentrik für eine immanente
Entstehungsgeschichte der copernicanischen Theorie beigelegt hat
[Blumenberg 1975, 272–299].
Für Clavius wird die Sonne deshalb zu Recht zwar nicht in die
Mitte der Welt, wohl aber in diejenige der Planeten gestellt, damit sie
alle in gleicher Weise an ihrem Licht teilhaben lassen kann:

in medio illorum commodissime est collocata,
am angemessensten ist sie in die Mitte von jenen versetzt worden:

es ist die "commoditas"-Argumentation Regiomontans.
4. Das astrologische Argument: Clavius führt Abū Maʿšars "Kitāb al-
madkhal al-kabīr ilā ʿilm aḥkām an-nuǧūm", die "Große Einführung
in die Wissenschaft der Astrologie" an, die ihm seit 1489 gedruckt in
Hermann von Carinthias lateinischer Übersetzung zur Verfügung
stand [Pingree 1970, 35]. Um maßvoll (*temperata*) und angemessen
(*commode*) auf die unteren Körper wirken zu können, durfte die
Sonne nicht über dem Saturn stehen, da sonst ihre Entfernung von
der Erde zu groß gewesen wäre. Alles würde gefrieren. Wäre sie
unmittelbar über dem Mond angesiedelt, wäre ihre Bewegung wegen
der zu großen Entfernung vom primum mobile von Ost nach West zu
langsam, zudem würde alles verbrennen.
5. Das Ordnungsargument: Die Bahn des Merkur ist viel unregelmäßiger
als die der Venus. Merkur erhielt von den Astronomen fünf
Bahnkreise und einen Epizykel zugeteilt, Venus nur drei Bahnkreise
und einen Epizykel. Es sei daher vernünftiger, eher Merkur als
Venus über dem Mond anzusetzen.

Aus diesen fünf Argumenten zieht Clavius den Schluß: Das ptolemäische
System ist wahrer und entspricht mehr den erfahrenen Astronomen.
Andersartige Systeme sind daher abzulehnen.
Clavius hätte mit dem Erreichten aus seiner Sicht zufrieden sein können.
Daß er nichts Überzeugendes zu bieten hatte, belegt sein Probabioris-

mus. Er wendet sich jedoch zudem mit den bekannten Argumenten von
Ptolemaios und Regiomontan (relatives Größenargument) gegen die alten
Einwände gegen die Lokalisierung von Venus und Merkur unter der
Sonne. Indem er wie Ptolemaios meint, es müsse zu keiner Kreuzung der
Planetenbahn mit dem Sehstrahl kommen, übersieht er Ğābirs Gegenbe-
weis, dessen Werk er sonst ausgiebig wegen des Sinussatzes und wegen
Spezialfällen des Seiten- und Winkelkosinussatzes der sphärischen Tri-
gonometrie zitiert.

Abschließend macht er eine aufschlußreiche Bemerkung [Clavius 1611,
47]:

> Extra hunc vero mundum, seu extra caelum Empyreum, nullum prorsus
> corpus existit, sed est spatium quoddam infinitum, (si ita loqui fas est) in quo
> etiam toto Deus existit sua essentia, in quo infinitos alios mundos perfec-
> tiores etiam hoc, fabricare posset, si vellet, ut Theologi asserunt.

> Außerhalb dieser Welt aber oder außerhalb des empyreischen Himmels gibt
> es jedenfalls keinen Körper, sondern es ist ein gewisser unendlicher Raum
> (wenn man so zu sprechen das Recht hat), in dem ebenfalls als Ganzem Gott
> durch sein Wesen existiert, in dem er, wenn er wollte, wie die Theologen
> versichern, unendlich viele andere Welten, selbst vollkommenere als diese,
> schaffen könnte.

Dies klingt wie eine vorsichtige—er verweist auf die Theologen—Anspielung auf Giordano Brunos Lehre von den unendlich vielen
Welten.

7. IBN RUŠD: *KOMMENTARE ZU ARISTOTELES, DE CAELO; METAPHYSICA; METEOROLOGICA*

Die tendenziöse Art, in der Clavius über die Vertreter der von ihm
abgelehnten Theorie der homozentrischen Sphären berichtet, zeigt sich
noch stärker im Falle von Ibn Rušd. Selbst wenn Carmody mit seiner
These Recht haben sollte, daß die Zeitgenossen Ibn Rušd und al-Biṭrūǧī
als gemeinsame Quelle für ihr homozentrisches System Ibn Ṭufail gehabt
haben sollten [Carmody 1952, 558f.], bleiben drei grundlegende Unterschiede zwischen den Systemen beider bestehen:

1. Ibn Rušd vertritt gemäß Aristoteles, *De caelo* II, 10, 291a, b die Lehre
von den gegenläufigen Bewegungen der Planeten zur äußerlichen
Umlaufschale des Himmels. Da die nächststehende Saturnbahn am

meisten von deren Bewegung mitgerissen wird, ist ihre (gegenläufige) Bewegung am langsamsten, die des am entferntest stehenden Mondes am schnellsten. Ibn Rušd vertritt also nicht al-Biṭrūǧīs Verzögerungstheorie mit der damit einhergehenden Ablehnung gegenläufiger Bewegungen.

2. Ibn Rušd lehnt ausdrücklich die Theorie der modernen Astronomen ab, die die Existenz einer neunten Sphäre vor dem ersten Himmel forderten [Arnaldez und Iskandar 1975, 5].

3. Ihre Anordnungen der inneren Planeten stimmen nicht überein.

Im Kommendar zum vierten Buch der "Sphaera" des Johannes von Sacrobosco, in dem Johannes mit der Erklärung der exzentrischen Sonnenbahn beginnt, bemerkt Clavius [1611, 292] zur Widerlegung der Homozentrik gemäß Fracastoro:

Averroes quoque in commentariis in Almagestum Ptolemaei asserit, dari Excentricos orbes, et Epicyclos in sphaeris coelestibus.

Auch Ibn Rušd versichert in den Kommentaren zum Almagest des Ptolemaios, es gäbe exzentrische Kreise und Epizykel in den himmlischen Sphären.

Lattis [1991] ist auf dies Zitat nicht eingegangen, das wie das copernicanische Ibn-Rušd-Zitat erhebliche Identifizierungsprobleme bereitet:

1. Copernicus kommt mit seinem Verweis auf eine averroistische Paraphrase des Almagest nicht in Frage, da sich sein Zitat nicht auf die Existenz von Epizykeln und Exzentern bezieht.

2. Es ist fraglich, ob Clavius auf die 1213 von Jakob Anatoli angefertigte hebräische Übersetzung des "Compendium des Almagest" anspielt, die Ibn Rušd zugeschrieben wird. Carmody [1952, 559] gestand seine Unkenntnis eines solchen "Compendium" ein. Richard von Wallingford spricht von einem "Almagestum abbreviatum" [Richard 1976, I, 250f., 273f.] bzw. einem "Commentum Almagesti" des Kommentators [Richard 1976 I, 290f.], das unter dem Namen "Parvum Almagestum" bekannt ist: Es endet nach dem sechsten Buch des Almagest. Lorch [1992, 430–434] hat jedoch gute Gründe dafür geltend gemacht, daß Ibn Rušd nicht der Autor war.

Es scheint mir aber überhaupt sehr zweifelhaft zu sein, daß Ibn Rušd eine solche Äußerung selbst getan hat. Clavius zitiert nämlich drei andere Kommentarstellen, um Ibn Rušd der Wankelmütigkeit zu überführen, um zu zeigen, daß dieser gezwungen gewesen sei, die Existenz von Exzentern und Epizykeln anzunehmen. Seine fast feindselige Haltung gegenüber Ibn

Rušd ist unübersehbar [Lattis 1994, 61]. Ein genaueres Studium der zitierten Stellen zeigt jedoch, daß diese Wankelmütigkeit so nicht nachweisbar ist, daß vielmehr Clavius tendenziös berichtet, wie er denn gelegentlich meint, die Averroisten hießen besser Erroisten [Clavius 1611, 303], "die sich irrenden", ein Wortspiel, das im Deutschen nicht nachzumachen ist. Es handelt sich um die folgenden drei Stellen:

1. Kommentar zu De caelo, II, 4, 287 b 14–21, comm. 32; [Averroes 1562–1574 V, 115v–116r; Clavius 1611, 296: Variable Sonnenentfernung]

Erhalten ist dieser 1171 angefertigte Kommentar nur in der lateinischen Übersetzung von Michael Scotus, dem al-Biṭrūǧī-Übersetzer. Der Mond durchläuft, heißt es dort, einen Erdschatten, der bald größer, bald kleiner ist, d.h. die Sonne ist bald näher an der Erde, bald entfernter. Clavius sagt dazu:

> Atque haec apparentia tantam etiam apud Averroem vim habuit, ut ingenue asseruerit lib. 2 de Coelo, comm. 32. fortasse non alia via defendi posse hanc apparentiam de Eclipsi Lunari, quam per orbem Eccentricum, quod tamen alibi negavit.

Und dieses Phänomen hat Ibn Rušd derart beeindruckt, daß er freimütig Buch 2 über den Himmel, Kommentar 32 bekannte, vielleicht könne diese Erscheinung zur Mondfinsternis auf keine andere Weise als mittels eines exzentrischen Kreises verteidigt werden, was er dennoch sonst verneinte.

In Wahrheit sagt Ibn Rušd:

> Et ex hoc apparet quod hoc, quod dicunt Mathematici de eccentricis, est impossibile.

Und daraus wird offenbar, daß das, was die Mathematiker über Exzenter sagen, unmöglich ist.

Etwas später fährt er fort:

> Et fortasse possibile est inveniri Astronomiam convenientem huic, quod apparet de Luna, sine orbe excentrico.

Und vielleicht ist es möglich, ohne einen exzentrischen Kreis eine Astronomie zu finden, die zu dem paßt, was hinsichtlich des Mondes erscheint.

Dies ist fast das Gegenteil von dem, was Clavius behauptet hatte.

2. Kommentar zu Metaphysica XII, 8, 1073b 10–22, comm. 45; [Averroes 1562–1574 VIII, 329v; Clavius 1611, 293: variable Mondentfernung]

1186 entstand Ibn Rušds Metaphysikkommentar. Nunmehr alt, räumt er erneut eigene Schwierigkeiten ein, die verschieden großen Mondentfernungen durch sein homozentrisches Sphärenmodell zu erklären. Aber er gesteht keineswegs ein, daß dies unmöglich ist. Er sagt sogar:

> Sed nihil de eis, quae apparent in motibus stellarum, cogit necessario dicere epicyclum esse, aut eccentricum.

> Aber nichts von dem, was in den Bewegungen der Sterne erscheint, zwingt notwendigerweise zu sagen, es gebe einen Epizykel oder einen Ekzenter.

Und etwas später:

> Et apparet ex hoc propinquitas et remotio, ut accidit in Luna. In iuventute autem mea speravi, ut haec perscrutatio compleatur per me: in senectute autem iam despero. Sed forte iste sermo inducet aliquem ad perscrutandum de hoc.

> Und hieraus erscheint die Nähe und die Entfernung, wie sie beim Mond auftritt. In meiner Jugend aber hoffte ich, daß diese Untersuchung von mir abgeschlossen wird, im Alter verzweifle ich nunmehr aber daran. Aber vielleicht wird jene Rede jemanden veranlassen, dies zu erforschen.

Regiomontanus [1972, 51] spielt auf diesen resignierenden Abschnitt in seiner Einleitung zur Vorlesung über al-Farġānīs Astronomie mit den Worten an:

> Quantam gloriam commentator adeptus se putaverit, si Astronomiam concentricam tradidisset testimonio suo docemur, qui totam ferme aetatem in ea re consumens desperare tandem se confitetur.

> Welch großen Ruhm der Kommentator glaubte zu erlangen, wenn er die Astronomie konzentrisch gemacht hätte, darüber werden wir durch sein eigenes Zeugnis belehrt. Er gesteht, daß er fast sein ganzes Leben für dieses Problem aufwende und endlich daran verzweifle.

Diese Rede wurde in der oben aufgeführten Ausgabe von 1537 gedruckt, die al-Farġānīs und al-Battānīs astronomische Werke enthielt, die Ausgabe, aus der Clavius die Astronomie dieser Araber kennenlernte. Er zitiert sogar in seiner Algebra diese Rede Regiomontans. Bis in unsere Tage ist dieses Eingeständnis von Ibn Rušd immer wieder erwähnt worden [Carmody 1952, 559; Arnaldez und Iskandar 1975, 3].

3. Kommentar zu Meteorologica I, Kap. 1, eingeschoben zwischen Meteorologica I, 5, 342 b 24 und I, 6, 342b 25; [Averroes 1562–1574 V, 404v–405r, Clavius 1611, 292f.]; nicht herangezogen von [Carmody 1952]: die variablen Entfernungen der Planeten, insbesondete der Sonne

Unter Bezugnahme auf die variable Sonnenentfernung zitiert Clavius die "expositio media" von Ibn Rušd zur aristotelischen Meteorologie mit folgenden Worten:

> Hanc apparentiam concedit Averroes ... lib. 1 Meteor. ubi ait videtur, quod
> Natura aequalizavit in hoc. Nam cum remittitur calor, qui est per refle-
> xionem, ut Sole existente in Capricorno accidit aequalitas in calefactione ex
> proprinquitate, et e contrario, quando accidit intensa caliditas propter refle-
> xionem ad angulos rectos, vel prope, dum Sol est in Cancro, distat tunc magis
> Sol a centro terrae, ut remittator calor.

> Diese Erscheinung gesteht Averroes ... Buch 1 Meteorologie zu, wo er sagt:
> Es scheint, daß die Natur darin ein Gleichgewicht bewahrte. Denn wenn die
> von der Reflexion herrührende Hitze nachläßt, wie es der Fall ist, wenn die
> Sonne im Steinbock steht, tritt Gleichstand bei der Erhitzung aufgrund der
> Nähe ein, und umgekehrt, wenn starke Hitze wegen der Reflexion in rechten
> oder fast rechten Winkeln auftritt, während die Sonne im Krebs steht, dann
> ist die Sonne vom Erdzentrum weiter entfernt, so daß die Hitze vermindert
> wird.

Auf den ersten Blick macht dieses "Zitat" mehr Schwierigkeiten als die
beiden vorangehenden. Wie sich Ibn Rušd wirklich die gleichbleibende
Wärmeentwicklung vorstellt, geht freilich aus diesem ungenauen und
verkürztem Zitat nicht hervor. Ibn Rušd sagt nämlich eigentlich:

> Videtur quod natura aequalizavit in hoc. Et, cum remittitur calor, qui est per
> reflexionem, fuit propinquitas ut accidat aequalitas in calefactione, quae est
> propter motum. Et, cum invenitur caliditas, quae est per reflexionem est
> distantia, ut remittatur calor, qui est per motum.

> Die Natur scheint in dieser Hinsicht ein Gleichgewicht zu bewahren. Wenn
> die von der Reflexion verursachte Hitze nachläßt, liegt Nähe vor, so daß
> Gleichstand bei der Erhitzung eintritt, die von der Bewegung herrührt. Und
> wenn Hitze vorgefunden wird, die von der Reflexion verursacht wird, liegt
> Entfernung vor, so daß die Hitze vermindert wird, die von der Bewegung
> herrührt.

Ibn Rušd hatte nämlich zunächst erklärt, daß die Sonne und die Sterne
auf zwei Weisen erhitzen, durch die Bewegung und durch die Reflexion
des Lichtes. Nach seinen Ausführungen über die Erhitzung der Sonne
betonte er nachdrücklich, es gebe nur die folgenden Alternativen: Entweder ist das Zentrum des Sonnenhimmels dasselbe wie das des Tierkreises,
oder die Sonne befindet sich auf einem Exzenter oder Epizykel. Die
Erläuterungen über das Gleichgewicht der Natur beziehen sich nur auf die
zweite Alternative, ohne daß sich Ibn Rušd für diese ausgesprochen hätte.
Er entschied sich an dieser Stelle nicht ausdrücklich für eine von den
beiden Möglichkeiten, dies ist in der Tat bemerkenswert. Clavius hat
freilich mit seiner parteiischen Stellungnahme den Sinn des Abschnittes
verfälscht.

SCHLUßWORT

Clavius verquickte naturwissenschaftliche Daten mit aristotelischer Natur-philosophie, christlichen Vorstellungen, teleologischen und ästhetischen Argumenten und arabischer Astrologie. Er als Gegner wie al-Biṭrūǧī und Ibn Rušd als Befürworter desselben Weltsystems homozentrischer Sphären bedienten sich aristotelischer Lehren (Negierung des Vakuums, einfacher Beweger, der einfache, in eine Richtung orientierte Bewegungen hervor-ruft usf.). So sehr das Denken also an alten Weltbildern hing, so unübersehbar scheint mir zu sein, daß al-Biṭrūǧīs und Ibn Rušds Kritik am ptolemäischen System das astronomische Denken insofern zukunfts-weisend beeinflussen konnte, als sie die Denkmöglichkeit anderer Weltsys-teme aufzeigte [Sezgin 1978, 36f.].

ANMERKUNGEN

[1] Ptolemaios nennt keine Namen. Dies ändert sich bei Peurbach und Regiomon-tan in der 1462 fertiggestellten "Epitome" des Almagest, die erstmalig 1496 in Venedig erschien, 1543 erneut in Basel, 1550 in Nürnberg, und von Copernicus herangezogen wurde [Rosen 1961].

[2] Zur Literatur über Galilei und die Venusphasen s. [Galilei 1989, 107–109].

[3] Al-Battānī setzt für den Erdradius die Größe 1^p an, für die Entfernung Mond–Erde 64^p 10', also etwas mehr als 64 Erdradien, für die Entfernung Sonne–Erde 1070 Erdradien.

[4] Die Kenntnis hiervon verdanke ich einem Gespräch mit Juliane Lay (Paris) vom 4.10.1993; s. [Goldstein 1969, 53].

LITERATURVERZEICHNIS

Aiton, E. J. 1978. Peurbach's Theoricae novae planetarum. A translation with commentary. *Osiris* **3** (2), 4–43.

Al-Battānī. 1899–1907. *Al-Battānī sive Albatenii Opus astronomicum. Ad fidem codicis Escurialensis Arabice editum. Latine versum, adnotationibus instructum a Carolo Alphonso Nallino*, 3 Bde. Mailand: Ulrico Hoepli. Nachdruck Hildesheim —New York: Olms, 1977.

Al-Biṭrūǧī. 1952. *De motibus celorum*, F. J. Carmody, Ed. Berkeley and Los Angeles: University of California Press.

Al-Biṭrūjī. 1971. *On the principles of astronomy. An edition of the Arabic and Hebrew versions with translation, analysis, and an Arabic–Hebrew–English glossary*, by Bernard Goldstein, 2 Vols. New Haven and London: Yale University Press.

Arnaldez, R., and Iskandar, A. Z. 1975. Ibn Rushd. *Dictionary of Scientific Biography*, vol. 12, pp. 1–9. New York: Scribner's.

Averroes. 1562–1574. *Aristotelis opera cum Averrois commentariis*. Venedig: Iunctae. Nachdruck Frankfurt/M.: Minerva, 1962.

Baldini, U. 1992. *Legem impone subactis. Studi su filosofia e scienza dei gesuiti in Italia 1540–1632*. Rom: Bulzoni Editore.

Blumenberg, H. 1975. *Die Genesis der kopernikanischen Welt*. Frankfurt: Suhrkamp.

Carmody, F. J. 1951. Regiomontanus' Notes on al-Biṭrūjī's astronomy. *Isis* **42**, 121–130.

— 1952. The planetary theory of Ibn Rushd. *Osiris* **10**, 556–586.

Clavius, C. 1611. Commentarius in sphaeram Ioannis de Sacro Bosco. In *Christophorus Clavius, Opera mathematica*, Bd. 3, S. 1–317 (1. Zählung). Mainz: A. Hierat und R. Eltz.

Copernicus, N. 1949. *De revolutionibus orbium caelestium libri sex*, F. Zeller und K. Zeller, Hrsgg. München: Oldenbourg.

— 1990. *Das neue Weltbild, Drei Texte Commentariolus, Brief gegen Werner, De revolutionibus I, Im Anhang eine Auswahl aus der Narratio prima des G. J. Rheticus, übersetzt, herausgegeben und mit einer Einleitung und Anmerkungen versehen von Hans Günter Zekl*. Hamburg: Meiner.

Gabrieli, G. 1924. Medici e scienziati Arabi. 'Alī ibn Ridwān. *Isis* **24**, 500–506.

Galilei, G. 1989. *Sidereus nuncius [The sidereal messenger]*. Translated with introduction, conclusion, and notes by A. van Helden. Chicago and London: University of Chicago Press.

Goldstein, B. R. 1969. Some Medieval Reports of Venus and Mercury Transits. *Centaurus* **44**, 49–59.

Hallyn, F. 1987. *La structure poétique du monde*. Paris: Edition du Seuil.

Houzeau, J. C., and Lancaster, A. B. M. 1887. *Bibliographie générale de l'astronomie*. Brüssel: F. Hayez. Nachdruck London: The Holland Press, 1964.

Knobloch, E. 1990. Christoph Clavius. Ein Namens- und Schriftenverzeichnis zu seinen Opera Mathematica. *Bollettino di Storia delle Scienze Matematiche* **10**, 135–189.

Krayer, A. 1991. *Mathematik im Studienplan der Jesuiten, Die Vorlesung von Otto Cattenius an der Universität Mainz (1610/11)*. Stuttgart: Steiner.

Lattis, J. 1994. *Between Copernicus and Galileo. Christoph Clavius and the collapse of Ptolemaic cosmology*. Chicago: The University of Chicago Press.

— 1991. Homocentrics, eccentrics and Clavius's refutation of Fracastoro. *Physis* [N. S.] **28**, 699–725.

Lorch, R. 1973. Jābir ibn Aflaḥ al-Ishbīlī. Dictionary of Scientific Biography, *Vol.* 7, 37–39.

— 1992. Some remarks on the Almagestum parvum. In *Amphora. Festschrift für Hans Wussing zu seinem 65. Geburtstag*, S. S. Demidov, M. Folkerts, D. E. Rowe, und C. J. Scriba, Hrsgg., S. 407–437. Basel: Birkhäuser.

Magini, G. A. 1589. *Novae caelestium orbium theoricae congruentes cum observationibus N. Copernici*. Venedig: D. Zenarius.

Pingree, D. 1970. Abū Ma'shar al-Balkhī. *Dictionary of Scientific Biography*, Vol. 1, 32–39. New York: Scribner's.

Regiomontanus, J. 1972. *Opera collectanea. Faksimiledruck von neun Schriften Regiomontans und einer von ihm gedruckten Schrift seines Lehrers Peurbach*, F. Schmeidler, Hrsg. Osnabrück: O. Zeller.

Rheticus, G. J. 1982. *Narratio prima, Edition critique, traduction française et commentaire par H. Hugonnard-Roche et J.-P. Verdet avec la collaboration de M.-P. Lerner et A. Segonds*. Wrocław: Maison d'Edition de l'Académie Polonaise des Sciences.

Richard of Wallingford. 1976. *An Edition of his Writings with Introductions, English Translation and Commentary by John D. North*, 3 vols. Oxford: Clarendon Press.

Rosen, E. 1961. Copernicus and Al-Bitruji. *Centaurus* **7**, 152–156.

Samsó, J. 1978. al-Bitruji al-Ishbili. *Dictionary of Scientific Biography*, Vol. 15, pp. 13–36. New York: Scribner's.

Sezgin, F. 1978. *Geschichte des arabischen Schrifttums*. Vol. VI: *Astronomie bis ca. 430 H*. Leiden: E. J. Brill.

— 1979. *Geschichte des arabischen Schrifttums*. Vol. VII: *Astrologie-Meteorologie und Verwandtes bis ca. 430 H*. Leiden: E. J. Brill.

Steinschneider, M. 1892. Die arabischen Bearbeiter des Almagest. *Bibliotheca Mathematica* **6** (2), 53–62.

Suter, H. 1900. *Die Mathematiker und Astronomen der Araber und ihre Werke*. Leipzig: Teubner. Nachdruck New York und London: Johnson, 1972.

Die Rückführung des allgemeinen auf den Sonderfall—Eine Neubetrachtung des Grenzwertsatzes für binomiale Verteilungen von Abraham de Moivre

Ivo Schneider

Institut für Geschichte der Naturwissenschaften der Universität München, D-80306 München, Germany

In 1733 de Moivre acquainted his friends with an approximation of the binomial distribution in the case of a great number of trials by the normal distribution. He justified his result only in the symmetric case of an event with probability $p = \frac{1}{2}$. His procedure in the general case of $p \neq \frac{1}{2}$ remained enigmatic. This paper proposes a solution for this enigma and, at the same time, throws some light on the different conceptions of 18th-century analysis in Great Britain and on the continent.

Daß der Grenzwertsatz von de Moivre weitgehend mit dem Laplaceschen Grenzwertsatz für die Binomialverteilung äquivalent ist, wurde erst in den 90er Jahren des 19. Jahrhunderts von Johann Eggenberger [Eggenberger 1893] und Emanuel Czuber [Czuber 1899] erkannt [1]. Abhängig davon hat Karl Pearson erneut auf die Leistung de Moivres in einem Artikel von 1924 hingewiesen [Pearson 1924]. In den genannten Arbeiten mußte zum Nachweis der Äquivalenz der Grenzwertsätze von de Moivre und Laplace die Darstellung von de Moivre in die auf den Kontinent übliche, der Leibnizschen Form des Kalküls entsprechende "übersetzt" werden. Diese Form der Übersetzung mit dem Ziel, die Äquivalenz von zwei mathematischen Aussagen aufzuzeigen, überdeckt aber auch wesentliche Auffassungsunterschiede, die die Äquivalenz zumindest einschränken. Solche Auffassungsunterschiede stehen auch in einem Zusammenhang mit der Frage nach der Äquivalenz der von Newton und Leibniz geschaffenen beiden Formen des Infinitesimalkalküls. Die von Newton bejahte und von Leibniz verneinte Äquivalenz ist historisch vollkommen zurückgetreten hinter der nach der Unabhängigkeit der Entdeckung, etwa weil die zwischen Newtonianern und Leibnizianern veranstalteten Tests in Form schwieriger Probleme, die die Überlegenheit des einen oder anderen Kalküls nachweisen sollten, jeweils von beiden Seiten bestanden wurden. Der im 19. Jahrhundert unübersehbar gewordene Rückstand der englischen Mathematik gegenüber der auf die Analysis gestützten kontinentalen vor allem französischen Mathematik ist weit weniger auf mögliche Auffassungsunterschiede in der Analysis bei den Newtonianern und Leibnizianern, sondern auf eine Verschiebung des Interessenschwerpunkts in England auf geometrische Methoden und Darstellungsformen zurückgeführt worden, für die Newton selbst durch die Darstellung seiner *Principia* in Form einer geometrisierten Analysis zumindest Mitverantwortung trägt.

De Moivre, der als einer der wichtigsten Repräsentanten der sich an Newtons Arbeiten zum Kalkül anschließenden englischen Analysis anzusehen ist, zeigt in seiner Wahrscheinlichkeitsrechnung insbesondere bei seinem Grenzwertsatz, daß z.B. die eingeschränkte Verwendung mathematischer Symbolik in der Praxis der englischen Analysis nicht als rein formaler Unterschied zur kontinentalen Analysis aufgefaßt werden darf, sondern unterschiedlichen Zielvorgaben entspricht, die auch die Möglichkeiten einer Problemlösung verändern.

Bei de Moivre wird in Anlehnung an Newton eine von den Anwendungsbedürfnissen her diktierte Ausrichtung der Analysis auf die Numerik und eine daraus erwachsende Konzentration auf numerisch verwertbare Potenzreihendarstellungen sichtbar. Konkrete Potenzreihen stehen bei ihm für einen noch nicht vorhandenen Funktionsbegriff, der andererseits bereits in der Leibnizschen Form des Kalküls grundgelegt ist. Das Fehlen eines Integrationssymbols behinderte zumindest die Anwendung von Transformationen bei de Moivre, die z.B. notwendig waren, um tatsächlich die Konvergenz der Binomialverteilung gegen die Normalverteilung zu zeigen.

Stattdessen kann de Moivre für den Grenzwert der Wahrscheinlichkeit dafür, daß sich eine binomial verteilte Zufallsgröße gegebenen Parameters zwischen bestimmten Grenzen bewegt, zunächst nur eine von dem Parameter und den Grenzen abhängige Potenzreihe angeben, die aber für beliebige Parameter bei geschickter Wahl der Grenzen als Vielfaches der heute sogenannten Standardabweichung auf den Fall mit dem Parameter $\frac{1}{2}$ zurückgeführt werden kann. Das soll im Folgenden gezeigt werden.

1733 hatte de Moivre eine siebenseitige lateinische Druckschrift mit dem Titel *Näherungswert für die Summe der Glieder des in eine Reihe entwickelten Binoms* $(a + b)^n$ an einige Freunde verteilt [de Moivre 1733]. Die Bemerkung in der zweiten Auflage seiner *Doctrine of Chances* von

1738 [de Moivre 1738, 235]:

> Ich werde hier eine Arbeit von mir übersetzen, die am 12. November 1733 gedruckt und an einige Freunde vermittelt, bisher aber nicht veröffentlicht worden war, indem ich mir das Recht vorbehielt, meine eigenen Überlegungen bei passender Gelegenheit zu erweitern.

zeigt, daß es sich dabei zunächst nicht um eine Veröffentlichung, sondern nur um eine Form der Prioritätssicherung handelte.

Ausgangspunkt ist ein Experiment wie das Münzwurfexperiment, bei dem die Möglichkeit einer beliebigen unabhängigen Wiederholung stillschweigend vorausgesetzt wird. Dabei interessiert man sich für das Eintreten oder Nichteintreten eines möglichen Ereignisses bei n-maliger Durchführung des Experiments.

Jakob Bernoulli hatte in seiner *Ars Conjectandi* von 1713 bereits die Binomialverteilung, wonach die Wahrscheinlichkeit für genau k-maliges Eintreten, $k \leq n$, des Ereignisses mit der Wahrscheinlichkeit p gerade $\binom{n}{k} p^k (1 - p)^{n-k}$ ist [2], und eine erste Form des Gesetzes der großen Zahlen, den von ihm so genannten Hauptsatz, abgeleitet [Bernoulli 1713, 45, 236]. Der Hauptsatz besagt, daß die Wahrscheinlichkeit P dafür, daß sich die relative Häufigkeit r_n des fraglichen Ereignisses von seiner Grundwahrscheinlichkeit um höchstens einen vorgegebenen, beliebig kleinen Betrag unterscheidet, mit wachsendem n dem Wert 1 beliebig annähert.

Genauer zeigte Jakob Bernoulli verbal, was sich in heutiger Schreibweise, wie folgt, wiedergeben läßt:

$$\forall c > 0, \ \varepsilon > 0 \ \exists n'(p, c, \varepsilon): \forall n \geq n' \ \frac{P(|r_n - p| \leq \varepsilon)}{P(|r_n - p| > \varepsilon)} > c$$

oder da

$$P(|r_n - p| > \varepsilon) = 1 - P(|r_n - p| \leq \varepsilon)$$

$$1 > P(|r_n - p| \leq \varepsilon) > \frac{c}{c + 1}.$$

Zum Beweis hatte Jakob Bernoulli gerade Teilsummen der Form

$$\sum_{k = l_1}^{l_2} \binom{n}{k} p^k (1 - p)^{n-k}$$

abschätzen müssen, die nach dem Summensatz für die Wahrscheinlichkeiten disjunkter Ereignisse die Wahrscheinlichkeit dafür darstellen, daß die relativen Häufigkeiten $r_n = k/n$ Werte von l_1/n bis l_2/n annehmen können. Diese Vorleistungen kommentierte de Moivre zu Beginn seiner Abhandlung von 1733 bzw. in leicht modifizierter Form zu Beginn des entsprechenden Abschnitts in der *Doctrine*, wie folgt [de Moivre 1733, 1; 1738, 235] [3]:

> Obwohl die Lösung von Glücksspielproblemen oft die Summierung mehrerer Glieder des Binoms $(a + b)^n$ erfordert, erscheint dies doch bei sehr hohen Potenzen so mühselig, daß sich nur wenige Leute dieser Mühe unterzogen haben; denn außer Jakob und Niklaus Bernoulli, zwei großen Mathematikern, kenne ich niemanden, der dies versucht hat. Obwohl sie dabei sehr große Geschicklichkeit bewiesen und die ihren Bemühungen angemessene Anerkennung gefunden haben, müssen doch einige Dinge darüberhinaus gefordert werden; denn sie waren weniger um einen Näherungswert als um die Bestimmung sehr weiter Grenzen bemüht, innerhalb von denen die Summe der Glieder, wie sie bewiesen, enthalten war.

Anders als Jakob und Niklaus Bernoulli interessierte sich de Moivre für die Größe des Intervalls, in dem sich die relative Häufigkeit um den erwarteten Wert der Grundwahrscheinlichkeit mit einer vorgegebenen Wahrscheinlichkeit p bewegen kann. Mit anderen Worten: De Moivre gab sich eine Wahrscheinlichkeit von 0,5 oder 0,7 vor und fragte nach der Länge $l_2/n - l_1/n$ des Intervalls, in dem sich die relativen Häufigkeiten r_n mit dieser Wahrscheinlichkeit bewegen können.

De Moivre ging aus vom einfachsten Fall des Münzwurfexperiments mit $p = {}^1\!/_2$. Dabei bestimmte er zunächst noch ähnlich wie Bernoulli die Wahrscheinlichkeit W dafür, daß die Häufigkeit h_n in das Intervall $[n/2 - l, n/2 + l]$ bzw. die relative Häufigkeit r_n in das Intervall $[1/2 - l/n, 1/2 + l/n]$ fällt, zu

$$W = \sum_{i=n/2-l}^{n/2+l} \binom{n}{i} \left(\frac{1}{2}\right)^i \left(\frac{1}{2}\right)^{n-i} = 2^{-n} \sum_{i=n/2-l}^{n/2+l} \binom{n}{i}.$$

Bezeichnet man, abweichend von de Moivre, den Binomialkoeffizienten $\binom{n}{i}$ mit T_i und den maximalen Binomialkoeffizienten $\binom{n}{[n/2]}$ mit $[n/2]$ $= :m$ entsprechend mit T_m, so gelten nach de Moivre für $n \gg l \gg 1$ die folgenden Abschätzungen:

$$\frac{T_m}{2^n} \approx \frac{2}{\sqrt{2\pi n}} \quad \text{und} \quad \ln\frac{T_{m\pm l}}{T_m} \approx -2\frac{l^2}{n}.$$

Der Weg zu diesen Abschätzungen war alles andere als einfach. De Moivre hatte sich nach seinen eigenen Angaben mit der Berechnung von Binomialkoeffizienten für sehr großes n und mit der damit zusammenhängenden Bestimmung von $n!$ schon in den 20er Jahren des 18. Jahrhunderts zu beschäftigen begonnen [de Moivre 1730a, 102–106, 226–229]. Er war mit James Stirling Ende der 20er Jahre in einen Wettbewerb getreten, in dessen Verlauf eine semikonvergente, aber als solche damals noch nicht erkannte Reihe für $\ln n!$ gefunden wurde, aus der sich unmittelbar die als Stirlingsche Formel bekannte Beziehung für $n!$ ergibt [4].

Wie es zu diesen Abschätzungen im Wettbewerb mit Stirling kam, schilderte [de Moivre 1733], wie folgt:

Es ist jetzt mindestens 12 Jahre her, seit ich folgendes Ergebnis gefunden hatte: Erhebt man das Binom $1 + 1$ zu einer sehr hohen Potenz n, so läßt sich das Verhältnis des mittleren Gliedes zu allen Gliedern, das heißt zu 2^n, durch den Bruch $\dfrac{2A \cdot (n-1)^n}{n^n \cdot \sqrt{n-1}}$ ausdrücken, wobei A die Zahl ist, deren hyperbolischer Logarithmus [5] der für mich beliebig fortsetzbaren Reihe $\frac{1}{12} - \frac{1}{360} + \frac{1}{1260} - \frac{1}{1680}$ usw. gleich ist. Da aber die Größe $\dfrac{(n-1)^n}{n^n}$ bzw. $\left(1 - \dfrac{1}{n}\right)^n$ für sehr großes n nahezu gegeben ist, wie man leicht zeigen kann, folgt, daß diese Größe für eine unendlich große Potenz absolut gegeben ist, nämlich durch die Zahl, deren hyperbolischer Logarithmus -1 ist. Daraus folgt, wenn B die Zahl bezeichnet, deren hyperbolischer Logarihmus gleich $-1 + \frac{1}{12} - \frac{1}{360} + \frac{1}{1260} - \frac{1}{1680}$ usw. ist, daß der obige Bruch durch $\dfrac{2B}{\sqrt{n-1}}$ bzw. durch $\dfrac{2B}{\sqrt{n}}$ ausgedrückt wird; wenn man deshalb die Vorzeichen dieser Reihe ändert und annimmt, daß B die Zahl bezeichnet, deren hyperbolischer Logarithmus gleich $1 - \frac{1}{12} + \frac{1}{360} - \frac{1}{1260} + \frac{1}{1680}$ usw. ist, wird dieser Ausdruck zu $\dfrac{2}{B\sqrt{n}}$.

Zu Beginn meiner Untersuchung begnügte ich mich mit einer groben Bestimmung von B über die Summierung einiger Glieder der obigen Reihe, die als Logarithmus dieser Größe angesehen werden kann. Aber die langsame Konvergenz schreckte mich davon ab, weiterzumachen, bis der mit mir befreundete hochgelehrte Herr James Stirling, der sich nach mir mit dieser Untersuchung auf eine von meiner völlig verschiedenen Weise beschäftigt hatte, herausfand, daß die Größe B die Quadratwurzel des Umfangs des Einheitskreises bezeichnet, so daß, wenn man diesen Umfang c nennt, das Verhältnis des mittleren Gliedes zur Summe aller Glieder durch $\dfrac{2}{\sqrt{nc}}$ ausgedrückt wird. [6]

Unter den gemachten Voraussetzungen konnte dann de Moivre die gesuchte Wahrscheinlichkeit dafür abschätzen, daß die Häufigkeit h_n um maximal l von $m = [n/2]$ abweicht:

$$W = \frac{T_m}{2^n} \sum_{i=-l}^{l} \frac{T_{m+i}}{T_m} \approx 2 \cdot \frac{2}{\sqrt{2\pi n}} \sum_{i=0}^{l} e^{-2i^2/n} \approx \frac{4}{\sqrt{2\pi n}} \int_0^l e^{-2x^2/n} \, dx.$$

Diese Darstellung weicht insofern von der de Moivres ab, als de Moivre weder über die hier benutzte Symbolik noch die dafür unerläßlichen begrifflichen Voraussetzungen verfügte. So stehen bei ihm statt der Summe über verschiedene Werte der e-Funktion nur die Reihenentwicklungen für die verschiedenen Argumente, die Summe ersetzt er ohne weitere Rechtfertigung durch die gliedweise bis l integrierte Reihe von $e^{-2x^2/n}$. Er ersetzte also

$$\frac{4}{\sqrt{2\pi n}} \sum_{i=0}^{l} e^{-2i^2/n} = \frac{4}{\sqrt{2\pi n}} \sum_{i=0}^{l} \sum_{v=0}^{\infty} \frac{(-2i^2/n)^v}{v!}$$

durch

$$\frac{4}{\sqrt{2\pi n}} \sum_{v=0}^{\infty} \frac{(-1)^v \cdot 2^v \cdot l^{2v+1}}{v!(2v+1) \cdot n^v},$$

wobei

$$\frac{4}{\sqrt{2\pi n}} \sum_{v=0}^{\infty} \frac{(-1)^v \cdot 2^v \cdot l^{2v+1}}{v!(2v+1) \cdot n^v} = \frac{4}{\sqrt{2\pi n}} \sum_{v=0}^{\infty} \int_0^l \frac{(-2x^2/n)^v}{v!} \, dx.$$

Er erkannte, daß sich diese Reihe für $l = s \cdot {}^1\!/_2 \sqrt{n}$ numerisch auswerten ließ. Insbesondere für $s = 1$ erhielt er die sehr schnell konvergierende Entwicklung

$$\frac{2}{\sqrt{2\pi}} \sum_{v=0}^{\infty} \frac{(-1)^v}{v!(2v+1) \cdot 2^v},$$

mit dem auf fünf Dezimalen genauen Wert O,682688. Dieser Wert entspricht nach den vorhergehenden Überlegungen dem Grenzwert für die Wahrscheinlichkeit einer Abweichung der Häufigkeit von $^1\!/_2 n$ um maximal $^1\!/_2 \sqrt{n}$ oder der relativen Häufigkeit von $^1\!/_2$ um maximal $^1\!/_2 n^{-1/2}$. De Moivre hatte also gefunden

$$\lim_{n \to \infty} W\left(|h_n - \tfrac{1}{2}n| \leq \tfrac{1}{2}\sqrt{n}\right) \cong 0,682688,$$

was er, wie folgt, ausdrückte [de Moivre 1733, 4]:

> Für den Fall unendlich vieler Experimente ist deshalb die Wahrscheinlichkeit dafür, daß ein Ereignis, für dessen Eintreten oder Nichteintreten die

Anzahl der Chancen gleich groß ist, weder häufiger als $\frac{1}{2}n + \frac{1}{2}\sqrt{n}$ noch seltener als $\frac{1}{2}n - \frac{1}{2}\sqrt{n}$ eintritt, ... 0,682688 und folglich wird die Wahrscheinlichkeit für das Gegenteil, häufiger oder seltener als in den oben genannten Verhältnissen einzutreffen, 0,317312 sein, wobei die beiden Wahrscheinlichkeiten zusammengenommen Eins, das Maß der Sicherheit, ergeben.

Da der Durchführung von unendlich vielen Experimenten unüberwindliche Schwierigkeiten entgegenstehen, hätte dieses Ergebnis nur theoretischen Wert. De Moivre hatte aber diesen Grenzwert im Interesse einer praktischen Anwendung auf den für ihn numerisch für eine genaue Bestimmung zu aufwendigen Fall sehr großer n abgeleitet. In diesem Sinn stellte er fest ([de Moivre 1733, Korollarien III, IV]; siehe [de Moivre 1756, 246]):

Obwohl die Durchführung unendlich vieler Experimente unmöglich ist, können die vorhergehenden Ergebnisse sehr gut im Fall endlicher Anzahlen, wenn sie nur groß sind, angewandt werden. Wenn z. B. 3600 Experimente gemacht werden, also $n = 3600$ und daher $\frac{1}{2}n = 1800$ sowie $\frac{1}{2}\sqrt{n} = 30$ ist, wird die Wahrscheinlichkeit für ein höchstens 1830maliges und mindestens 1770maliges Eintreten des Ereignisses 0,682688 sein ... .

Doch muß man sich n dabei nicht als unermeßlich groß vorstellen; denn selbst für ein n, das nicht größer als 900, ja nicht einmal größer als 100 ist, wird die hier angegebene Regel, wie ich durch Versuche bestätigt habe, einigermaßen genau sein.

Nach dieser empirischen Überprüfung der Brauchbarkeit des Grenzwerts als Näherungswert für große n interessierte sich de Moivre dafür, welche Masse der Wahrscheinlichkeit der Binomialverteilung in einem größeren Intervall um das maximale Glied enthalten ist. Er vergrößerte dazu die zulässige maximale Abweichung der Häufigkeit von $^1/_2\sqrt{n}$ auf $\sqrt{n}$, bestimmte also

$$\lim_{n \to \infty} W\left(|h_n - \tfrac{1}{2}n| \le \sqrt{n}\,\right).$$

Die gliedweise bis l integrierte Reihe von $e^{-2x^2/n}$ konvergierte für $l = \sqrt{n}$ bei weitem nicht mehr so schnell wie für $l = {}^1/_2\sqrt{n}$. Von dem dabei zu berechnenden Integral

$$\frac{4}{\sqrt{2\pi n}} \int_0^{\sqrt{n}} e^{-2x^2/n}\, dx = \frac{4}{\sqrt{2\pi n}} \int_0^{(1/2)\sqrt{n}} e^{-2x^2/n}\, dx$$

$$+ \frac{4}{\sqrt{2\pi n}} \int_{(1/2)\sqrt{n}}^{\sqrt{n}} e^{-2x^2/n}\, dx$$

war das erste Teilintegral der rechten Seite schon vorher berechnet worden. Zur Bestimmung des zweiten Teilintegrals bediente sich de Moivre einer mechanischen Quadratur mit den vier äquidistanten Stützstellen bei $\frac{1}{2}\sqrt{n}$, $\frac{2}{3}\sqrt{n}$, $\frac{5}{6}\sqrt{n}$ und $\sqrt{n}$. Für das Integral über das Gesamtintervall und damit die obige Wahrscheinlichkeit fand de Moivre so den Wert 0,95428. Schließlich bestimmte de Moivre noch die Wahrscheinlichkeit einer Abweichung der Häufigkeit von $\frac{1}{2}n$ um maximal $\frac{3}{2}\sqrt{n}$ ebenfalls mit Hilfe einer mechanischen Quadratur numerisch zu 0,99874.

All dies zeigt hinlänglich, daß es de Moivre ausschließlich um die numerische Bestimmung der Wahrscheinlichkeit für eine Abweichung der Häufigkeit vom Erwartungswert in der Größenordnung von Vielfachen von $\frac{1}{2}\sqrt{n}$ ging.

Das ausschließliche Interesse an numerischen Entwicklungen ist mit dafür verantwortlich, daß de Moivre für den allgemeinen Fall von Ereignissen, deren Eintrittswahrscheinlichkeit auch von $\frac{1}{2}$ verschiedene Werte $p = a/(a + b)$ annehmen kann, nicht mehr als einige Andeutungen geben konnte. Aufgrund der durch die Glücksspielrechnung nahegelegten Vorstellungen ging de Moivre davon aus, daß sich zumindest die meisten Ereignisse als aus jeweils endlich vielen gleichwahrscheinlichen Elementarereignissen zusammengesetzt darstellen lassen. Insofern ist auch der allgemeine Fall $p = a/a + b$ eingeschränkt auf natürliche a und b und damit auf rationale p. De Moivres Andeutungen beschränkten sich auf eine Abschätzung für große n einmal des maximalen Gliedes [De Moivre 1733, Corollarium VIII, 6] in der Entwicklung von $(p + q)^n$ zu $(a + b)/\sqrt{2\pi nab}$ bzw. zu $1/\sqrt{2\pi npq}$ und des logarithmierten Verhältnisses eines Gliedes mit dem Indexabstand l vom maximalen Glied zu diesem maximalen Glied zu [De Moivre 1733, Corollarium IX, 6f.]

$$-\frac{(a + b)^2}{2abn}l^2 \quad \text{oder} \quad -\frac{l^2}{2pqn},$$

wobei l von der Größenordnung $c\sqrt{n}$ sein sollte.

Folgt man de Moivres Vorgehen zur Bestimmung des Grenzwerts im Fall des Münzwurfexperiments, so ist

$$\lim_{n \to \infty} W(|h_n - pn| \le l) \quad \text{als das Äquivalent von}$$

$$\frac{2}{\sqrt{2\pi pqn}} \int_0^l e^{-x^2/2pqn}\, dx$$

durch gliedweise Integration in der Reihenentwicklung von $e^{-x^2/2pqn}$ zu bestimmen.

Laplace, der analytisch in der Tradition von Leibniz stand, konnte dieses Integral unmittelbar durch die Substitution $x = t \cdot \sqrt{2npq}$ für $l = c \cdot \sqrt{2npq}$ in die nur noch von c, nicht aber von p und q abhängige Form $(2/\sqrt{\pi}) \cdot \int_0^c e^{-t^2} dt$ bringen [7], die unmittelbar in die übliche Form der Normalverteilung überführt werden kann. Eine allgemeine Formulierung seines Grenzwertsatzes in der obigen, auf Leibniz zurückgehenden Form war de Moivre aus zwei Gründen versagt. Einmal kannte die auch für de Moivre verbindliche Form des Newtonschen Infinitesimalkalküls kein dem Leibnizschen vergleichbares Integralsymbol [8]. Außerdem verbaute die Konzentration auf numerisch verwertbare Reihenentwicklungen und das Fehlen eines Funktionsbegriffs sowie, damit zusammenhängend, das geringe Interesse an geschlossenen Ausdrücken für die benutzten Reihenentwicklungen die Möglichkeit für de Moivre, seinen Grenzwertsatz in einer der obigen äquivalenten Form darstellen zu können.

Wie weit de Moivre trotz dieser Einschränkungen einem Laplace vergleichbaren Verständnis nahegekommen war, muß einer Interpretation der sehr lapidaren Feststellung des Korollars 10 überlassen bleiben, wo es unmittelbar an die bereits referierten Abschätzungen heißt [9]:

> Wenn die Wahrscheinlichkeiten des Eintretens oder Nichteintretens sich beliebig voneinander unterscheiden, werden die sich auf die Summierung von Gliedern des Binoms $(a + b)^n$ beziehenden Probleme genauso leicht gelöst werden wie die, bei denen die Wahrscheinlichkeiten des Eintretens und Nichteintretens gleich sind.

Stiegler hat z.B. in seiner ausführlichen Analyse des de Moivreschen Grenzwertsatzes sicherlich zu Recht festgestellt, daß de Moivre keinerlei Hinweis darauf hinterlassen habe, wie man zu der in Korollar 10 behaupteten leichten Lösung kommen könne, und offengelassen, ob de Moivre eine solche Lösung überhaupt zuzutrauen ist [Stiegler 1986, 84].

Einem Mathematiker, der wie De Moivre sein Brot hauptsächlich durch Privatunterricht von gut betuchten, aber nicht notwendig besonders begabten Schülern verdienen mußte, sollte man nicht leichtfertig Leichtfertigkeit unterstellen. Es ist einigermaßen wahrscheinlich, daß ihn der eine oder andere seiner Schüler, der es wie die Mathematikhistoriker des 20ten Jahrhunderts in keiner Weise als "genauso leicht" ansah, die erforderliche Summe zu bilden, beim Wort nahm und gegen Entgeld entsprechende Unterweisung einforderte. Es wäre de Moivres Reputation nicht zuträglich gewesen, wenn sich der von ihm angebotene Weg nicht als ziemlich einfach erwiesen hätte. Tatsächlich ergibt eine dem Vorgehen beim Münzwurfexperiment analoge Berechnung, die die von de Moivre in den

Korollarien 8 und 9 angegebenen Abschätzungen in moderner Schreibweise benutzt:

$$W(|h_n - pn| \leq l) \approx 2\frac{1}{\sqrt{2\pi pqn}} \sum_{i=0}^{l} e^{-i^2/2pqn}$$

$$= \frac{2}{\sqrt{2\pi pqn}} \sum_{i=0}^{l} \sum_{v=0}^{\infty} \frac{(-i^2/2pqn)^v}{v!}$$

$$\approx \frac{2}{\sqrt{2\pi pqn}} \sum_{v=0}^{\infty} \int_0^l \frac{-(x^2/2pqn)^v}{v!} dx$$

$$= \frac{2}{\sqrt{2\pi pqn}} \sum_{v=0}^{\infty} \frac{(-1)^v \cdot l^{2v+1}}{v!(2v+1) \cdot (2pqn)^v}.$$

Billigt man de Moivre zu, gesehen zu haben, daß sich diese Reihe für $l = c \cdot \sqrt{npq}$ vereinfacht zu der folgenden

$$\frac{2}{\sqrt{2\pi}} \sum_{v=0}^{\infty} \frac{(-1)^v \cdot c^{2v+1}}{v!(2v+1) \cdot 2^v},$$

die für $c = 1$ mit der für $\lim_{n \to \infty} W(|h_n - \frac{1}{2}n| \leq \frac{1}{2}\sqrt{n})$ gefundenen übereinstimmt, dann hat de Moivre tatsächlich den Grenzwert für die Binomialverteilung im Fall $p \neq \frac{1}{2}$ auf den für $p = \frac{1}{2}$ zurückführen und damit seine Behauptung einer ebenso einfachen Bestimmung bestätigen können. Dabei entsprechen die für $p = \frac{1}{2}$ berechneten Werte von $l = \sqrt{n}$ und $l = \frac{3}{2}\sqrt{n}$ den Werten von $c = 2$ und $c = 3$ im Fall $p \neq \frac{1}{2}$.

Das bedeutet, daß die für den Fall $p = \frac{1}{2}$ berechneten drei Werte einer unmittelbar auf den allgemeinen Fall $p \neq \frac{1}{2}$ übertragbaren Tabulierung der Normalverteilungsfunktion entsprechen. Das von de Moivre benutzte Verfahren der mechanischen Quadratur könnte grundsätzlich zur Berechnung beliebig vieler Werte dieser Funktion und damit zu einer systematischen Tabulierung eingesetzt werden.

Mit der hier vorgelegten Deutung, die dem Denken und dem mathematischen Rüstzeug de Moivres gut entspricht, läßt sich erklären, was bis in jüngste Zeit als ein nicht gerechtfertigter Anspruch de Moivres erschien.

Gleichzeitig wird deutlich, daß die allgemein als der den englischen weit überlegen eingestuften analytischen Darstellungsmittel der französischen Mathematik zur Zeit von Laplace im konkreten Fall für die Bedürfnisse und Erwartungen de Moivres keinen Vorteil brachten. Laplace's Fähigkeit, die später so genannte Normalverteilung in geschlossener Form als Grenzwert einer Wahrscheinlichkeit angeben zu können war irrelevant für das Bedürfnis, über einen in die Wettpraxis übertragbaren konkreten Wert einer solchen Wahrscheinlichkeit verfügen zu können. De Moivres Berechnungen waren hingegen geeignet, Wettern, die bei vielmaligem Werfen von Münzen oder Würfeln auf oder gegen eine vorgegebene maximale Abweichung der Häufigkeit des interessierenden Ereignisses von seinem Erwartungswert setzten, vernünftige Hinweise auf die relative Höhe ihrer Einsätze zu geben. Laplace hätte zur Befriedigung eines solchen Bedürfnisses nur dasselbe machen können wie de Moivre, nämlich die Normalverteilung tabulieren. Seit 1798 stand eine Tabulierung der Funktion

$$f(x) = \int_0^x e^{-t^2}\, dt$$

von Christian Kramp zur Verfügung [Kramp 1798], mit deren Hilfe sich dann weit mehr als die drei von de Moivre angegebenen numerischen Werte der Grenzverteilung bestimmen ließen.

ANMERKUNGEN

[1] Siehe dazu [Schneider 1968, 179].

[2] Jakob Bernoulli benutzte für die Binomialkoeffizienten, abweichend von der hier gewählten Form, eine explizite Produktdarstellung, und für p bzw. q die Brüche b/a und c/a, mit $a = b + c$ und a, b, c natürlich.

[3] Übersetzung nach [de Moivre 1738].

[4] [de Moivre 1730b] (mit 22 gesondert paginierten Seiten sowie zwei Seiten Errata, die sich sowohl auf die *Miscellanea* als auch auf das *Supplementum* beziehen); für eine Zusammenfassung der von De Moivre und Stirling getragenen Entwicklung der so genannten Stirlingschen Formel siehe [Schneider 1968, 266–276].

[5] Die Bezeichnung hyperbolischer Logarithmus entspricht der der Zeit de Moivres geläufigen Erkenntnis, daß die Fläche der Hyperbel $\xi y = 1$ zwischen den Abszissen $\xi = 1$ und $\xi = x$ die Eigenschaften eines Logarithmus besitzt, daß also insbesondere gilt, die Fläche von $\xi = 1$ bis $\xi = x_1 x_2$ ist der Summe der Flächen von $\xi = 1$ bis $\xi = x_1$ und $\xi = 1$ bis $\xi = x_2$ gleich. Der hyperbolische Logarithmus entspricht dem erst später so bezeichneten natürlichen Logarithmus.

[6] [de Moivre 1733, 1f.]. Die spätere englische Übersetzung des letzten Abschnitts in [de Moivre 1738, 1756] weicht in einigen Formulierungen geringfügig von der ursprünglichen lateinischen Fassung von 1733 ab.

[7] Am leichtesten ist der Grenzwertsatz für binomiale Verteilungen und die angesprochene Substitution bei Laplace in dessen *Théorie analytique des probabilités*, Paris 1812 und öfter, zu finden; siehe [Laplace 1886, 280–284] und deutsche Übersetzung in [Schneider 1988, 145–149].

[8] Newton hatte seine Analysis auf der Grundlage formaler Potenzreihen aufgebaut. Solche formalen Potenzreihen können differenziert und integriert werden, indem man sie gliedweise differenziert und integriert. Bei der gliedweisen Integration muß nur beachtet werden, daß für $r \neq -1$ das Integral von x^r gleich $x^{r+1}/(r + 1)$ ist. Das Integral von x^{-1} kennzeichnete Newton symbolisch durch ein dem Integranten x^{-1} umbeschriebenes Quadrat. Dieses von Newton nur dem Sonderfall $r = -1$ vorbehaltene Symbol kann nicht mit dem von Leibniz eingeführten Integralsymbol verglichen werden. Die unmittelbar auf Newton folgende Generation englischer Mathematiker übernahm auch dieses Quadratsymbol nicht, sondern bezeichnete die Stammfunktion von x^{-1} als Logarithmus Hyperbolicus.

[9] [de Moivre 1733, 7]. Entsprechende englische Fassungen des Korollars finden sich in [de Moivre 1738, 242] und [de Moivre 1756, 250]. Stiegler verglich die Beweiskraft und Durchsichtigkeit dieses Korollars mit der oft beklagten und gerne benutzten Formulierung heutiger Mathematiker "wie leicht zu sehen ist" [Stiegler 1986, 84].

LITERATURVERZEICHNIS

Bernoulli, J. 1713. *Ars Conjectandi*. Basel: Gebrüder Thurneisen.

Czuber, E. 1899. Die Entwicklung der Wahrscheinlichkeitstheorie und ihrer Anwendungen. *Jahresbericht der D.M.V.* **7**.

de Moivre, A. 1730a. *Miscellanea Analytica de Seriebus et Quadraturis*. London: J. Tonson & J. Watts.

de Moivre, A. 1730b. *Miscellaneis Analyticis Supplementum*. London: o.A.

de Moivre, A. 1733. *Approximatio ad Summam Terminorum Binomii $(a + b)^n$ in Seriem expansi*, Autore A. D. M. R. S. S. Der Privatdruck ist datiert vom 12. November 1733.

de Moivre, A. 1738. *Doctrine of Chances*, 2. Aufl., London: H. Woodfall.

de Moivre, A. 1756. *Doctrine of Chances*, 3. Aufl. London: A. Millar in the Strand.

Eggenberger, J. 1893. Beiträge zur Darstellung des Bernoullischen Theorems, der Gammafunktion und des Laplaceschen Integrals. *Mittheilungen der Naturforschenden Gesellschaft Bern*, S. 110–182.

Kramp, C. 1798. *Analyse des réfractions astronomiques et terrestres*. Leipzig: E. B. Schwikkert.

Laplace, P. S. 1886. *Oeuvres complètes de Laplace*, Bd. 7. Paris: Gauthier–Villars.

Pearson, K. 1924. Note on the origin of the normal curve of errors. *Biometrika* **16**, 402–404.

Schneider, I. 1968. Der Mathematiker Abraham de Moivre (1667–1754). *Archive for History of Exact Sciences* **5**, 177–317.

—, Hrsg. 1988. *Die Entwicklung der Wahrscheinlichkeitstheorie von den Anfängen bis 1933—Einführungen und Texte*. Darmstadt: Wissenschaftliche Buchgemeinschaft.

Stiegler, S. M. 1986. *The History of Statistics—The Measurement of Uncertainty before 1900*. Cambridge, MA: Harvard Univ. Press.

The Evolution of Notation

Zur Geschichte der negativen Zahlen

Helmuth Gericke

Sonnenbergstraße, D-79117 Freiburg, Germany

Numbers, which were subtracted from other numbers, were often written even in the pre-Greek period independently of the numbers from which they were to be taken away, without being thought of as numbers less than zero. Rules of computation for these "subtractive" numbers result from calculations with binomials, i.e., $(a - b)(c - d) = ac - ad - bc + bd$. Later they were applied as binomials of the form $(0 - b)$. In practical calculations negative or subtractive numbers appear in systems of linear equations. They do not arise in quadratic equations, as long as these are solved geometrically. Only when a general theory of equations arose in modern times (Cardano) were negative numbers as "fictitious solutions" unavoidable. For a long time the subtractive numbers were taken to be those juxtaposed with the positive numbers. Later these were understood to be the negative values that arise in the equation $a + (-a) = 0$. The construction of negative numbers by means of the extension of the additive half-group of positive (integer, rational, or real) numbers to a group will not be considered in this paper.

1. VORWORT

In einem Didaktik-Seminar wurde in einem Vortrag über die Einführung der negativen Zahlen im Schulunterricht gesagt: es sei wenig hilfreich, dabei auf die Geschichte zurückzugreifen, denn daß Schulden mit Schulden multipliziert ein Vermögen ergäben, sei den Schülern ja doch unverständlich. Dazu wäre zu sagen, daß negative Zahlen früher durchaus nicht immer als Schulden aufgefaßt wurden. Und: man kann negative Zahlen mit negativen Zahlen nur dann multiplizieren, wenn man eine geeignete Definition der Multiplikation zugrunde legt, z.B. mittels des Verhältnisses, wie es durch die Gleichung $1 : a = b : ab$ ausgedrückt wird. Diese Erklärung der Multiplikation steht z.B. in den *Regulae ad directionem ingenii* von Descartes (Regel XVIII) und am Anfang der "Géométrie".

Die vorliegende Arbeit entstand anläßlich der dänischen Übersetzung meiner "Geschichte des Zahlbegriffs" [Gericke 1970], sie ist also keine Forschungsarbeit, sondern eine lehrbuchartige Übersicht. Ich danke Frau Kirsti Andersen für Hinweise auf Mängel der alten Ausgabe und für wertvolle Anregungen.

2. SUBTRAKTIVE ZAHLEN

Nicht jede Zahl, vor der ein Minus-Zeichen steht, ist eine negative Zahl in dem Sinne, daß sie < 0 ist. Manchmal ist das Zeichen kein Vorzeichen, sondern ein Operationszeichen und bedeutet, daß die dahinter stehende (positive) Zahl von einer anderen subtrahiert werden soll, die manchmal, z.B. in babylonischen Tabellen, an anderer Stelle steht. Wir wollen dann von *subtraktiven* Zahlen sprechen, obwohl es sich nicht um eine besondere Art von Zahlen handelt.

Am Anfang von Diophants Arithmetik steht die Regel Λεῖψις ἐπὶ λεῖψιν πολλαπλασιασθεῖσα ποιεῖ ὕπαρξιν, λεῖψις δὲ ἐπὶ ὕπαρξιν ποιεῖ λεῖψιν. Wenn man wissen will, wie Diophant sich das gedacht hat, muß man vorsichtig übersetzen, aber eine rein umgangssprachliche Deutung der Fachausdrücke ist auch nicht angebracht. ὕπαρξις bedeutet das Vorhandensein, λεῖψις, das Weglassen oder Wegnehmen.

Weggenommenes mit Weggenommenem vervielfacht ergibt Vorhandenes, Weggelassenes mit Vorhandenem ergibt Wegzulassendes.

Ein Scholion von Maximus Planudes (Byzanz 1255 – 1310) besagt: Οὐχ ἁπλῶς λεῖψιν λέγει, μὴ καὶ ὑπάρξείως τινος οὔσης, ἀλλὰ ὕπαρξιν ἔχουσαν λεῖψιν [Diophant 1893 – 1895, II, 139]. "Er spricht nicht einfach von λεῖψις, als ob keine ὕπαρξις da wäre, sondern von etwas Vorhandenem, das eine Wegnahme enthält." Diophant meint also Rechenregeln wie die folgende

$$(a - b)(c - d) = ac - bc - ad + bd. \tag{1}$$

Die Regeln gelten also für subtraktive Zahlen; echt negative Zahlen lehnt Diophant ab. Die Gleichung [Diophant 1893–1895, I, Buch V, Aufg. 2]

$$4 = 4x + 20$$

erklärt er als ἄτοπος, sinnlos (ohne Ort — sei es in der Algebra oder in der Wirklichkeit; "utopisch" würde dem Wort nahekommen).

Algebraisch konnten damals solche Regeln wie (1) nicht bewiesen werden, weil für die Arithmetik und Algebra keine Axiome formuliert waren.

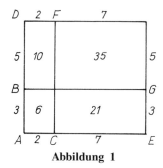

Abbildung 1

Euklid beweist solche Regeln in Buch II geometrisch, und für Diophants Rechenregeln beschreibt Maximus Planudes geometrische Beweise, die sich jahrhundertelang — man kann wohl sagen: bis heute — erhalten haben. Simon Stevin gibt für die Gleichung

$$(8 - 5) \cdot (9 - 7) = 8 \cdot 9 - 8 \cdot 7 - 5 \cdot 9 + 5 \cdot 7$$

nach einer Demonstration durch Ausrechnen eine *Autre demonstration geometrique* mittels der Figur in Abb. 1. Das kleine Rechteck 6 entsteht, wenn von dem großen Rechteck 8 · 9 die beiden Rechtecke 8 · 7 und 5 · 9 weggenommen werden und das doppelt weggenommene Rechteck 5 · 7 wieder hinzugefügt wird [Stevin 1585, Livre II, S. 167; Stevin 1958, 561].

Es ist für mich nicht klar erkennbar, ob Stevin nur subtraktive oder auch negative Zahlen meint. Aus (1) entsteht ja die Regel für negative Zahlen, wenn man $a = c = 0$ wählt. Dieser Gedanke steht in der Arithmetik von Stifel [1544], die Stevin bekannt war.

Im folgenden halte ich mich ungefähr an die historische Reihenfolge. Eine vollständige und ausgewogene Darstellung der Entwicklung ist nicht beabsichtigt, vielmehr werde ich solche Fragen und Aussagen hervorheben, die mir persönlich interessant erscheinen.

3. MINUS BEI DEN BABYLONIERN

Natürlich wird seit den ältesten Zeiten subtrahiert. Daß eine größere Zahl nicht von einer kleineren abgezogen werden kann, ist dabei selbstverständlich und wird nicht als störend empfunden.

Ein Minus-Zeichen kommt schon in altsumerischen Wirtschaftstexten um 2500 v. Chr. vor: 𒆷 transkribiert *là* oder *lal*, bedeutet "weniger sein". Es wird auch zur Darstellung von Zahlen benutzt, z.B. [Selz 1989, Text 47]

$$17 = \underset{\text{⟨symbols⟩}}{} = 20 - 3 \ (\ \bigcirc = 10,\ D = 1).$$

In astronomischen Texten aus der Seleukidenzeit kommt es vor, daß einzelne Glieder einer Zahlenreihe von einem an anderer Stelle angegebenen Wert subtrahiert (*lal*) oder zu ihm addiert (*tab* ⊨) werden sollen. Das sieht dann so aus, als ob negative oder positive Zahlen angeschrieben sind. Als Beispiel seien einige Spalten aus einer Tafel aus Uruk wiedergegeben ([Neugebauer 1955, Text Nr. 1]; zur Erläuterung siehe auch [Neugebauer 1957, Kap. V] und [van der Waerden 1966, Kap. IV]).

2.5	C	G	J		C′		K
I	3,16,05	4,19,59,...,44	57, 4	*lal*	9,22,30	*lal*	3,13,33
II	3,18,54	4,45,47,39,15	57, 4	*lal*	6,24,30	*lal*	3,42,19
III	3,34,43	4,55, 2,13,20	57, 4	*lal*	2,54,30	*lal*	3,55, 4
IV	3,33,32	4,32,29, 8, 9	57, 4	*lal*	.,35,30	*tab*	3,36
V	3,25,21	4, 6,40,29,37	57, 4	*lal*	4, 5,20	*tab*	3,13,41
VI	3, 9,38,40	3,40,51,51, 6	33,13,23	*lal*	7,51,10	*tab*	3,15,29
VII	2,49,38,40	3,15, 3,12,35			10	*tab*	3,25, 3
VIII	2,33,47,12	2,49,31,51, 6,40			7,55,44	*tab*	2,57,26
IX	2,25,55,44	2,40			3,55,44	*tab*	2,43,55
X	2,26, 4,16	2,40			., 4,16	*lal*	2,39,56
XI	2,34,12,48	2,42,23,42,13			4, 4,16	*lal*	2,38,19
XII	2,50,21,20	3, 3,59,30,22			8, 4,16	*lal*	2,55,55
I	3, 9,10	3,29,48, 8,53	54,16,22	*lal*	9,24,20	*lal*	2,26, 8

(Ausschnitte aus [Neugebauer 1955, Nr. 1. Rev.]. Spalten- und Monatsbezeichnungen sind geändert.)

Die Tafel enthält die Vollmonddaten der Jahre 124/125 der Seleukidenzeit (187/186 v. Chr.). Die römischen Ziffern geben die Nummern der babylonischen Mondmonate an. Natürlich stehen im Text die babylonischen Namen. I = *bar* = *Nisannu* ist der Monat des Frühlingsanfangs. Der Monat beginnt mit dem Tag, an dem nach dem Neumond die erste schmale Mondsichel am Morgenhimmel sichtbar wird, um bald im Tageslicht zu verschwinden. Alle Daten gelten für den Tag des Vollmonds in dem betreffenden Monat, hier für das Jahr 2.5 = 125 S.E.

Die Spalte C war nicht erhalten, die Rekonstruktion ist aber durch die aus anderen Tafeln ersichtlichen Daten und Rechnungen gesichert. Diese

Spalte gibt die Sonnenscheindauer an, in "Großstunden" (1 Großstunde = 4 normale Stunden) und den entsprechenden Minuten und Sekunden.

Die Spalte G enthält die Monatslänge nach einer vorläufigen Rechnung, bei der angenommen ist, daß die Sonne in jedem Monat den gleichen Weg von 30° zurücklegt, die Geschwindigkeit des Mondes aber veränderlich ist. Angegeben ist der Überschuß über 29 Tage, in Großstunden und deren Teilen.

Tatsächlich legt die Sonne nicht in jedem Monat den gleichen Weg zurück. In dem "System A", das der besprochenen Tafel zugrunde liegt, wird angenommen, daß die Sonne in einem Halbjahr jeden Monat 30° zurücklegt, in dem anderen Halbjahr aber nur 28°7′30″. In diesen Monaten braucht auch der Mond etwas weniger Zeit für einen Umlauf, von der Monatslänge ist 57′4″ abzuziehen (Spalte J). In den Monaten, in denen der Wechsel der Geschwindigkeit der Sonne angenommen wird, sind Zwischenwerte anzusetzen.

Die Spalte C′ enthält weitere Korrekturwerte, und zwar die halbe Differenz der Sonnenscheindauer zweier aufeinanderfolgender Monate, z.B.

Sonnenscheindauer im Monat I   3,16,05
                       im Monat II 3,28,54
halbe Differenz   6,24,30.

Diese halbe Differenz gibt also an, um wieviel Minuten der Sonnenaufgang im Monat II früher stattfand als im Monat I.

Wenn der Mondmonat zwischen Sonnenaufgang und Sonnenaufgang gerechnet wird, ist diese halbe Differenz von der Monatslänge abzuziehen (*lal*) oder, wenn die Sonnenscheindauer geringer geworden ist, zu addieren (*tab*). Die korrigierten Werte stehen in Spalte K. Z.B. ist im Monat II

die vorläufige Monatslänge	G =	29 Tage + 4,45,47,39,15
davon ist zu subtrahieren	J =	57, 4
und	C′ =	6,24,20
die korrigierte Monatslänge ist	K =	29 Tage + 3,42,19.
Im Monat IX ist	G =	29 Tage + 2,40
	J	entfällt
dazu ist zu addieren	C′ =	3,55,44 *tab*
also	K =	29 Tage + 2,43,55.

Die Zeichen *tab* und *lal* können hier also stets als Aufforderung zum Addieren und Subtrahieren aufgefaßt werden.

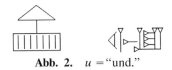

Abb. 2. $u =$ "und."

Der gleiche Text enthält auch Angaben über die Breite des Mondes, d.h. den Abstand des Mondmittelpunktes von der Ekliptik, in der Einheit $\check{s}e = 1/2$ Grad. Es sind für das Jahr 124 der Seleukidenzeit die folgenden

I	6,22;56,12	*u*	*lal*
II	4,24;10,30	*u*	*lal*
III	2,25;24,48	*u*	*lal*
IV	1,30;41,48	*lal*	*lal*
V	3,56;6,36	*lal*	*lal*
VI	6,..;46,18	*lal*	*lal*
VII	6,16;58	*lal*	*u*
VIII	4,10;42,18	*lal*	*u*
IX	1,44;53,12	*lal*	*u*
X	2,25;49, 6	*u*	*u*
XI	4,32; 4,48	*u*	*u*
XII	6,35;56,30	*u*	*u*

An der ersten Stelle bedeutet *lal* = nördliche Breite, *u* = südliche Breite. An der zweiten Stelle bedeutet *lal* = Weiterbewegung nach Norden, *u* = Weiterbewegung nach Süden. *u* bedeutet in sprachlichen Texten "und".

Ptolemaios gibt in einer entsprechenden Tabelle die Zahlen ohne Vorzeichen an und sagt im Text, in welchen Zeilen (bei ihm sind es die Zeilen 1–15) die Abweichung nach Norden und in welchen Zeilen sie nach Süden zu rechnen ist.

Die Verwendung des Vorzeichens in diesem Sinne als Richtung nach oben oder nach unten kommt wohl erst bei den Temperaturskalen von Réaumur und Celsius am Anfang des 18. Jahrhunderts wieder vor.

4. SYSTEME LINEARER GLEICHUNGEN BEI DEN CHINESEN

Die Chinesen benutzten mindestens seit dem 4. Jahrhundert v. Chr. Rechenstäbe (Zählstäbchen, counting rods), und zwar rote für positive

Zahlen (*cheng*), schwarze für negative Zahlen (*fu*) [Needham 1959, 70, 90; Lam und Ang 1987, 102].

cheng: aufrecht, genau, positiv
fu: auf dem Rücken tragen, den Rücken kehren, negativ.

Rechenregeln dafür finden sich in dem Rechenbuch *Jiuzhang suanshu* (früher geschrieben *Chiu-chang suan shu*), zusammengestellt im 1. Jahrhundert n. Chr., in Buch VIII, bei Aufg. 3 [Vogel 1968]:

> Die Plus-Minus-Regel lautet: Für die Subtraktion gilt: Bei gleichen Benennungen wird voneinander subtrahiert, bei verschiedenen Benennungen wird zueinander addiert. Positives, ohne daß etwas dazukommt, mache es negativ, Negatives, ohne daß etwas dazukommt, mache es positiv. Für die Addition gilt: Sind die Benennungen verschieden, wird voneinander subtrahiert, bei gleichen Benennungen wird zueinander addiert. Positives, ohne daß etwas dazukommt, mache es positiv, Negatives, ohne daß etwas dazukommt, mache es negativ.

Gebraucht werden diese Regeln bei der Lösung von Systemen linearer Gleichungen, bei denen es sich um den Ertrag verschiedener Getreidesorten oder um An- und Verkauf von verschiedenen Tierarten handelt.
Die Aufgabe 8 lautet:

> Jetzt hat man 2 Rinder und 5 Schafe verkauft und damit 13 Schweine gekauft, wobei ein Rest von 1000 Geldstücken übrig blieb. Man hat 3 Rinder und 3 Schweine verkauft und damit 9 Schafe gekauft; das Geld reichte gerade. Man hat 6 Schafe und 8 Schweine verkauft und damit 5 Rinder gekauft, aber das Geld reichte nicht um 600 Geldstücke.

Setzen wir (!) den Preis eines Rindes $= x$, den eines Schafs $= y$, den eines Schweins $= z$, so erhalten wir das Gleichungssystem

$$
\begin{aligned}
2x + 5y - 13z &= 1000 \\
3x - 9y + 3z &= 0 \\
-5x + 6y + 8z &= -600
\end{aligned}
$$

Der Text sagt:

> Lege hin die 2 Rinder und die 5 Schafe als positiv, die 13 Schweine als negativ, die Anzahl des restlichen Geldes als positiv. Als nächstes lege hin die 3 Rinder als positiv, die 9 Schafe als negativ, die 3 Schweine als positiv. Als nächstes lege hin die 5 Rinder als negativ, die 6 Schafe als positiv, die 8 Schweine als positiv, das Geld, um das es nicht reicht, als negativ.

Aus dem Kommentar von Liu Hui zu den "Neun Büchern" (3. Jahrhundert n. Chr.) geht hervor, daß die Anzahlen (die Koeffizienten der Gleichungen) in Spalten auf einem Rechenbrett niedergelegt wurden, und zwar als schwarze und rote Stäbchen. Die Spalten werden von rechts nach links gezählt. So entsteht das Bild

	(C_3)	(C_2)	(C_1)
Rinder	-5	3	2
Schafe	6	-9	5
Schweine	8	3	-13
Geldwert	-600	0	1000

Positive und negative Zahlen bezeichnen also zunächst Verkauf und Kauf, Einnahmen und Ausgaben, aber in der Darstellung auf dem Rechenbrett werden sie zu reinen Zahlen. Liu Hui betrachtet Zahlen, die verschiedene "Namen" haben, als verschiedene "Kategorien" von Zahlen [Lam und Ang 1987, 138 und 140].

Das Lösungsverfahren ist bei einer früheren Aufgabe genau geschildert; hier wird nur darauf verwiesen. Es könnte folgendermaßen ausgesehen haben: Die Spalte C_2 wird mit 2 multipliziert und davon dreimal die Spalte C_1 abgezogen:

-5	0	2
6	-33	5
8	45	-13
-600	-3000	1000

Die folgenden Schritte wären

$C_3 \rightarrow 2C_3 + 5C_1$

0	0	2
37	-33	5
-49	45	-13
3800	-3000	1000

$C_3 \rightarrow C_3 + C_2$

0	0	2
4	-33	5
-4	45	-13
800	-3000	1000

$C_3 \rightarrow 33C_3 + 4C_2$

0	0	2
0	-33	5
48	45	-13
14400	-3000	1000

Ein Schwein kostet also 14400 : 48 = 300 Geldstücke. Ferner wird berechnet: Ein Schaf kostet 500, ein Rind 1200.

Regeln für die Multiplikation mit negativen Zahlen oder die Division durch negative Zahlen sind nicht erforderlich.

Die negativen Zahlen treten also hier in zweifacher Weise auf: 1) bei der Zusammenstellung von Einnahmen und Ausgaben, 2) im Verlauf des Rechenverfahrens, aber nicht im Resultat. Ein Tier kann ja keinen negativen Preis haben. Wie mag der Hersteller der Aufgaben das erreicht haben? Die Aufgabe 8 könnte er vom Resultat her konstruiert haben. Aber z.B. in der Aufgabe 11 ergibt sich als Preis eines Pferdes $5454\frac{6}{11}$ Geldstücke und als Preis eines Rindes $1818\frac{2}{11}$ Geldstücke.

5. SYSTEME LINEARER GLEICHUNGEN BEI DIOPHANT

Sicher haben auch die griechischen Kaufleute und Steuerbeamten Rechenaufgaben gelöst, doch ist davon nichts überliefert. Diophant fragt nicht nach Einnahmen oder Ausgaben, sondern nach Zahlen, und eine Zahl ist [Euklid VII, Def. 2] "eine aus Einheiten zusammengesetzte Menge"   da kommt der Gedanke an andere als positive Zahlen gar nicht auf. Eine Aufgabe, die keine positive Lösung hat, ist "unmöglich"

Diophants Aufgaben sind oft schematisch konstruiert, und da kann es bei Systemen linearer Gleichungen vorkommen, daß eine oder mehrere der gesuchten Zahlen negativ werden, was später bei Leonardo von Pisa und bei Chuquet tatsächlich vorgekommen ist. Kann man das vielleicht der Aufgabe von vornherein ansehen?

In Buch I, §16 verlangt Diophant, "drei Zahlen von der Beschaffenheit zu finden, daß die Summen zu je zweien gegebenen Zahlen gleich sind", also in unserer Schreibweise

$$y + z = a \quad \text{(bei Diophant 20)}$$
$$x + z = b \quad \text{(bei Diophant 30)}$$
$$x + y = c \quad \text{(bei Diophant 40)}.$$

"Es ist dabei notwendig, daß die halbe Summe der drei gegebenen Zahlen größer ist als jede der gegebenen Zahlen." Daß diese Bedingung nicht nur notwendig, sondern auch hinreichend ist, ist leicht zu sehen: Es ist

$$S = x + y + z = (a + b + c)/2 \quad \text{und}$$
$$x = S - a, \quad y = S - b, \quad z = S - c.$$

6. WIE KAMEN DIE NEGATIVEN ZAHLEN INS ABENDLAND?

Wurden sie aus dem Orient überliefert oder selbständig entdeckt? Die Kontakte zwischen China und Indien sind noch nicht vollständig geklärt. Die Inder arbeiteten mindestens seit Brahmagupta (*Brāhmasputasiddhānta*, 628 n. Chr.) mit negativen Zahlen [Brahmegupta and Bhascara 1817, 329]. Sie benutzten die Fachwörter *dhana* = Vermögen, *rina* = Schulden, Verlust, die vermutlich abstrakt gemeint waren. Die negativen Zahlen werden durch einen darübergesetzten Punkt bezeichnet. Brahmagupta gibt außer den Regeln für Addition und Subtraktion auch die Regeln für Multiplikation und Division mit negativen Zahlen und mit Null, auch Aussagen über das doppelte Vorzeichen der Quadratwurzel. Die negativen Zahlen treten besonders bei quadratischen Gleichungen auf; sie gestatten eine einheitliche Normalform; bei den Lösungen werden sie nur anerkannt, wenn es sachlich sinnvoll ist.

Die Araber haben zwar die indischen Ziffern übernommen, aber nicht die negativen Zahlen. Nur in einem einzigen Rechenbeispiel von Abū-l-Wafā (940–997/8), nämlich

$$3 \cdot 5 = (10 - 5) \cdot (10 - 3) - [(10 - 5) - (10 - 3)] \cdot 10,$$

kommt eine negative Zahl vor, nämlich

$$(10 - 5) - (10 - 3) = -2;$$

sie wird mit *dain* = Schuld bezeichnet [Juschkewitsch 1964, 256].

Auf diesem Weg können die negativen Zahlen wohl nicht ins Abendland gekommen sein.

7. NEGATIVE ZAHLEN IM ABENDLAND IM FRÜHEN MITTELALTER

Im Abendland erscheinen negative Zahlen zum erstenmal in einer Schrift, die fälschlich Beda zugeschrieben wurde, aber jedenfalls vor dem 10. Jahrhundert in Westdeutschland oder Nordfrankreich verfaßt wurde [Folkerts 1972, 41]. Drei Aufgaben behandeln das Erraten von Zahlen, dann folgen Regeln für das Addieren ("Verbinden") von positiven und negativen Zahlen, ohne Zusammenhang mit den Aufgaben und ohne

Hinweis auf Anwendungen. Eine positive Zahl wird als *verum*, eine negative als *minus* bezeichnet. Der Text beginnt

Verum cum vero facit verum. Minus cum vero facit verum.
Verum cum minus facit minus. Minus cum minus facit minus.

Es wird also vorausgesetzt, daß der zweite Summand der größere ist. Es folgt die Erklärung

Verum essentiam, minus nihil significat.
("Wahr" (positiv) bedeutet Sein, minus bedeutet das Nichts.)

Dann werden Rechnungen vorgeführt und erklärt:

Iunge III et VII, fiunt X;
iterum iunge III minus et VII, fiunt IIII;
iunge III et VII minus, fiunt IIII minus;
iunge III minus et VII minus, fiunt VII minus.

Die dritte dieser Operationen wird so erläutert:

... si iungantur III veri nomine et VII minus, quia maior est nihili quam essentiae summa, vincit septenarius non existens ternarium existentem et consumit eum sua non essentia, et remanent de ipso illi IIII numeri non existentes.
Wenn 3 als wahr benannte (Zahl) und 7 minus zusammengefaßt werden, dann überwindet, weil der Betrag des Nichts größer ist als der Betrag des Seins, die nicht-existierende 7 die existierende 3 und verbraucht (verzehrt) sie durch ihr Nicht-Sein, und es bleiben von ihr selbst jene 4 nicht-existierende Zahlen übrig.

Man möchte sagen: Gerade dadurch, daß die negativen Zahlen als nicht-existierend bezeichnet werden, wird ihnen eine gewisse Existenz zugesprochen. Immerhin sind diese nicht-existierenden Größen imstande, existierende Größen zu besiegen und aufzuzehren.

8. LEONARDO VON PISA (1202): SCHULDEN

Der Beda zugeschriebene Text fand in der Folgezeit keine Beachtung. Erst als im 12. Jahrhundert der Fernhandel im Mittelmeerraum aufblühte, die arabische Wissenschaft ins Abendland einströmte und in Kaufmannskreisen Rechnen und Algebra gepflegt und geübt wurde, kamen auch negative Zahlen wieder ins Gespräch. Sie begegnen uns zuerst als Schulden bei Leonardo von Pisa. Er hatte auf Reisen in den Orient Rechnen und Algebra gelernt und hat in seinem *Liber abaci* (1202) eine Fülle von Lern- und Übungsmaterial zusammengestellt.

Übungsaufgaben wurden sicher meistens von der Lösung her konstruiert, und dann kamen natürlich keine negativen Lösungen vor. Aber es gab auch schematisch konstruierte Aufgaben, bei denen es Überraschungen geben konnte. Eine Aufgabe von Leonardo von Pisa (*Liber abaci*, [Boncompagni 1857–1862, I, 349f.]; auch *Flos*, [Boncompagni 1857–1862, II, 238f.]) handelt "Von vier Personen und einer Börse". "Der erste besitzt mit der Börse die Hälfte des zweiten und dritten". In dieser Weise werden die folgenden Gleichungen beschrieben (x_i die Vermögen der Personen, b der Inhalt der Börse):

(1) $$x_1 + b = 2(x_2 + x_3)$$

(2) $$x_2 + b = 3(x_3 + x_4)$$

(3) $$x_3 + b = 4(x_4 + x_1)$$

(4) $$x_4 + b = 5(x_1 + x_2).$$

Der Text fährt fort: "Ich werde zeigen, daß diese Aufgabe unlösbar ist, wenn nicht zugestanden wird, daß der erste Partner Schulden hat." (*Hanc quidem questionem insolubilem esse monstrabo, nisi concedatur, primum hominem habere debitum*). Leonardo beweist das, und zwar so: Er setzt $x_1 = 1$ Drachme. Da das Gleichungssystem homogen ist, da auch b nicht vorgegeben ist, kann tatsächlich eine Unbekannte beliebig angenommen werden. Wir wollen jedoch $x_1 = a$ setzen. Ferner bezeichnet Leonardo den Anteil des zweiten Partners als *res*, das ist die damals aufgekommene Bezeichnung für die Unbekannte; wir schreiben $x_2 = r$.

Nun werden x_3 und x_4 eliminiert, so daß zwei Gleichungen für a, b, r entstehen: Aus (1) folgt

$$x_2 + x_3 = \frac{a+b}{2}, \quad \text{also} \quad x_3 = \frac{a+b}{2} - r,$$

Hiermit wird aus (2)

$$\frac{r+b}{3} = \frac{a+b}{2} - r + x_4, \quad x_4 = \frac{4}{3}r - \frac{1}{6}b - \frac{a}{2}.$$

Trägt man diese Werte ein, so erhält man

aus (3) $$b = \frac{38}{13}r + \frac{9}{13}a,$$

aus (4) $$b = \frac{22}{5}r + \frac{33}{5}a.$$

Wenn r und a positiv sind, widersprechen sich diese beiden Gleichungen, denn es ist $\frac{22}{5} > \frac{38}{13}$ und $\frac{33}{5} > \frac{9}{13}$.

Die Aufgabe ist unbestimmt; Leonardo ermittelt eine ganzzahlige Lösung:

$$x_1 = -1, \quad x_2 = 4, \quad x_3 = 1, \quad x_4 = 4, \quad b = 11.$$

Sesiano [1985] berichtet über eine Reihe weiterer Stellen bei Leonardo von Pisa, er berichtet auch über ein in der Provence um 1430 geschriebenes Manuskript, das eine Aufgabe über den gemeinsamen Kauf eines Stückes Tuch durch mehrere Partner enthält, wobei der Anteil eines Partners sich als $-10\frac{3}{4}$ (*10 he tres quarts, que le prumier ha me(n)s de non res*).

9. CHUQUET: ZAHLEN MIT VORZEICHEN

Im 13.–15. Jahrhundert, der Zeit, in der es vor allem in Italien zahlreiche Rechenmeister und Rechenschulen gab, hat man sich vielleicht allmählich an das Auftreten negativer Zahlen gewöhnt. Sie erschienen als Schulden, und man konnte mit ihnen rechnen. Jetzt begann eine neue Phase der Entwicklung mit der Frage: Was für Objekte sind das?

In der *Triparty* von Chuquet (verfaßt 1484) erscheinen negative Zahlen als reine Zahlen — anscheinend im Abendland erstmals nach der Schrift von Pseudo-Beda.

Chuquet spricht von *Nombres composez par plus et par moins*, d.h., er sieht diese "mit plus und mit minus zusammengesetzten Zahlen" als eine neue Art von Objekten an, für die die Rechengesetze erklärt — eigentlich: definiert — werden müssen. Chuquet erklärt [Chuquet 1880, 641 und 715] die Addition, womit die Subtraktion mit erfaßt ist (*Plus et plus, moins et moins, adioustons, plus et moins soustrayons*). Er bespricht die verschiedenen Fälle, die davon abhängen, welche der beiden Zahlen die größere ist.

Für die Lösung linearer Gleichungssysteme reicht das aus, denn dabei werden die zusammengesetzten Zahlen nur mit positiven rationalen Zahlen multipliziert oder dividiert. Und die meisten Aufgaben des kaufmännischen Rechnens, auch der Unterhaltungsmathematik, sind lineare Gleichungen mit komplizierten Koeffizienten und lineare Gleichungssysteme. Nur bei Zins- und Zinseszinsaufgaben kommen Gleichungen höheren Grades vor.

Ein Beispiel von Chuquet ist: *Je veulx trouuer cinq nombres de telle nature que tous ensemble sans le p'mier facent .120. Sans le second .180. Sans le tiers*

.240. Sans le quart .300. et sans le quint .360. In unserer Schreibweise:

$$S = x_1 + x_2 + x_3 + x_4 + x_5$$
$$S - x_1 = 120$$
$$S - x_2 = 180$$
$$S - x_3 = 240$$
$$S - x_4 = 300$$
$$S - x_5 = 360$$

Jceulx trouuer Je assemble tous les cinq nombres et montent .1200. que Je divise par .4. et men vient .300. desquels Je soustraitz les cinq nombres ci dessus cestas .120.180.240.300. et 360. et me restent .180. 120. 60. 0. et moins 60. qui sont les cinq nombres Je desiroye.

Diophants Nebenbedingung würde hier lauten: 1200 : 4 = 300 muß kleiner sein als jede der gegebenen Zahlen; das ist hier nicht der Fall.

10. GLEICHUNGEN ZWEITEN UND HÖHEREN GRADES. VORBEMERKUNG

Außer bei den Indern sind bisher quadratische Gleichungen im Zusammenhang mit negativen Zahlen nicht vorgekommen. Bis zu Stevin wurden die negativen Lösungen quadratischer Gleichungen überhaupt nicht bemerkt. Der Grund ist, daß die Lösungen geometrisch gefunden oder die Lösungsregeln geometrisch bewiesen wurden. Da bei Strecken und Flächen nur die absolute Größe, nicht aber die Richtung oder der Umlaufsinn berücksichtigt wurden, ist die Fläche eines Quadrats stets positiv, und ebenso die Länge der Seite, die Quadratwurzel. Die beiden Vorzeichen einer Quadratwurzel erscheinen als zwei Möglichkeiten des Zeichnens.

11. EUKLID

Schon die Babylonier konnten quadratische Gleichungen lösen. Ihre algebraischen Lösungsverfahren entsprachen den geometrischen Verfahren Euklids und sind vielleicht auf diesem Wege gefunden worden.

Euklid gibt Aufgaben und Lösungen rein geometrisch, ohne algebraische Gesichtspunkte auch nur anzudeuten. Ausgangspunkt ist der Satz II,5: "Teilt man eine Strecke ($AD = a$ in Abb. 3) sowohl in gleiche als

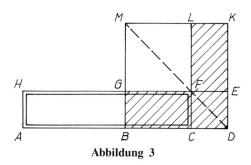

Abbildung 3

auch in ungleiche Abschnitte $\langle AB = BD = \frac{a}{2},\ AC = x,\ CD = y \rangle$, so ist das Rechteck aus den ungleichen Abschnitten der ganzen Strecke $\langle xy \rangle$ zusammen mit dem Quadrat über der Strecke zwischen den Teilpunkten $\langle BC^2 = (\frac{x-y}{2})^2 \rangle$ dem Quadrat über der Hälfte gleich." Algebraisch geschrieben:

(1)
$$xy + \left(\frac{x-y}{2} \right)^2 = \left(\frac{x+y}{2} \right)^2 .$$

Der Beweis beruht darauf, daß das doppelt umrandete Rechteck gleich dem schraffierten Gnomon ist.

Daraus folgt (VI,27): "Von allen Rechtecken, die man an eine feste Strecke so anlegen kann, daß ein Quadrat fehlt, ist das Quadrat über der Hälfte das größte." ⟨Etwas vereinfacht; Euklid spricht nicht von Rechtecken, sondern von Parallelogrammen.⟩

Dieser Satz ist Vorbedingung für die Lösbarkeit der Aufgabe VI,28: "An eine gegebene Strecke $\langle a \rangle$ ein einer gegebenen Fläche $\langle b \rangle$ gleiches Rechteck so anzulegen, daß ein Quadrat fehlt. Hierbei darf die gegebene Fläche $\langle b \rangle$ nicht größer sein als das über der Hälfte der Strecke zu zeichnende Quadrat."

Algebraisch ausgedrückt: Verlangt ist die Lösung der Gleichungen

(2)
$$x + y = a, \qquad xy = b$$

mit der Nebenbedingung $(\frac{a}{2})^2 \geq b$. Wenn diese Bedingung nicht erfüllt ist, ist die Aufgabe nach VI,27 unmöglich.

Wir wollen Euklids geometrische Lösung algebraisch erläutern. Wenn $(\frac{a}{2})^2 = b$ ist, dann ist $x = y = \frac{a}{2}$ die Lösung. Wenn $(\frac{a}{2})^2 > b$ ist, zeichne man die Strecke $a = AD$ und errichte über ihrer Hälfte BD das Quadrat.

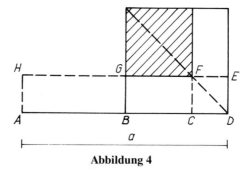

Abbildung 4

Dann muß man sich

$$\frac{x-y}{2} = \sqrt{\left(\frac{a}{2}\right)^2 - b}$$

verschaffen. Geometrisch hat man dazu die Differenz der beiden Flächen $\left(\frac{a}{2}\right)^2$ und b in ein Quadrat zu verwandeln. Wie man das machen kann, lehrt Euklid in Buch 1. Dieses in einer Nebenzeichnung gezeichnete Quadrat hat man in das Quadrat über $\frac{a}{2}$ oben links einzutragen. Seine Seite $GF = BC$ ist zu $AB = \frac{a}{2}$ zu addieren. Man erhält

$$x = \frac{x+y}{2} + \frac{x-y}{2} = AB + BC = \frac{a}{2} + \sqrt{\left(\frac{a}{2}\right)^2 - b}\,.$$

y, die zweite Seite des Rechtecks, ergibt sich aus der Figur (Abb. 4) zu

$$y = DE = CD = BD - BC = \frac{a}{2} - \sqrt{\left(\frac{a}{2}\right)^2 - b}\,.$$

In den Gleichungen (1), (2) sind x und y vertauschbar. Geht man zu

$$x(a - x) = b$$

über, so ist außer

$$x = \frac{a}{2} + \sqrt{\left(\frac{a}{2}\right)^2 - b} \quad \text{auch} \quad y = a - x = \frac{a}{2} - \sqrt{\left(\frac{a}{2}\right)^2 - b}$$

eine Lösung.

12. AL-ḪWĀRIZMĪ

Al-Ḫwārizmī teilte die Gleichungen 1. und 2. Grades in sechs Typen ein, mit nur positiven Koeffizienten, und zwar drei einfache

$$x^2 = ax, \quad x^2 = b, \quad x = b,$$

und drei zusammengesetzte

I. $x_2 + ax = b$ mit der Lösung $x = \sqrt{b + (\frac{a}{2})^2} - \frac{a}{2}$

II. $x^2 + b = ax$ mit der Lösung $x = \frac{a}{2} \pm \sqrt{(\frac{a}{2})^2 - b}$

III. $x^2 = ax + b$ mit der Lösung $x = \sqrt{b + (\frac{a}{2})^2} + \frac{a}{2}$.

Er beschreibt die Gleichungen und die Lösungen in Worten und beweist die Lösungen geometrisch.

Die hier noch fehlende Gleichungsform

$$x^2 + ax + b = 0$$

wäre den damaligen Mathematikern sinnlos vorgekommen; sie hätte bedeutet, daß eine positive Zahl = 0 werden soll.

Wenn (oder: Da) a und b positiv sind, haben die einfachen Gleichungen je eine Lösung. Bei der ersten Gleichung wird die Lösung 0 nicht beachtet.

Von den zusammengesetzten Gleichungen haben I und III je eine positive Lösung. Da die Wurzel als positiv angesehen wird, wird die negative Lösung gar nicht bemerkt.

Der Typ II entspricht der Aufgabe VI,28 bei Euklid. Al-Ḫwārizmī stellt fest: Wenn $(\frac{a}{2})^2 < b$ ist, ist die vorgelegte Frage nichtig (*quaestio tibi proposita nulla est*), und wenn $(\frac{a}{2})^2 = b$ ist, hat die Gleichung nur eine Lösung. Für den Fall $(\frac{a}{2})^2 > b$ beschreibt er die Lösungsregel an dem Beispiel [Karpinski 1915, 74]

Substantia et 21 drachmae 10 radices coaequantur.
Ein Vermögen und 21 Drachmen sind gleich 10 Wurzeln.

$$x^2 + 21 = 10x.$$

Halbiere zuerst die Zahl der Wurzeln, das ist in diesem Falle 5; diese mit sich multipliziert ergeben 25. Von diesen ziehe die 21 Drachmen ab, es bleiben 4. Davon nimm die Quadratwurzel, es sind 2. Diese ziehe von der Hälfte der Wurzeln ab, es bleibt 3 als eine Wurzel. Du kannst auch 2 zur Hälfte der Wurzeln addieren; es kommt 7 heraus.

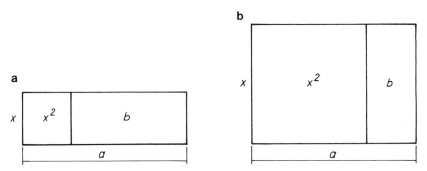

Abb. 5. Die Größenverhältnisse entsprechen dem Beispiel $x^2 + 21 = 10x$. (a) $x < a/2$. (b) $x > a/2$.

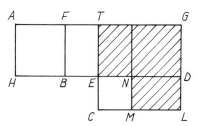

Abb. 6. Die Buchstaben entsprechen denen von Robert von Chester. $MBFA = x^2$; $BDGF = b$; $HD = a$.

Diese Lösungsregel bestätigt al-Hwarizmi geometrisch. Er beweist, und zwar für (im Rahmen der Aufgabe) beliebige x, a, b:

$$\text{Wenn } x^2 + b = ax \text{ ist, dann ist } x = \frac{a}{2} \pm \sqrt{\left(\frac{a}{2}\right)^2 - b}\,.$$

Dazu zeichnet er das aus x^2 und b zusammengesetzte Rechteck, dessen Fläche $= ax$ sein soll, von dem also eine Seite $= a$ ist. x kann nicht ganz willkürlich angenommen werden. Aus der Gleichung ergibt sich, daß $x < a$ sein muß. Aber die Zeichnung fällt verschieden aus, je nachdem $x < \frac{a}{2}$ oder $x > \frac{a}{2}$ angenommen wird. Das zeigen die Abb. 5a,b.

Wenn $x < \frac{a}{2}$ angenommen wird, dann wird die Abb. 5a zur Abb. 6 ergänzt: Über $TG = \frac{a}{2}$ wird das Quadrat $TGCL$ gezeichnet, ferner wird $ND = DG = x$ gemacht.

Dann ist $EN = EC = BE = \frac{a}{2} - x$ und das Rechteck $BDGF = b$ gleich dem schraffierten Gnomon, somit

$$\left(\frac{a}{2}\right)^2 - b = EN^2 = \left(\frac{a}{2} - x\right)^2.$$

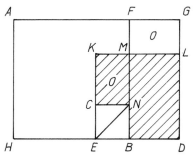

Abb. 7. $MBFA = x^2$; $BDGF = b$; $HD = a$; $BD = ML = FG = a - x$.

Die Seite dieses Quadrats ist

$$\sqrt{\left(\frac{a}{2}\right)^2 - b} = \frac{a}{2} - x, \quad \text{w.z.b.w.}$$

Wenn $x > \frac{a}{2}$ angenommen wird, ist aus der Abb. 5b die Abb. 7 zu entwickeln.

Wir verwandeln wieder das Rechteck $BDFG = b$ in einen Gnomon, indem wir das Quadrat über $ED = \frac{a}{2}$ zeichnen ($EDLK$) und $EC = EB = x - \frac{a}{2}$ machen. Dann ist $CK = x - 2 \cdot (x - \frac{a}{2}) = a - x = FG$. Daher sind die mit O bezeichneten Rechtecke einander gleich, und b ist gleich dem schraffierten Gnomon:

$$b = \left(\frac{a}{2}\right)^2 - \left(x - \frac{a}{2}\right)^2,$$

$$x - \frac{a}{2} = \sqrt{\left(\frac{a}{2}\right)^2 - b}, \quad \text{w.z.b.w.}$$

Daß eine Gleichung vom Typ II zwei positive Lösungen haben kann, wird also nicht aus dem doppelten Vorzeichen der Wurzel abgeleitet, sondern ergibt sich daraus, daß die Zeichnungen für $x < \frac{a}{2}$ und für $x > \frac{a}{2}$ verschieden ausfallen.

Für den Fall $x > \frac{a}{2}$ wird in älteren Handschriften nur die Lösung $x = 7$ genannt. Die ausführliche Erläuterung der Konstruktion erscheint erst in einer Ausgabe aus dem 16. Jahrhundert [Hughes 1989].

13. CARDANO: GLEICHUNGSTHEORIE

Wie damals üblich war, so verteilt auch Cardano (*Ars magna*, 1545) [1968] die Glieder einer Gleichung so auf die beiden Seiten, daß nur positive Koeffizienten auftreten. Infolgedessen muß er bei kubischen Gleichungen

22 Typen behandeln. Aber es gelingt ihm, diese Typen durch Umfor-
mungen der Gleichungen (Einführung einer neuen Unbekannten) auf die
beiden Typen

$$x^3 + bx = c$$
$$x^3 = bx + c$$

zurückzuführen. Ferner stellt er allgemeine Aussagen über Anzahl und
Wert der Lösungen voran.

Hierzu sind nun negative Zahlen unentbehrlich. Cardano nennt die
aestimatio rei (Wert der Unbekannten) oder auch die *aequatio: vera* oder
ficta, sic enim vocamus eam, quae debiti est seu minoris (fingiert, so nennen
wir diejenige ⟨Lösung⟩, die von Schulden oder von Minus ⟨handelt⟩).

Die folgenden Beispiele von Cardanos allgemeinen Sätzen sollen belegen, daß hier negative Zahlen gebraucht werden.

Wenn in einer Gleichung nur gerade Potenzen der Unbekannten
vorkommen, gehört zu jeder wahren Lösung eine fingierte [Cardano 1968,
Kap. I,3].

Wenn beliebig viele ungerade Potenzen der Unbekannten (mit positiven
Koeffizienten) einer Zahl gleich sind, hat die Gleichung genau eine wahre
und keine fingierte Lösung [Cardano 1968, Kap. I,4].

Ferner kann es vorkommen, daß zwei Gleichungen einander in der
Weise entsprechen, daß jeder wahren Lösung der einen Gleichung eine
fingierte Lösung der anderen entspricht und umgekehrt. Ein Beispiel von
Cardano ist [Cardano 1968, Kap. I,5]: Die Gleichung $x^3 + 16 = 12x$ hat
die Lösungen $+2, +2, -4$, und die Gleichung $x^3 = 12x + 16$ hat die
Lösungen $-2, -2, +4$. Cardano weiß auch, daß in diesem Falle die
Summe der wahren Lösungen (dem Betrage nach) gleich der Summe der
fingierten Lösungen ist.

Für Stevin war es ein Argument für die Anerkennung der *solutions
par* − , die er als *solutions songées* bezeichnet, daß man durch sie die
wahren Lösungen einer anderen Gleichung finden kann [Stevin 1585,
Livre II, S. 332; Stevin 1958, 642].

Wenn man nun erwartet, daß Cardano alle positiven und negativen
Lösungen der von ihm behandelten Aufgaben angibt, wird man enttäuscht.
Die quadratischen Gleichungen behandelt er in Kapitel V wie al-Ḫwārizmī,
d.h. mit geometrischen Beweisen; negative Lösungen werden nicht erwähnt.
Auch bei kubischen Gleichungen gibt er nur selten die fingierten Lösungen
an.

Ein besonderer Fall findet sich in Kap. XVIII. Die Gleichung

(1) $$x^3 + 21x = 9x^2 + 5$$

wird durch die Substitution $x = y + 3$ umgeformt zu

(2) $y^3 + 4 = 6y$.

Huius aestimationes sunt tres, prima est 2, secunda R.3.m.1, tertia ficta m.
R.3.p.1; quas adde ad 3, habebis veras aestimationis illas quas a latere vides.
Deren Lösungen sind drei: die erste ist 2, die zweite $\sqrt{3} - 1$, die dritte
eine fingierte $-(\sqrt{3} + 1)$. Diese addiere zu 3, dann hast du die wahren
Lösungen ⟨der Gleichung (1)⟩, nämlich 5, $2 + \sqrt{3}$, $2 - \sqrt{3}$.

⟨Die Lösungen von (2) kann man z.B. finden, indem man die Gleichung
durch $y - 2$ dividiert. Man erhält die quadratische Gleichung $y^2 + 2y = 2$.⟩
Hier wird also eine fingierte Lösung einer Hilfsgleichung angegeben, weil
man daraus eine wahre Lösung der Ausgangsgleichung erhält.

Cardano hat die Unterscheidung zwischen wahren und fingierten
(erfundenen, uneigentlichen) Lösungen sehr ernst genommen[1]: Als Koeffizienten der Ausgangsgleichungen dürfen "fiktive" Zahlen nicht auftreten.
Das ist eigentlich erst möglich, wenn auch die bekannten Größen durch
Buchstaben bezeichnet werden. Dann kann z.B. die Normalform einer
quadratischen Gleichung als

$$x^2 = ax + b$$

geschrieben werden, wobei für a, b positive oder negative Zahlen eingesetzt werden können.

Stifel [1544, Buch III, Kap. 5] muß noch drei Formen angeben:

$1_{\mathcal{Z}}$ *aequatus* $1_{\mathfrak{x}} + 35156$ $1x^2 = 1x + 35156$

$1_{\mathcal{Z}}$ *aequatus* $18_{\mathfrak{x}} - 72$ $1x^2 = 18x - 72$

$1_{\mathcal{Z}}$ *aequatus* $84 - 8_{\mathfrak{x}}$ $1x^2 = 84 - 8x$.

Er denkt auch nicht daran, etwa die dritte Gleichung in der Form

$1_{\mathcal{Z}} = (-8)_{\mathfrak{x}} + 84$ $1x^2 = (-8)x + 84$

zu schreiben. Man könnte sagen: Er läßt das Minuszeichen als Operationszeichen in der Normalform der Gleichungen zu, aber keine negativen
Koeffizienten.

Jedoch: Bei der Lösung der Gleichungen ist der zweite Schritt: Multipliziere die Hälfte der Zahl der Wurzeln mit sich. Und da rechnet Stifel
bei der dritten Gleichung:

scilicet $- 4$ *in* -4 *facit* $+ 16$.

Stifel kann die Lösungen der drei Gleichungen einheitlich angeben, aber
mit zweimaliger Alternativvorschrift der Form "Addiere oder subtrahiere,
wie es das Zeichen erfordert."

Stevin zieht ausdrücklich das Operationszeichen als Vorzeichen zur Zahl und kann damit Alternativvorschriften vermeiden [1958, II, Problème LXVIII].

Viète, der ja Buchstaben auch zur Bezeichnung der bekannten Größen eingeführt hat, denkt geometrisch und läßt daher negative Zahlen weder als Lösung noch als Koeffizienten zu, gestattet aber das Auftreten von Minuszeichen als Operationszeichen in der Normalform der Gleichungen. Wenn das Glied mit der höchsten Potenz der Unbekannten negativ ist, stellt er es nicht an die erste Stelle, z.B. [Viète 1646, 130]

$$D\ 2\ in\ A\ -\ A\ quad.,\ aequetur\ Z\ plano.$$

(A ist die Unbekannte, D und Z sind die bekannten Größen.)

14. RECHNEN MIT BINOMEN

Das Rechnen mit den fingierten Größen war unproblematisch, weil die mindestens seit Diophant bekannten Rechenregeln für subtraktive Zahlen benutzt werden konnten. Grundlage war das Rechnen mit Binomen, dem in vielen Rechenbüchern ein besonderes Kapitel gewidmet war. Schon al-Ḫwārizmī bringt 13 Beispiele [Hughes 1989, 45–49]. Das vorletzte lautet: *10 dragmata et rei medietas in medietate dragmatis, 5 rebus abiectis, multiplicata, quantum procreant?* In gleicher Art werden die Lösungsschritte und das Resultat angegeben. In unserer Schreibweise ist das ($dragma = x^0 = 1$)

$$\left(10 + \frac{x}{2}\right) \cdot \left(\frac{1}{2} - 5x\right) = 5 - 2\frac{1}{2}x^2 - 49\frac{3}{4}x.$$

In einer Schrift von Initius Algebras, die von Adam Ries vor 1525 abgeschrieben worden ist, steht eine Multiplikationstabelle, von der ein Beispiel so aussieht (Abb. 8).

	$-1d$ $-1x$
$+1d$ $+1x$	$-1d$ $-1x^2$ $-2x$

Abb. 8. d steht für $\phi = dragma = x^0$, x für $\underset{\smile}{r}$, x^2 für $\mathfrak{z}$ [Kaunzner 1992, 207].

Stifel hat das Rechnen mit Binomen systematisch auf Binome der Form $(0 - a)$ übertragen. Er nennt die negativen Zahlen *numeri absurdi* und *numeri ficti infra nihil* [Stifel 1544, Kap. VI, fol. 248v/249r] und sagt dabei, daß diese Fiktion von größtem Nutzen für die Mathematik ist (*fitque haec fictio summa utilitate pro rebus mathematicis*).

15. ENTGEGENGESETZTE GRÖSSEN

Die Einführung "fingierter Zahlen unterhalb von Null" ist also möglich, weil man mit ihnen wie mit subtraktiven Zahlen rechnen kann, und sie ist für die Algebra nützlich, weil bei manchen Aufgaben Nebenbedingungen wegfallen und allgemeine Aussagen über Gleichungen möglich werden. Für den Übergang der Algebra von der Lösung einzelner Gleichungen zu einer Gleichungstheorie ist das geradezu notwendig. Dabei spielt auch die Einführung der komplexen Zahlen eine wichtige Rolle. Diese Entwicklung der Algebra geschah in mehreren Schritten in der Zeit etwa von Cardano bis Descartes.

Zunächst galten also die negativen Lösungen von Gleichungen nicht als wahre, sondern nur als fingierte Lösungen. Konnte man sie in irgendeiner Weise als wirklich ansehen?

Das ist z.B. möglich, wenn man sie als Schulden im Gegensatz zu Vermögen oder als Ausgaben im Gegensatz zu Einnahmen ansieht, oder auch als Strecken in einer zu einer anderen entgegengesetzten Richtung. Bhaskara (um 1250) [Brahmegupta and Bhascara 1817, Lilavati VI, 166] fragt nach den Höhenabschnitten eines Dreiecks mit der Basis 9 und den Seiten 10 und 17 (Abb. 9).

Dafür kennt schon Brahmagupta die Regel

$$p, q = \frac{1}{2}\left(c \pm \frac{(a + b)(a - b)}{c}\right).$$

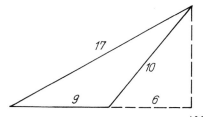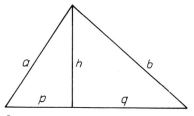

Abbildung 9

Sie läßt sich aus $p + q = c$ und $h^2 = a^2 - p^2 = b^2 - q^2$ ableiten
[Brahmegupta and Bhascara 1817, 297. Chap. XII, Sect. IV,22.]. In der
Aufgabe von Bhaskara ist $[(a + b)(a - b)]/c = 21$. "Das kann von der
Basis nicht abgezogen werden, also zieht man die Basis hiervon ab. Die
Hälfte des Restes ist der Abschnitt: 6; und das ist negativ, d.h. in der
entgegengesetzten Richtung."

In der Neuzeit wurde der Begriff der entgegengesetzten Größen
mehrfach zur Erklärung der negativen Zahlen herangezogen. Eine Definition steht allerdings — wie sollte es anders sein — schon bei Aristoteles
[Kateg. Kap. 5: 6a17 und gleichlautend in Metaphys. 10: 1018a27]: "Entgegengesetzt heißt, was innerhalb derselben Gattung am weitesten
voneinander absteht."

Wallis sagt in seiner Arithmetik [1685, vermehrte lateinische Ausgabe
1693, Kap. 66]: "Es ist unmöglich, daß eine Größe weniger sei als nichts.
Trotzdem ist die Annahme einer negativen Größe (*suppositio quantitatis
Negativae*) weder nutzlos, noch absurd, wenn sie nur richtig verstanden
wird.... -3 Schritte vorwärts zu gehen ist dasselbe wie 3 Schritte
zurückgehen (*processisse passibus -3, tantundem est ac 3 passibus retroces-
sisse*)."

Kant sagt in seiner Schrift "Versuch, den Begriff der negativen Größen
in die Weltweisheit einzuführen" im Vorwort: "Es sind die negativen
Größen nicht Negationen von Größen, wie die Ähnlichkeit des Ausdrucks
hat vermuten lassen, sondern etwas an sich selbst wahrhaftig Positives, nur
was dem andern entgegengesetzt ist." Der 1. Abschnitt beginnt: "Einander
entgegengesetzt ist: wovon eines dasjenige aufhebt, was durch das andre
gesetzt ist."

Dabei sagt Kant aber auch [im Vorwort]: "Ich habe vorjetzo die Absicht,
einen Begriff, der in der Mathematik bekannt genug, allein der
Weltweisheit noch sehr fremde ist, in Beziehung auf diese zu betrachten."

Bolzano hat in seiner "Einleitung zur Größenlehre" [Bolzano 1975–1976,
Bd. 7, §70] die Definitionen von Aristoteles und Kant besprochen und als
zu eng abgelehnt. Die Definition von Aristoteles passe z.B. nicht für
Vermögen und Schulden, und bei der Definition von Kant könne er sich
nicht vorstellen, daß zwei von einem Punkt aus in entgegengesetzten
Richtungen abgetragene gleiche Strecken einander aufheben. Er selbst
definiert: "A und B heißen einander entgegengesetzt, wenn es irgendeine
durch bloße Begriffe darstellbare Regel gibt, vermöge deren sich B aus A
ableiten läßt, und wenn nach eben derselben Regel auch wieder aus B sich
A ableiten läßt."

In der Einleitung zur Größenlehre bemerkt Bolzano, daß nicht alle Arten von Größen die Beschaffenheit haben, daß es zu jeder Größe eine entgegengesetzte gibt. In der Zahlenlehre erklärt er (positive ganze) Zahlen als Mengen von Einheiten gleicher Art. Die Erklärung der negativen Zahlen findet sich in einer sehr umfangreichen Randbemerkung [Bolzano 1975–1976, Bd. 7, §19]:

"In Übereinstimmung mit dem, was wir schon in der allgemeinen Größenlehre erklärt, setzen wir auch in dieser Zahlenlehre fest, daß wir bei allen Zahlenausdrücken die Einheit, auf welche sie sich beziehen, als eine des Gegensatzes fähige Einheit betrachten wollen. Wir setzen überdies fest, daß uns, wo ein auf verständliche Weise gebildeter Zahlenausdruck wegen der eigenthümlichen Beschaffenheit der in ihm vorkommenden Zahlen etwas Unmögliches fordert, erstattet sein solle, das Gesetz, nach welchem die durch ihn vorgestellte Zahl aus den gegebenen gewonnen werden soll, auf eine solche Art zu verändern, welche sich aus der Verschiedenheit der beyden Fälle, des hier obwaltenden u. desjenigen, wo die in dem Ausdrucke vorkommenden Forderungen verglichen wären, auf das Genügendste erklären läßt."

Dies steht also am Rande einer von einem Kopisten angefertigten Kopie eines Entwurfes und erweckt bei mir den — vielleicht ganz falschen — Eindruck, als habe Bolzano beim Lesen der Kopie eine Erklärung der negativen Zahlen vermißt und nun eine vorläufige Bemerkung notiert, die noch genauer auszuarbeiten gewesen wäre. Bolzano hat ja diese Schrift nicht mehr vollenden können.

Bolzano postuliert also die Existenz einer negativen Einheit. Ferner verlangt er, die Rechengesetze der neuen Zahlenart anzupassen.

Weierstraß hat in den Jahren 1861 bis 1885 ungefähr alle zwei Jahre eine Vorlesung "Einleitung in die Theorie der analytischen Funktionen" gehalten und in einem ersten Abschnitt den Begriff der Zahl behandelt. Eine (positive) Zahl ist für ihn ein Aggregat aus verschiedenen Anzahlen verschiedener Einheiten. Dann erklärt er,

...daß es, um die Subtraktion stets möglich zu machen, notwendig ist und genügt, den Bereich der Größen, mit denen bisher operiert wurde, durch ein neues Größengebiet zu erweitern, derart daß sich äquivalente Mengen beider Größen in irgendeinem Aggregat aufheben. Die neu eingeführten Größen nennen wir den ursprünglichen, für die wir die Subtraktion allgemein durchzuführen suchten, *entgegengesetzt*; die einer Zahlgröße des ursprünglichen Gebietes, *a*, entgegengesetzte *a'* definieren wir durch die Gleichung

$$A + a + a' = A,$$

wo *A* eine beliebige Summe bedeutet." [Weierstraß 1880/81, 18–19]

Die negativen Zahlen sind also auch für Weierstraß fingierte Größen, eingeführt, um die Subtraktion allgemein möglich zu machen, also zur Vereinfachung der Theorie. Er definiert sie, wie es zur Zeit von Kant schon bekannt war, als zu den positiven Zahlen entgegengesetzt, und "entgegengesetzt" erklärt er nicht nur durch "einander auslöschen", sondern drückt es auch in einer mathematischen Formel aus [Weierstraß 1878].

In der Folgezeit ist dann der Begriff "entgegengesetzt" aus der Erklärung verschwunden und durch die Gleichung

$$a + (-a) = 0$$

ersetzt worden ([Weber 1895, 16]. Allerdings schreibt Weber: "Das zweite System ⟨das der negativen Zahlen⟩ ordnen wir nun dem ersten ⟨dem der positiven Zahlen⟩ gerade entgegengesetzt, so daß überall, wo in dem System x "größer" steht, in dem System $-x$ "kleiner" gesetzt wird, und umgekehrt.).

ANMERKUNGEN

[1] Tatsächlich ist es wohl ein entscheidender Schritt in der Entwicklung der Mathematik, daß der Mathematiker sich neue Objekte verschafft (erfindet), die in der Wirklichkeit (anscheinend) nicht vorkommen. So mag es verständlich sein, daß man mit diesen neuen Objekten zunächst sehr vorsichtig umgeht, sie nur dort verwendet, wo es nicht anders geht, und daß man dann doch versucht, sie in der Wirklichkeit vorzufinden. Erst im 19. Jahrhundert hat man auf den Bezug zur Wirklichkeit grundsätzlich verzichtet.

LITERATURVERZEICHNIS

Bolzano, B. 1975−1976. *Bernard Bolzano Gesamtausgabe*. Reihe II A (nachgelassene Schriften). Stuttgart-Bad Cannstadt: Friedrich Frommann. Bd. 7. Einleitung zur Größenlehre und Erste Begriffe der allgemeinen Größenlehre. Jan Berg (Ed.). Bd. 8. Reine Zahlenlehre. Jan Berg (Ed.).

Boncompagni, B. (Hrsg.). 1857−1862. *Scritti di Leonardo Pisano*. 2 Bde. Rom: Tipografia delle scienze matematiche e fisiche.

Brahmegupta and Bhascara. 1817. *Algebra with Arithmetic and Mensuration*. Translated by H. Th. Colebrooke. London: John Murray. Nachdruck Walluf bei Wiesbaden: Sändig 1973.

Cardano, H. 1663. *Hieronymi Cardani...Operum tomus quartus*. C. Spon (Ed.) Lyon. Nachdruck: New York, London: Johnson 1967.

Cardano, H. 1968. *The great Art or the Rules of Algebra*. Witmer, T. R. (Übersetzung ins Englische). Cambridge, Mass., London: M.I.T. Press.

Chuquet, N. 1880. La triparty en la science des nombres. A. Marre (Ed.). *Bollettino di bibliografia e di storia delle scienze matematiche e fisiche* 13, 555–659, 693–814.

Diophant. 1893–1895. *Diophanti Alexandrini opera omnia cum Graecis commentariis*. P. Tannery, Hrsg. Bd. I, II. Leipzig: B. G. Teubner.

Folkerts, M. 1972. Pseudo–Beda: De arithmeticis propositionibus. Eine mathematische Schrift aus der Karolingerzeit. *Sudhoffs Archiv* 26, 22–43.

Gericke, H. 1970. *Geschichte des Zahlbegriffs*. Mannheim: Bibliographisches Institut.

Hughes, B. B. 1989. *Robert of Chester's Latin Translation of al-Khwarizmi's Al-Jabr. A New Critical Edition*. Stuttgart: Steiner.

Juschkewitsch, A. 1964. *Geschichte der Mathematik im Mittelalter. Deutsche Übersetzung von V. Ziegler*. Basel: Pfalz-Verlag.

Karpinski, L. C. (Ed.) 1915. *Robert of Chester's Latin Translation of the Algebra of al-Khowarizmi*. New York, Macmillan.

Kaunzner, W. 1992. Über das wissenschaftliche Umfeld und die mathematischen Handschriften von *Adam Ries*. *In Adam Rieß vom Staffelstein*. Staffelsteiner Schriften, Bd. 1, Staffelstein: Verlag für Staffelsteiner Schriften, 157–279.

Lam, Lay-Yong, und Ang, T. S. 1987. The earliest negative numbers. How they emerged from a solution of simultaneous linear equations. *Archive for History of Exact Sciences* 37, 222–267.

Li Yan und Du Shiran. 1987. *Chinese Mathematics. A concise history*. Oxford: Clarendon Press.

Needham, J. 1959. *Science and Civilisation in China*. Vol. 3: *Mathematics and the Sciences of the Heavens and the Earth*. Cambridge: Cambridge University Press.

Neugebauer, O. 1955. *Astronomical Cuneiform Texts*. London: Lund Humphreys. Nachdruck: Berlin, Heidelberg: Springer 1983.

Neugebauer, O. 1957. *The Exact Sciences in Antiquity*. Providence, Rhode Island: Brown University Press. 2. Aufl.

Selz, G. 1989. Die altsumerischen Wirtschaftsurkunden der Eremitage zu Leningrad. Stuttgart: Steiner.

Sesiano, J. 1985. *The Appearance of Negative Solutions in Mediaeval Mathematics*. Archive for History of Exact Sciences 32, 105–150.

Stevin, S. 1585. *Arithmetique*. Leiden: Christophe Plantin.

Stevin, S. 1958. *Principal Works*. II B. D. Struik, Hrsg. Amsterdam: Swets u. Zeitlinger.

Stifel, M. 1544. *Arithmetica integra*. Nürnberg: apud Iohan. Petreium.

van der Waerden, B. L. 1966. *Anfänge der Astronomie*. Groningen: Noordhoff.

Viète, F. 1646. *Opera mathematica*. F. Schooten (Ed.) Leiden: Ex Officina Bonaventurae et Abrahami Elzeviriorum. Nachdruck Hildesheim: Olms. J. E. Hofmann (Ed.) 1970.

Vogel, K. 1968. *Chiu-chang suan-shu*, deutsche Übersetzung "Neun Bücher arithmetischer Technik" von K. Vogel. Braunschweig: Vieweg.

Weber, H. 1895. *Lehrbuch der Algebra*. Bd. 1.

Weierstraß, K. 1878. *Einleitung in die Theorie der analytischen Funktionen. Vorlesung Berlin 1878 in einer Mitschrift von Adolf Hurwitz*. Bearbeitet von Peter Ullrich. Braunschweig, Wiesbaden: Vieweg. 1988.

Weierstraß, K. 1880/81. *Einleitung in die Theorie der analytischen Funktionen*. Vorlesung vom W.S. 1880/81, Mitschrift von Adolf Kneser. Sein Sohn, Hellmuth Kneser, hat es mir seinerzeit ermöglicht, mir eine Photokopie anfertigen zu lassen.

Begriffs- und Zeichenkonzeptionen in der Mathematik des lateinischen Mittelalters und der Renaissance

Wolfgang Kaunzner

Zoller Straße 9, D-93053 Regensburg, Germany

In the 15th and 16th centuries a structural change occurred in Western European mathematics which finally offered the appearance of a mathematical statement, e.g. of an algebraic equation, with the same content as before, but in a new shape. Until then, numbers had not always been ciphered and all arithmetic and algebraic technical terms were traditionally expressed in words. But from about 1460 on scientists and later reckoners, too, tried to render mathematical texts more concisely and to make their meaning easier for readers to understand. Without any reasons being given, new signs appeared. For example, Johannes Widmann, in his well-known arithmetic-book, Leipzig 1489, f. 88^r, only writes: "was − ist dz ist minus" und "das + das ist mer" ("what − is, is minus" and "that +, that is more").

1. EINLEITUNG

Das Bestreben der Fachleute, die sich im lateinischen Mittelalter und in der Renaissance (500−1550) mit Mathematik beschäftigten, galt großenteils der Übernahme des überkommenen Wissens—*ex oriente lux*—, vor allem aber der Umformung dieser Kenntnisse. Der Inhalt der Mathematik zu Beginn dieses Zeitraums gliedert sich in Anlehnung an früher in einen anwendungsbezogenen Teil, der sich um 500 auf Ziffern und einfache praktische Geometrie der Römer, ferner auf von alters her übermittelte Methoden wie Kerbhölzer, Abacus mit unbezifferten Rechensteinen (*calculi*) und Fingerzahlen vorwiegend zur Zahlendarstellung hätte stützen können, sowie in einen theoretischen Abschnitt, wie er durch Boetius (ca. 480−524/525) aus der prinzipiell zahlentheoretischen Arithmetik des Nikomachos (um 100) überliefert war, und einzelne Stücke aus den "Elementen" I-V Euklids (365?−300?).

Die Problematik in der Mathematik des genannten Zeitabschnitts blieb in dem Bereich gleich, wo sie sich an den Bedürfnissen aus dem Alltagsleben der damaligen Menschen orientierte.

Nach der Völkerwanderung führten elementare Veränderungen im politischen Leben, im Leben der Stämme und Völker überhaupt, zu Umwandlungen in Gewerbe und Handel. Eine Verdichtung der zwischenmenschlichen Beziehungen verlangte nach Güteraustausch bzw. -absatz, der auf Maß und Zahl, Schätzung und Rechnung angelegt war. Man mußte sich neuer Methoden bedienen, um mit dem Partner rascher "ins Geschäft" zu kommen.

Auch in der Mathematik dieser Periode stand am Anfang das Wort, denn bei schriftlicher Formulierung wurde alles ausgeschrieben. Von dieser rhetorischen Darstellung eines mathematischen Inhalts ging man über zur synkopierten, wo bereits sinnvolle Kürzungen angewandt wurden, wie sie sich bei wiederholtem Anschreiben von selbst anboten. Der nächste Schritt war die symbolische Schreibart, welche manchmal überregionale, schließlich in einigen Fällen allgemeine Anerkennung fand. Die Zeichen im mittelalterlichen Rechnen erfüllten mehrere Funktionen: Abkürzungen von Wörtern; Abkürzungen von Zahlwörtern, somit Darstellung von Zahlen; Verbindungen von Zahlen; Abkürzungen von mathematischen Operationswörtern; Abkürzungen algebraischer Operationswörter; Verdeutlichung und Vereinfachung des Zahlbegriffs; Abkürzung von mathematischen Operationen.

2. ZUR ENTWICKLUNG DER ARITHMETIK UND IHRER METHODEN

Die Impulse, welche auf die Mathematik damals einwirkten, kamen von fremden Völkern.

2.1.

Das erste Zahlensystem mit der Grundzahl 10, das prinzipmäßig wie unser Positionssystem aufgebaut und als Vorläufer unserer Zahlschrift anzusehen ist, entstand in Nordindien. Im 9. Jahrhundert dürfte die indische Arithmetik, die auf den dortigen neuen Zahlzeichen aufbaute, bei den Westarabern Eingang gefunden haben. In Spanien wurden die Gobar-Ziffern verwendet, namentlich hergeleitet vom indischen Verfahren, Zeichnungen oder Rechnungen im Sand durchzuführen. Diese Zeichen wurden den christlichen Völkern Mitteleuropas bekannt. Von 976 datiert

I	ʓ	ʒ	ɣ	Ʋ	Ь	7	8	9	
I	ʊ	ʓ	⌐ᶜ	ꟼ	Ƅ	⋀	8	9	⊙
I	♀	⊠	ᴄᴄ	Ꙋ	Λ	⤳	8	ꟼ	
ı	ꟼ	ſ	℞	ꓯ	6	7	8	9	0
ι	ꜱ	ʓ	℞	5	Ɠ	ꝯ	8	9	0
ι	Ꙁ	3	ᵘꟼ	ꓯ	Ɠ	Ʋʃ	8	9	0
I	2	ꝯ	ꟼ	ꓯ	6	7	ꝯ	9	0
1	ι	ꜱ	ℓ	ɣ	Ɛ	⋀	8	9	0
1	2	ꜱ	4	5	6	7	ꝸ	9	0
I	℥	ꜱ	℞4	54	6	⋀	8	9	0
I	2	3	4	5	6	7	8	9	0

Abb. 1. Entwicklung der indisch-arabischen Ziffern in Europa zwischen 976 und Ende des 15. Jahrhunderts (entnommen aus [Tropfke 1980]).

die älteste Darstellung im Abendland ohne die Null, den Gobar-Ziffern sehr ähnlich (Abb. 1, Zeile 1). Vom 6. bis zum 10. Jahrhundert ging es in der Arithmetik großenteils darum, die kirchliche Festtagsrechnung zu bewältigen. Hierbei scheint man sich vorwiegend der Fingerzahlen bedient zu haben, die von Beda Venerabilis (672/673–735) erstmals schriftlich überliefert wurden. An die 1000 Jahre lang findet man ab dann diese Methode beschrieben, wo mittels der 28 Fingergelenke und verschiedener Handstellungen am Körper Zahlwerte ausgedrückt wurden (Abb. 2). In der 2. Hälfte des 10. Jahrhunderts läßt sich ein neuer Abacus nachweisen —wohl eine hölzerne Tafel mit einer Anzahl von senkrecht zum Rechner hin verlaufenden Spalten von rechts nach links steigender Zehnerpotenzen —, auf dem mit markierten Rechensteinen gearbeitet wird: zuerst mit griechischen Buchstabenziffern, am Ende des 11. Jahrhunderts nur mehr mit Gobar-Ziffern. Statt mehrerer unbezifferter *calculi* früherer Systeme

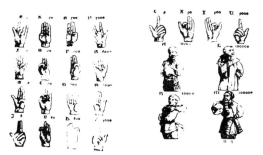

Abb. 2. Fingerzahlen des Beda bei Turmair, Johannes (Aventinus): *Abacus atque vetustissima... consuetudo.* Regensburg 1532.

setzte man nun einen einzigen "gemerkten"—also bezifferten—Rechen-
stein in eine Spalte, Leerstellen blieben frei (Abb. 1, Zeile 2). Auf diese
Weise wurden die indisch-arabischen Ziffern durch sogenannte *apices*
oder *caracteres* ohne die Null im 11. und 12. Jahrhundert im Abendland
eingeführt. Hiermit beherrschte man die vier Grundrechenarten.

Über Spanien war man um 1150 auch mit Muḥammad ibn Mūsā
al-Ḫwārizmīs (783?–850?) "Rechenbuch" vertraut geworden, in dem er das
Rechnen mit indischen Ziffern bis zum Quadratwurzelziehen und den
Gebrauch des Stellenwertsystems lehrt. Leonardo von Pisa (ca.
1170–1250?) schuf 1202 mit seinem "Liber abbaci" ein mathematisches
Sammelwerk, in dem erstmals indisch-arabisches Rechnen für den
kaufmännischen Gebrauch auftritt.

Bald folgten Kommentare, in denen aus dem Namen al-Ḫwārizmī die
Begriffe Algorismus oder Algorithmus hervorgingen. Die Null bürgerte
sich im Abendland ab dem 12. Jahrhundert ein. In den Algorithmen
behandelte man etwa 250 Jahre lang das Rechnen mit den neuen indisch-
arabischen Zahlzeichen, anfangs freilich völlig mißverstanden, weil man
römisch verzifferte. Das indische *śūnya* (leer) wurde wörtlich arabisch als
ṣifr übersetzt, dann lateinisch *cephirum*, schließlich *zefiro, zero, chiffre*,
und wurde letztlich zur Bezeichnung der Zahlzeichen allgemein. Im Laufe
des 15. Jahrhunderts setzten sich die neuen Ziffern in Multiplikations-
tafeln, Kalendern und großen Tabellenwerken zur Trigonometrie
schließlich durch, später begünstigt durch Buchdruck und Papierherstel-
lung, und erlangten ihr heutiges Aussehen (Abb. 1). Die Methoden des
täglichen Rechnens wurden jetzt übersichtlich im Positionssystem in der
heute üblichen schriftlichen Form als "Rechnen mit der Feder" gelehrt;
dies ist nicht als Folge verstärkter menschlicher Intelligenz zu werten,
sondern es geschah dank gezielter Symbolisierung.

2.2.

Die Einführung arithmetischer Operationszeichen verlief parallel. Addieren drückte man durch Nebeneinanderstellen der Summanden aus, das
Minuszeichen wird aus schnell geschriebenem *minus* hervorgegangen sein,
bei Übergewicht durchkreuzte man diesen langen waagrechten Strich und
hatte das Pluszeichen. Ab etwa 1480 waren diese beiden Symbole normiert.
Bezeichnungen für Multiplikation wurden ausgeschrieben, gewöhnlich *in*
oder *per*, der Bruchstrich erscheint ab dem 13. Jahrhundert. Die Brüche
wurden auch weiterhin nach dem in fortlaufende Zwölftel unterteilten

römischen Pfund, dem *as librarius*, bezeichnet, so daß etwa mit *uncia* nicht nur $\frac{1}{12}$ Pfund, sondern allgemein auch $\frac{1}{12}$ dargestellt wurde. Eine Vermengung von Dezimal- und Sexagesimalsystem findet sich im Radius 60.000, 6.000.000, 100.000, 10.000.000 bei trigonometrischen Rechnungen in den Sinustafeln Johannes Regiomontans (1436–1476).

2.3.

Das Mittelalter wurde der Zeitabschnitt, in dem in unserem Lebensraum erstmals Maß-, Gewichts- und Münzeinheiten, auch in ihren Umrechnungsbeziehungen, schriftlich überliefert sind. Die römische *libra* (327,45 g) zu zwölf Unzen wurde zur Grundlage des mittelalterlichen Gewichts- und Münzsystems. Karl d. Gr. stellte das Münzwesen auf Silberwährung um. Pfenning (Pfennig, Denar) wurde in weiten Teilen Europas zur Bezeichnung der silbernen Münzeinheit, so daß 1 Pfund (*pondus Caroli*, zwischen 367 und 491 g) = 20 Schilling = 240 Pfennig; Pfund und Schilling waren keine Münzen, sondern nur Rechnungsgrößen. Der mittelalterliche *solidus* (Schilling) wurde im 12. Jahrhundert in Italien als *grosso* geprägt, in der 2. Hälfte des 13. Jahrhunderts in Frankreich als Groschen zu 12 Pfennig; es gab auch andere Stückelungen. Die Münzbezeichnung richtete sich oft nach dem ersten Prägungsort. Für gut zwei Jahrhunderte waren Groschen die größten Silbernominale, bei großen Beträgen wurde oft in Schock Groschen verhandelt. Im 13. Jahrhundert verlangte der Handel größeres Zahlgeld in Gold; in Florenz erschien 1252 der *fiorino* (*florenus*, Gulden), in Venedig 1284 der Dukaten. In der Mitte des 12. Jahrhunderts wurde die kölnisch-erfurtische Mark (zwischen 233 und 237 g) maßgeblich für den europäischen Handel. Bedingt durch vielerlei Privilegien, entwickelte sich besonders im deutschsprachigen Raum eine unüberblickbare Vielfalt an Münz-, Maß- und Gewichtseinheiten mit meist dualer, dezimaler oder duodezimaler Unterteilung, wobei man oft auf frühere natürliche Etalons zurückgriff, deren Kürzel zum Teil heute noch verständlich sind: Fuß (Schuh), Elle, Klafter (6 oder 10 Fuß), Meile (Doppelstundenweg); Morgen, Tagwerk (Joch); Faß, Saum, Fuder, Metze, Scheffel, Eimer, Kanne, Maß, Ster; Pfund, Schilling, Groschen, Plappart, Pfennig, Heller; Groß, Schock, Dutzend, Paar usw. Auch die Buchführung wurde, von Italien ausgehend, von den Methoden, die das Gedächtnis nicht beanspruchten (Kerbholz, Rechensteine), zu den später üblichen entwickelt, wobei 1521 erstmals deutsch gedruckt die Terme *Zornal* (Journal), *Kaps* (Kassabuch des in einer Kapsel verwahrten Bargeldes), Schuldbuch erscheinen.

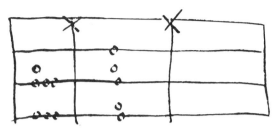

Abb. 3. Duplation von 83 ergibt 166. Handschrift, um 1500 (Staats- und Stadtbibliothek Augsburg, 8° Cod 1).

2.4.

Eine dem Abacus ähnliche Methode auf Rechentischen mit parallel zum Rechner verlaufenden Linien tritt schriftlich ab dem 15. Jahrhundert in Deutschland, Frankreich und England auf, die Tausenderlinie mit einem *x* markiert; unbezifferte Rechensteine stellen auf den Linien Einer, Zehner, Hunderter usw. dar, in den Zwischenräumen Fünfer, Fünfziger usw. Fünf Einer ergeben beim Linienrechnen einen Fünfer, zwei Fünfer einen Zehner usw. (Abb. 3).

3. ZUR ENTWICKLUNG DER ALGEBRA

Die abendländische Gleichungslehre geht auf al-Ḫwārizmīs *Al-Kitāb al-muḥtaṣar fī ḥisāb al-ǧabr wa-l-muqābalah* zurück mit den Formen $ax^2 = bx$; $ax^2 = c$; $bx = c$; $ax^2 + bx = c$; $ax^2 + c = bx$; $ax^2 = bx + c$. Über Spanien wurde dieses Werk durch Robert von Chester (um 1150) und Gerhard von Cremona (ca. 1114–1187) lateinisch vermittelt. Aus *al-ǧabr wa-l-muqābalah* (Ergänzen und Ausgleichen) wurde "Algebra". Gemäß muslimischer Tradition hieß die Unbekannte meist *radix* oder *res*, das Quadrat *census*, die Einheit *dragma* oder *numerus*. Jordanus Nemorarius (um 1220) arbeitete bereits mit allgemeinen Buchstaben *a*, *b* usw., seine komplizierte Methode setzte sich aber nicht durch. Um 1320 findet man in Italien in einer Bearbeitung der "Algebra" eine gezielte Symbolik für *dragma*, *radix* und *census*, nämlich *d*, *r* und *c* (Abb. 4). In Italien kennt man Manuskripte aus dem 14. und 15. Jahrhundert, wo die Terminologie in der algebraischen Gleichung 2. und höheren Grades samt Operationszeichen stark verkürzt erscheint, ehe Regiomontanus ab 1456 mit einer nahezu symbolischen Form der quadratischen Gleichung hervortritt (Abb. 5). In der "deutschen Coß" (*cosa = res = x*) zwischen etwa 1460 und 1550 wird der äußere

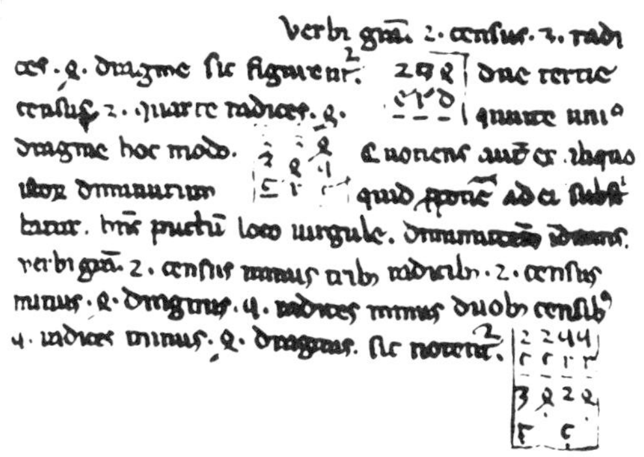

[handschriftlicher Text, Abb. 4]

Abb. 4. $2x^2 - 3x$, $2x^2 - 4$, $5x - 2x^2$, $5x - 4$, Beispiel unten. Handschrift, um 1320 (Bodleian Library Oxford, Lyell 52).

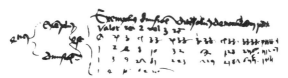

Abb. 5. $250x - 25x^2 = 2x^2 + 100 - 20x$. Handschrift von Regiomontanus, 1463 (Stadtbibliothek Nürnberg, Cent 5 app 56c).

Rechengang in eine immer knappere Form gebracht, und das Rechnen mit Potenzen wird als gesonderte Disziplin behandelt (Abb. 6). Hier wurde die Proportion $x^0 : x^1 = x^1 : x^2 = x^2 : x^3$ usw. in den Vordergrund der Überlegungen gestellt. Daneben bezog man sich in der Namensgebung lange auf die Anschaulichkeit; Michael Stifel (1487?–1567) etwa versuchte *radix* ᴦ, *census* ᴣ, *cubus* ᴄ, *census de censu* ᴣᴣ usw. für x, x^2, x^3, x^4 usw. geometrisch über den Grad drei hinaus zu deuten. Waren mehrere Unbekannte nötig, dann nannte er sie $1ᴦ$, $1A$, $1B$,...,$1F$.

In der Algebra traten auch irrationale Zahlwerte auf. Eine Formulierung von 1461 "Gib mir ain censum vnd zuech dar von sin wurcz, vnd von dem, daz vber belyb an dem censu, zuech och vß dye wurcz. Dye czwo wurcz tue zesamen, daz 2 zal dar auß werden" für $x + \sqrt{x^2 - x} = 2$ hätte man um 1550 wahrscheinlich geschrieben als "ᴦ $+ \sqrt{ᴣ \cdot ᴣ - ᴦ} \cdot facit$ 2". Mit

[handschriftlicher Text, Abb. 6]

Abb. 6. Cossische Schreibweise für Potenzen x^0, x, x^2 bis x^9. Handschrift, um 1486 (Universitätsbibliothek Leipzig, Ms. 1470).

Logarithmenrechnen waren um 1600 Grundrechenarten und algebraische Begriffe weitgehend symbolisiert erstmals im Besitz einer Epoche.

4. ÜBER DEN EINFLUSS DER GEOMETRIE

Anscheinend wurden innerhalb des angesprochenen Zeitraumes die bei uns ab dem 9. Jahrhundert nachweisbaren Geometrien nicht in den praktischen Ablauf des alltäglichen Geschehens einbezogen, denn bis um 1500 wurden diese Texte fast unverändert für den Unterricht weitergegeben. So wie die Proportion ein Kennzeichen der mittelalterlichen Betrachtungsweise in der Arithmetik war und im 15./16. Jahrhundert die Gleichungslehre mit fundierte, entsprechend wurde der goldene Schnitt das ideale Maßverhältnis der Geometrie, schließlich die ästhetische Norm der bildenden Künste. Mit der Erkenntnis des Fluchtpunkts in der Perspektive anstatt mehrerer Augpunkte und Horizonte wurden im 15. Jahrhundert neue zeichnerische Maßstäbe gesetzt, zum anderen wurde die Darstellung im Grund- und Aufriß hervorgehoben. In die mittelalterlichen Dombauhütten hatte die Geometrie insofern Einzug gehalten, als man dort geübt war in der Konstruktion verschiedenartigster geometrischer Figuren, die man in die Praxis umsetzte. Rechnerische Statik betrieb man nicht, man stützte sich auf Erfahrungen.

5. ZUR WECHSELBEZIEHUNG ZWISCHEN MATHEMATIK, PHILOSOPHIE UND PHYSIK

Durch die unmittelbare Beschäftigung mit den übermittelten antiken Texten, meist Euklid, Aristoteles (384–322) und Archimedes (287?–212), wurden ab dem 12. Jahrhundert mathematisch-philosophische und physikalische Fragen angerührt, die diskutiert und kommentiert, aber nicht gelöst wurden: Einheit, Wesen der Zahlen, Teil-Ganzes-Problem, Stetigkeit, Kontinuum, Funktion, Endliches, Unendliches, Parallelenpostulat, Verhältnis Geschwindigkeit zu (Kraft)/(Widerstand).

Vorläufig herrschte die Doktrin vom nicht atomistischen Aufbau des Kontinuums vor. Mittels einer Arithmetisierung des Kontinuums versuchte man, Arithmetik und Geometrie, die Prototypen für diskrete und kontinuierliche mathematische Betrachtungsweise, unter gemeinsamem Aspekt zu erfassen. Für Robert Grosseteste (ca. 1168–1253), der alle Ursachen natürlicher Wirkungen auf solche Fakten zurückzuführen ver-

suchte, die durch Linien, Winkel und Figuren darzustellen sind, war die
Länge einer Strecke ein Maß für die Anzahl ihrer unendlich vielen
Punkte. So gab es verschiedenartige unendliche Zahlen, die in ihrer Art
aber nur Gott, nicht der Vorstellung der Menschen zugänglich waren.
Prinzipiell wurden die Darstellungen zum Kontinuum ohne Symbole behandelt, also verbal. Bezeichnend für die mathematische Auffassung seit
dem Mittelalter ist der Versuch, Unendliches durch Endliches darzustellen.
Roger Bacon (ca. 1219–ca. 1292) argumentierte so, daß die Welt nur
endlich groß ist. Thomas Bradwardinus (ca. 1290/1300–1349) setzte voraus, daß die Geschwindigkeit nach einer arithmetischen Folge wächst,
wenn (Kraft)/(Widerstand) gemäß einer geometrischen Folge zunimmt,
also

$$\cdots \quad 3v \quad 2v \quad v \quad \frac{v}{2}$$

$$\cdots \quad (F/R)^3 \quad (F/R)^2 \quad F/R \quad (F/R)^{1/2}$$

einander gegenüberstehen. Nikolaus Oresme (ca. 1320–1382) ließ wei-
tergehend als Bradwardinus Brüche und gar irrationale Zahlverhältnisse
zu. Oresmes Fortschritt liegt vor allem in der Schreibart $1/2\ 2^p$ für $2^{1/2}$;
$1/3\ 9^p$ für $9^{1/3}$; $1/3\ 6^p\ 3/4$ für $(\frac{27}{4})^{1/3}$ usw. begründet, durch die die
Bruchpotenzen in die Mathematik eingeführt wurden; p ist *proportio*.
Hierdurch erfolgte sowohl ein großer Schritt in Richtung Algebraisierung
des Rechenganges und Vorbereitung der Logarithmen, als auch im Rechnen mit irrationalen Zahlen (Abb. 7). Die Gegenüberstellung rational-irra-
tional erscheint im lateinischen Mittelalter als *rationabilis, audibilis* einerseits und *irrationabilis, surdus, alogus* andererseits.
    Der Kontingenzwinkel an der Berührstelle zwischen Tangente und Kreis
nach Euklid III,16 wurde nach Campanus von Novara (gest. 1296) zwar
kleiner als jeder spitze Winkel, aber nicht gleich Null. So verneinte er die
Frage, ob man aus der Existenz kleinerer und größerer Werte hinsichtlich
einer bestimmten Größe auf die Existenz einer zu ihr gleichen schließen
darf. Die Quadratur des Kreises, aus der sich, vom Dreieck ausgehend, das
Studium isoperimetrischer Figuren weiterentwickelte, wurde ebenfalls

Abb. 7. $(2 \cdot \sqrt[3]{10}):(3 \cdot \sqrt[3]{5})$. Handschrift, um 1500 (Österreichische Nationalbiblio-
thek Wien, Ms. 5277).

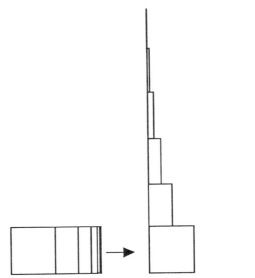

Abb. 8. Geometrische Zerlegung von 2 in $1 + \frac{1}{2} + \frac{1}{4} + \frac{1}{8} + \cdots$.

unter dem Gesichtspunkt des Zwischenwertsatzes—es gibt ein umbeschriebenes und ein einbeschriebenes Quadrat—behandelt.

Der wirkliche Fortschritt in der Mathematik des angesprochenen Zeitraums lag also in der Arbeit der Fachleute begründet, die sich um die Systematisierung der einzelnen Stoffgebiete Arithmetik, Algebra, Geometrie, Trigonometrie bemühten. Hierher gehört der Versuch, das Kontinuum aus nicht gleichartigen Indivisibilien aufzubauen, nämlich aus zueinander proportionalen Teilen. Oresme ordnete z.B. zwei Quadrate durch jeweiliges Halbieren des rechten Rechtecks und Auftürmen dieser Teile so an, daß eine endliche Fläche aus unendlich vielen Rechtecken entstand (Abb. 8). Hierdurch gelangte er geometrisch zur Summation unendlicher geometrischer Reihen.

Anschaulicher als etwa im "Liber calculationum" von Richard Swineshead (wirkte um 1340–1355), behandelte Oresme im "Tractatus de configurationibus qualitatum et motuum" gleichförmige und ungleichförmige Veränderungen der Intensität von Qualitäten. Mittels geometrischer Figuren, der Formlatituden, versuchte er, zeitliche Veränderungen einer physikalischen oder sonstigen Größe (Qualität) anzugeben. Seine *configurationes* nennt er *uniformis, uniformiter difformis incipiens a non gradu, uniformiter difformis incipiens a gradu*, ansonsten *difformiter difformis* auf vielerlei Arten (Abb. 9).

Abb. 9. Darstellungen Oresmes: *uniformis, uniformiter difformis, difformiter difformis*.

"Unitas est principium numeri et non est numerus", woraus wohl ein erheblicher Teil der Zahlenmystik des besprochenen Zeitabschnitts herzuleiten ist, zieht sich als Faden durch die mathematische Entwicklung der angesprochenen Zeit. So waren vom Erkennen her die Einheit und das Unendliche durch das ganze Mittelalter hindurch außermathematisch geblieben.

In der Renaissance änderte sich die Einstellung gegenüber den Werken der Antike; Regiomontanus z.B. sagte, daß—im Gegensatz zu Aristoteles—Geradliniges und Krummliniges gleichartige Gebilde, also kommensurabel sind, d.h. rektifiziert werden können.

LITERATURVERZEICHNIS

Bergmann, W. 1985. *Innovationen im Quadrivium des 10. und 11. Jahrhunderts.* Stuttgart: Franz Steiner. (Sudhoffs Archiv, Beiheft 26.)

Breidert, W. 1983. Zum maßtheoretischen Zusammenhang zwischen *indivisibile* und Kontinuum. In *Mensura. Maß, Zahl, Zahlensymbolik im Mittelalter* (Miscellanea Mediaevalia 16/1), 145–152. Berlin und New York: de Gruyter.

Bubnov, N. 1899. *Gerberti Opera mathematica.* Berlin: R. Friedländer & Sohn. Reprint Hildesheim: Olms, 1963.

Busard, H. L. L. 1971. Die Quellen von Nicole Oresme. *Janus* **58**, 161–193.

Clagett, M. 1968. *Nicole Oresme and the Medieval Geometry of Qualities and Motions.* Madison: University of Wisconsin Press.

Dictionary of Scientific Biography. 1970–1980. C. Gillespie, Ed. New York: Scribner's. 16 Bde.

Ernst, U. 1983. Kontinuität und Transformation der mittelalterlichen Zahlensymbolik in der Renaissance. *Euphorion* **77**, 247–325.

Folkerts, M. 1970. *"Boethius" Geometrie II. Ein mathematisches Lehrbuch des Mittelalters.* Wiesbaden: Franz Steiner.

Franci, R., und Toti Rigatelli, L. 1985. Towards a History of Algebra from Leonardo of Pisa to Luca Pacioli. *Janus* **72**, 17–82.

Gericke, H. 1980. Wie dachten und denken die Mathematiker über das Unendliche? *Sudhoffs Archiv* **64**, 207–225.

Hughes, B. 1981. *Jordanus de Nemore*: *De numeris datis*. Berkeley: University of California Press.

Kahnt, H., und Knorr, B. 1987. *Alte Maße, Münzen und Gewichte*. Mannheim, Wien und Zürich: Bibliographisches Institut.

Luschin von Ebengreuth, A. 1926. *Allgemeine Münzkunde und Geldgeschichte des Mittelalters und der Neueren Zeit*, 2. Aufl. München und Berlin: Oldenbourg.

Menninger, K. 1979. *Zahlwort und Ziffer*, 3. Aufl. Göttingen: Vandenhoeck & Ruprecht.

Molland, A. G. 1983. Continuity and measure in medieval natural philosophy. In *Mensura. Maß, Zahl, Zahlensymbolik im Mittelalter* (Miscellanea Mediaevalia 16/1), 132–144. Berlin und New York: de Gruyter.

von Naredi-Rainer, P. 1984. *Architektur und Harmonie. Zahl, Maß und Proportion in der abendländischen Baukunst*, 2. Aufl. Köln: DuMont.

Tropfke, J. 1980. *Geschichte der Elementarmathematik*. Bd. 1, 4. Aufl. Berlin und New York: de Gruyter.

Vogel, K. 1963. *Mohammed ibn Musa Alchwarizmi's Algorismus*. Aalen: Zeller.

Die Rolle Arnold Sommerfelds bei der Diskussion um die Vektorrechnung, dargestellt anhand der Quellen im Nachlaß des Mathematikers Rudolf Mehmke

Karin Reich

Fachhochschule für Bibliothekswesen, Feuerbacher Heide 38-42, D-70192 Stuttgart, Germany

Felix Klein induced Arnold Sommerfeld to take over the editorship of volume V, "Physics," of the *Encyklopädie der mathematischen Wissenschaften*. In 1901 Sommerfeld sent a directive of 8 pages to all contributors of this volume. In his note he proposed a special way of denoting vectors, vector calculus and the electromagnetic magnitudes which became obligatory for all contributors. In September 1903 Klein founded a so-called "vector commission" consisting of Sommerfeld, Ludwig Prandtl, and Rudolf Mehmke. Its aim was to create a unified vector symbolism and calculus. Contrary to Sommerfeld and Prandtl, Mehmke supported Grassmann's terminology, which he thought was the very best. A major point in the conflict was the term "bivector." According to Mehmke this was absolutely necessary, while Sommerfeld did not need it for his interpretation of electromagnetism. Sommerfeld, Mehmke, and Prandtl wanted to convince each other and, in reality, did not look for a compromise. As a consequence, the commission disbanded without reaching a final result.

Sommerfelds Beitrag zur Diskussion um die Vektorrechnung war bislang kein Thema in der Wissenschaftsgeschichte. So fehlt der Name Sommerfeld in Crowes "A History of Vector Analysis." Und in Werken über Sommerfeld war wiederum seine Bedeutung für die Konsolidierung der verschiedenen Richtungen innerhalb der Vektorrechnung und -analysis kein Gegenstand der Darstellung. So fehlt dieses Thema beim Sommerfeld-Biographen Ulrich Benz [1974, 1975] wie auch in der Dokumentation aus dem Sommerfeldnachlaß [Eckert u.a. 1984]. Und auch der Name Mehmke taucht in der genannten Literatur erst gar nicht auf.

HISTORY OF MATHEMATICS: STATES OF THE ART

319

Bis Dezember 1992 waren nur einige, wenige Dokumente des Mehmke-Nachlasses bekannt. Doch es stellte sich heraus, daß im Mathematischen Institut B der Universität Stuttgart, Mehmkes früherer Wirkungsstätte, noch zwei Kisten mit Dokumenten verborgen waren. Der Nachlaß ist sehr vielfältig, bislang listet die Universitätsbibliothek, d.h. das dort angesiedelte Universitätsarchiv, vier Bereiche auf: I. Werkmanuskripte (48 Nummern), II. Korrespondenzen (104 Nr.), III. Lebensdokumente (32 Nr.), IV. Sammlungen (6 Nr.). Aber es gibt noch eine Reihe von Dokumenten, die bislang noch nicht eingeordnet wurden. Besondere Bedeutung kommt der Korrespondenz zu, sie enthält sowohl Briefe an Mehmke als auch Briefe von Mehmke an seine Kollegen. Diese Briefe sind meistens nur als Konzepte vorhanden, die fast ausschließlich in Kurzschrift abgefaßt sind. Nur dank der finanziellen Hilfe von Herrn Menso Folkerts gelang es, einen Teil der Briefe umschreiben zu lassen. Herr Hans Gebhardt in Eckersdorf (bei Bayreuth), Fachmann in allen Kurzschriften, sorgte für eine sorgfältige und korrekte Umschrift der Dokumente. Er verwandte sicherlich mehr Zeit darauf, als ihm bezahlt wurde. Sowohl Herrn Folkerts als auch Herrn Gebhardt sei herzlichst gedankt. Erst diese Hilfe machte die folgenden Ausführungen möglich.

1. DIE ENCYKLOPÄDIE DER MATHEMATISCHEN WISSENSCHAFTEN

Im September 1894 wurde auf der Naturforscherversammlung in Wien ein erster Beschluß gefaßt, eine *Encyklopädie der mathematischen Wissenschaften* zu erarbeiten. Hinter diesem Mammutunternehmen stand Felix Klein (1849–1925), dessen Assistent in demselben Jahr Arnold Sommerfeld (1868–1951) wurde: Sommerfeld gab im Herbst 1894 seine Göttinger Assistentenstelle bei Theodor Liebisch (1842–1922) auf und wechselte ins mathematische Lesezimmer zu Felix Klein [Eckert u.a. 1984, 17]. Klein wollte Sommerfeld zunächst mit der Redaktion der "Zeitschrift für Mathematik und Physik" betrauen, aber Sommerfeld lehnte ab [Tobies 1986–1987, II, 38]. Am 11.3.1895 wurde Sommerfeld Privatdozent, 1897 konnte er Göttingen mit Clausthal vertauschen, wo er mit Hilfe einer Empfehlung von Klein Professor der Mathematik an der Bergakademie wurde. 1898 übertrug Klein Sommerfeld die Redaktion des Bandes V der *Encyklopädie der mathematischen Wissenschaften*: "Physik" [von Meyenn 1993, 245f.]. Zunächst hoffte Sommerfeld, diese Bürde an Willy Wien (1864–1928) abgeben zu können, doch dieser lehnte ab [Eckert u.a. 1984,

26f.]. So blieb Sommerfeld nichts anderes übrig, als die "Physik" selbst zu redigieren; das Werk erschien zwischen 1903 und 1926 in drei Teilbänden. Auf der Suche nach qualifizierten Mitarbeitern reisten Klein und Sommerfeld 1898 nach Holland und 1899 nach England. Der Erfolg dieser Reisen war, daß unter anderen Hendrik Antoon Lorentz (1853–1928) (Leiden) für die Maxwellsche Theorie und George Hartley Bryan (1864–1928) (Bangor) für die Thermodynamik gewonnen werden konnten [Sommerfeld 1903–1921, V]; siehe auch [Lorentz 1903a, b, 1909, Bryan 1903]. Gleichzeitig übernahm Klein zusammen mit Conrad Müller (1878–1953) die Redaktion des Bandes IV der *Encyklopädie*: "Mechanik", der zwischen 1901 und 1935 in vier Teilbänden herauskam.

Sowohl für die "Mechanik" als auch für die "Physik" war die Vektorrechnung und -analysis von Bedeutung. Das Problem war, daß bislang noch keine Vereinheitlichung der verschiedenen Tendenzen in der Vektorrechnung stattgefunden hatte. Klein und Müller lösten das Problem anders als Sommerfeld: Im Band IV erschienen gleich zwei Artikel über Vektorrechnung, einer von Max Abraham [1901] "Geometrische Grundbegriffe," abgeschlossen im Februar 1901, und einer von Heinrich Emil Timerding "Geometrische Grundlegung der Mechanik eines starren Körpers," abgeschlossen im Februar 1902, der mit einem Kapitel über "Geometrische Grundbegriffe" beginnt [Timerding 1902, 125–141]. Abraham (1875–1922) legte seine Darstellung möglichst breit an und schilderte alle möglichen Systeme. Als Literatur nannte er u.a. die einschlägigen Werke seiner Göttinger Kollegen August Föppl (1854–1924) und Woldemar Voigt (1850–1919), sowie J. W. Gibbs' "Elements of Vector Analysis." Timerding (1873–1945) dagegen war viel stärker von Hermann Graßmann (1809–1877) beeinflußt; er behandelte in Anlehnung an diesen Plangrößen und Linienteile, während Abraham wie Voigt zwischen polaren und axialen Vektoren unterschied.

Sommerfeld war im Gegensatz zu Klein und Müller von Anfang an um eine Vereinheitlichung der Bezeichnungsweise bemüht. Dies zeigt ein 8 Seiten umfassendes Papier im Nachlaß Mehmkes.

Sommerfeld und Rudolf Mehmke (1857–1944) hatten sich spätestens im September 1900 anläßlich der Tagung der deutschen Naturforscher und Ärzte vom 16.–23.9. in Aachen kennengelernt. Daß Mehmke anwesend war, beweist ein Brief Mehmkes an Klein [Tobies 1986–1987 I, 23f., 31 Anm. 31]. Sommerfeld hatte im Januar 1900 seine Professur in Clausthal aufgegeben und war einem Ruf an die TH Aachen gefolgt. Klein benützte die Gelegenheit dieser Tagung, um über den Stand der Arbeiten an den Bänden IV und V der *Encyklopädie der mathematischen Wissenschaften* zu berichten; dabei erwähnte er insbesondere Sommerfeld als Herausgeber des Bandes V [Klein 1901, 163]. Bereits im Jahr 1901 veröffentlichte Sommerfeld eine erste Arbeit in der nun neuerdings von Mehmke und Carl Runge zusammen herausgegebenen "Zeitschrift für Mathematik und Physik" [Sommerfeld 1901], der noch viele weitere folgen sollten [Tobies 1986–1987, I, 32 Anm. 65]. Daß Sommerfeld Mehmke als Mathematiker schätzte, zeigt z.b. die Tatsache, daß er ihn als Nachfolger von Hans von Mangoldt in Betracht zog, denn am 20.6.1904 schrieb Mehmke an Sommerfeld: "Haben Sie herzlichen Dank für Ihr freundliches Schreiben vom 15. d.M. ... Wie ich wohl nicht zu versichern brauche, hat es mich mit großer Freude erfüllt, daß Sie und die Abteilung resp. Berufungskommission bei der Wiederbesetzung des Lehrstuhles, den Prof. Mangoldt verlassen, an mich gedacht haben, wenn ich mich auch umsonst frage, wie es kommt, daß Sie mich für würdig halten, diese Stellung einzunehmen, trotzdem ich auf den Herrn v. Mangoldt vertretenen Gebieten fast nichts veröffentlicht habe. Nun sitze ich zwar hier nicht so fest, daß ich mich nicht entschließen könnte, bei passender Gelegenheit an eine andere Hochschule zu gehen, aber im vorliegenden Fall wäre eine Verminderung meines Gehaltes zu befürchten, die es mir von vorneherein verbietet, der Sache näherzutreten, trotzdem es mich locken würde, mit Ihnen zusammen an derselben Hochschule wirken zu können. Wohl sind nach Ihren freundlichen Mitteilungen die durchschnittlichen Bezüge eines Professors bei Ihnen ungefähr dieselben wie hier, aber es ist mir schon bei meiner Berufung von Darmstadt nach Stuttgart eine bedeutende Personalzulage gewährt worden, die jetzt rund 2000 M beträgt." Der erwähnte Brief Sommerfelds vom 15.6. ist im Mehmke-Nachlaß nicht vorhanden. Hans von Mangoldt (1854–1925), an der Technische Hochschule Aachen tätig, wechselte 1904 nach Danzig.

1.1. Sommerfelds Papier: "Encyklopädie der mathematischen Wissenschaften, Bd. V, Physik"

Diese Schrift Sommerfelds stammt aus dem Jahre 1901. Sie beginnt mit einem Brief Sommerfelds: "Den Herren Bearbeitern der Artikel über Elektricität und Optik erlaube ich mir, Vorschläge für eine einheitliche Bezeichnung der elektromagnetischen Grössen vorzuschlagen. Dieselben sind bereits von den Bearbeitern der elektrischen und optischen Hauptartikel, den Herren H. A. Lorentz und W. Wien, gebilligt, bez. nach deren Wünschen abgeändert und erweitert worden". Sommerfelds Ausführungen betrafen nicht nur die Bezeichnungen der elektromagnetischen Größen, sondern auch die Darstellung der Vektoren und die Vektorrechnung und -analysis. Sein Papier umfaßt 12 Grundsätze und 4 Fragen, die er zu beantworten bat. Sommerfeld bemerkte anfangs, daß "eine Einigung über Zeichenwahl und Namengebung nur möglich ⟨sei⟩, wenn jeder an seinem Theile zu Compromissen bereit ist und nöthigenfalls die eigene Gewöhnung und den persönlichen Geschmack hintansetzt". Im folgenden sind nur die die Vektorbezeichnung und -rechnung betreffenden Punkte ausgeführt:

1. Grundsatz: Bedeutung der Vereinheitlichung. "Eine einheitliche Bezeichnung der den elektrischen und optischen Theorien gemeinsamen Begriffe ist für die einschlägigen Theile der mathematischen Encyklopädie unerlässlich, sowohl im Interesse der leichten Lesbarkeit, wie zur Vermeidung der durch Definition neuer Symbole entstehenden Längen ... Es ist nicht möglich und wünschenswerth, die Bezeichnungen der Autoren, über die referiert wird, beizubehalten ... ".

2. Die Bezeichnung der Vektoren. "Eine einheitliche Bezeichnung wäre aus Mangel an Symbolen unmöglich, wenn man mit dem gewöhnlichen (lateinischen) Alphabet auskommen oder jede Componente einer gerichteten Grösse mit einem eigenen Buchstaben bedenken wollte. Deshalb wird vorgeschlagen, die gerichteten Grössen durch gothische Buchstaben und ihre Componenten durch Indices zu unterscheiden. Auf diese Weise wird das Wesentliche, der Begriff des Vectors, in der Bezeichnung hervorgekehrt, und das Unwesentliche nämlich die zufällige Wahl des Coordinatensystems, gebührend zurückgedrängt, bez. wo es auf den Vector selbst ankommt, überhaupt nicht zum Ausdruck gebracht. Man schreibe also $\mathfrak{E}$ (elektr. Kraft), $\mathfrak{E}_x, \mathfrak{E}_y, \mathfrak{E}_z, \mathfrak{E}_r, \mathfrak{E}_\varphi$ (Componenten nach rechtwinkligen oder Polar-Coordinaten etc), allgemein $\mathfrak{E}_l, \mathfrak{E}_s, \mathfrak{E}_n$ (Componenten nach irgend welchen Richtungen l, s, n). Dieser Grundsatz entspricht der eigentlichen Meinung Maxwell's besser wie der von ihm selbst gebrauchten, recht regellosen Bezeichnungen der Componenten $P, G, R, f, g, h, a, b, c$ etc.

Die Punkte 3. bis 8. betrafen die Bezeichnungen bzw. Definitionen der im Elektromagnetismus wichtigen physikalischen Größen. Im Grundsatz 9 wird das rechtwinklige Koordinatensystem mittels einer Rechtsschraube festgelegt.

10. Dieser Grundsatz betrifft die Vektoranalysis: "Für die unter dem Worte "curl" bekannte Vectoroperation wird die Be-

zeichnung "rot" (sprich Rotation) vorgeschlagen... Der Gleichförmigkeit wegen wolle man ferner div (nicht Div) schreiben. Statt des Vectorsymbols $\nabla\varphi = i\,\partial\varphi/\partial x + j\partial\varphi/\partial y + k\partial\varphi/\partial z$ wolle man nach dem sehr glücklichen Vorschlage von H. Weber die Bezeichnung grad (sprich "Gradient" oder Anstieg) benutzen. Das Zeichen $\Delta\varphi$ wird man für den sog. 2ten Differentialparameter (in rechtwink. Coordinaten $= \partial^2\varphi/\partial x^2 + \partial^2\varphi/\partial y^2$ bez. $= \partial^2\varphi/\partial x^2 + \partial^2\varphi/\partial y^2 + \partial^2\varphi/\partial z^2$) reservieren. Der sog. 1te Differentialparameter (in rechtw. Coordinaten $= \sqrt{(\partial\varphi/\partial x)^2 + (\partial\varphi/\partial y)^2}$ bez. $\sqrt{(\partial\varphi/\partial x)^2 + (\partial\varphi/\partial y)^2 + (\partial\varphi/\partial z)^2}$ kann eventuell zum Unterschiede $\Delta_1\varphi$ genannt werden.... ".

Im 11. Grundsatz legte Sommerfeld die Produktbildungen wie folgt fest: "Das Vectorielle Product zweier Vectoren $\mathfrak{A}$ und $\mathfrak{B}$ werde mit $[\mathfrak{A}\mathfrak{B}]$, das skalare mit $(\mathfrak{A}\mathfrak{B})$ bezeichnet...".

Der Betrag eines Vektors und die von Hamilton dafür eingeführte Bezeichnung Tensor sind das Thema von Punkt 12:

Handelt es sich nicht um die Richtung sondern nur um die Grösse eines Vectors, so nehme man den zugehörigen lateinischen Buchstaben oder das Zeichen $|\,|$ des absoluten Betrages. Wo keine Verwechslung zu befürchten kann dies aber auch fortbleiben... Die Bezeichnung "Tensor" für die Grösse eines Vectors bitte ich, obwohl sie ziemlich verbreitet ist und von Hamilton herrührt, zu vermeiden, da in der Encyclopädie (Art. Abraham) dies Wort für eine andere Art gerichteter Grössen, die in der Elasticität principiell sind, in Anspruch genommen ist. Ich schlage vor, wenn eine besondere Bezeichnung erforderlich scheint, "Grösse des Vectors" oder "absoluter Betrag" oder "Betrag".

Was die Fragen anbelangt, so steht die 4. Frage in Zusammenhang mit dem 10. Grundsatz:

Soll man an dem zwar sehr verbreiteten aber nicht sehr geschmackvollen Worte "curl" (Maxwell–Heaviside und viele andere) festhalten, oder statt dessen das Wort rot (H. A. Lorentz) einzuführen suchen. Es kommen ferner in Betracht Quirl (Wiechert, Ebert), vort (Vortex oder Wirbel, bis auf den Factor 1/2, Voigt), P (von E. Cohn als Ausdruck des Wortes Rotation benutzt.) Der Unterzeichnete war ursprünglich für die Bezeichnung curl; da aber die Herren H. A. Lorentz und W. Wien entschieden gegen curl und für rot sind, so empfiehlt der Unterzeichnete nun ebenfalls die allgemeine Annahme von "rot".

Sommerfelds Papier schließt mit einer Liste der für elektromagnetische Größen zu verwendenden Buchstaben.

Die vorliegende Schrift ist keine Reinschrift, sondern ein Konzept, vieles ist durchgestrichen oder wurde durch Randbemerkungen ergänzt. An welchen Personenkreis Sommerfeld diese verschickt hat, läßt sich nicht mehr nachvollziehen, sicher an Rudolf Mehmke und doch wohl an alle Mitarbeiter der Encyklopädiebandes "Physik."

Im Jahre 1903 veröffentlichte die Deutsche Physikalische Gesellschaft Vorschläge für eine einheitliche Bezeichnungsweise, an denen Sommerfeld nicht beteiligt war, die aber durch die Bezeichnungen der *Encyklopädie* angeregt worden war (Brief Sommerfelds an Prandtl und Mehmke vom 7.12.1904). Was die Vektoren anbetraf, so wurde hier lediglich deren Bezeichnungsweise durch sog. große deutsche Buchstaben und deren Komponenten durch Indizes $\mathfrak{H}_x, \mathfrak{H}_y, \mathfrak{H}_z$ erwähnt [Vorschläge 1903, 71]. Ein Jahr später veröffentlichte Sommerfeld seine Vorschläge unter dem Titel "Bezeichnung und Benennung der elektromagnetischen Grössen in der *Enzyklopädie der mathematischen Wissenschaften* V" (Sommerfeld 1904). Dieser Arbeit liegt sicherlich das Papier von 1901 zugrunde, sie ist aber mit diesem durchaus nicht identisch. So fehlen z.B. die Ausführungen über den Betrag von Vektoren, dafür erwähnte Sommerfeld noch den sogenannten und von vielen Mathematikern favorisierten Bivektor:

Man sage "Vektorprodukt" und nicht "äusseres Produkt," weil die letztere Grassmannsche Bezeichnung für den "Bivektor" zu reservieren ist, welche direkt das Parallelogramm aus $\mathfrak{A}$ und $\mathfrak{B}$ nach Inhalt und Stellung im Raume darstellt. Das Vektorprodukt ist im Grassmannschen Sinne die "Ergänzung" des äusseren Produktes. In der Mechanik ist eine Unterscheidung zwischen Bivektor und Vektor, wie sie durch die scharfe und folgerichtige Begriffsbildung der Grassmannschen Theorie vorgezeichnet ist, am Platze. Z.B. wird man das Moment einer Kraft um einen Aufpunkt als äusseres Produkt der Entfernung des Aufpunktes vom Angriffspunkt der Kraft und der Kraft selbst bezeichnen; von diesem äusseren Produkt nimmt man die Ergänzung und geht zum Vektorprodukte über, wenn es sich um die Zusammensetzung mehrerer Momente handelt. In der Elektrizitätslehre dagegen schien uns durch eine Unterscheidung zwischen Vektor und Bivektor, wie sie von Wiechert konsequent durchgeführt ist (Vektor und Rotor), die Kürze des Ausdrucks beeinträchtigt zu werden, da wir hier über den vektoriellen Charakter der Zustandsgrössen (polaren und axialen Charakter nach der Ausdrucksweise von Voigt) nicht genau unterrichtet sind.

In der Tat wurden im Band V der *Encyklopädie* Sommerfelds Vorschläge berücksichtigt. Besondere Erwähnung verdienen die Arbeiten von H. A. Lorentz, er gehörte ja mit Willy Wien zum Kreis derjenigen, mit denen Sommerfeld seine Vorschläge abgestimmt hatte. So beginnt Lorentz' erster Artikel "Maxwells elektromagnetische Theorie" mit einer Liste der

Bezeichnungen sowie einem Kapitel über "Vorbereitende Begriffe und Rechnungsmethoden" [Lorentz 1903a, 65–67, 71–78], in dem die Vektorrechnung und -analysis im Stile Sommerfelds eingeführt wird [1]. In diesem Sinne verfuhr Lorentz auch in seinen weiteren Encyklopädieartikeln [Lorentz 1903b und 1909].

Erst im Jahre 1910 griff Sommerfeld das Thema Vektorrechnung wieder auf. Diesmal ging es um deren vierdimensionale Gestaltung als Folge der speziellen Relativitätstheorie [Reich 1993b, 178–181].

2. DIE VEKTORKOMMISSION

Im Jahre 1901 wurde Sommerfeld Mitglied der Gesellschaft deutscher Naturforscher und Ärzte, ein Jahr später ständiges Mitglied. Bei der im folgenden Jahr in Kassel stattfindenden Tagung der Gesellschaft (20.9.–26.9.1903) trafen sich Sommerfeld und Mehmke abermals. Ludwig Prandtl (1875–1953) hielt am 24.9. einen Vortrag "Über eine einheitliche Bezeichnungsweise der Vektorenrechnung im technischen und physikalischen Unterricht", der im Jahresbericht der Deutschen Mathematiker-Vereinigung veröffentlicht wurde [Prandtl 1904a]. Prandtl begann seine Ausführungen, indem er den Nutzen einer Vereinheitlichung betonte und berichtete, daß er erst vor kurzem erfahren hatte, daß auch von seiten der Encyklopädie derartige Bestrebungen ausgingen. Prandtl favorisierte im folgenden ausdrücklich nicht die Schreib- und Bezeichnungsweise seines Lehrers August Föppl, sondern die von Josiah Willard Gibbs (1839–1903), nämlich: Vektoren: kleine fette Buchstaben,

a · b : inneres Produkt,
a × b : äußeres Produkt,
∇· : Divergenz,
∇× : Curl.

An der anschließenden Diskussion beteiligten sich Sommerfeld, Arthur von Oettingen (1836–1920), Georg Hamel (1877–1954), Mehmke, Ernst Zermelo (1871–1953), Klein, Heinrich Burkhardt (1861–1914), Ludwig Boltzmann (1844–1906), und Eugen Jahnke (1861–1921). Burkhardt hatte 1901 einen historischen Abriß über Vektoranalysis veröffentlicht, in dem er zu dem Schluß kam: "Es kann keine allumfassende, geometrische Symbolik geben, wie sie Grassmann und Hamilton sich dachten. Alles in Quaternionen zwängen zu wollen, ist zwecklos. Man erhält das für physikalische Zwecke geeignetste System, wenn man Grassmann's System nach der Seite der Infinitesimalrechnung hin ausbaut" [Burkhardt 1901, 52]. Auch Jahnke lehnte sich, wie seine Veröffentlichungen aus dem Jahre 1904 [Jahnke 1904a, 278; 1904b, 488ff.] und sein Lehrbuch [Jahnke 1905] zeigen, sehr stark an Graßmann an, ohne aber auf eigene Weiterentwicklungen zu verzichten.

Im Anschluß daran wurde gemäß einem Vorschlag von Felix Klein eine Vektorkommission gegründet, die aus Prandtl, Sommerfeld und Mehmke bestand. Diese Kommission hatte die Aufgabe der weiteren Erörterung der Angelegenheit und das Recht, weitere Mitglieder hinzuzuwählen [Verhandlungen 1904, 31, Prandtl 1904a, 40]. Das Ziel war eine Vereinheitlichung der verschiedenen Tendenzen innerhalb der Vektorrechnung.

Am 25.10.1903 schrieb Heinrich Weber (1842–1913) an Sommerfeld: "Aus den Zirkularen der D.M.V ersehe ich, daß in Cassel eine Kommission eingesetzt ⟨wurde⟩, die Vorschläge über eine konsequente Bezeichnung in der Vektorrechnung machen soll. Ich hätte gewünscht, daß ich auch in die Kommission gewählt worden wäre, da ich an dieser Frage doch sehr interessiert bin. Vielleicht genügt es aber, daß ich Ihnen, der Sie doch die Seele der Kommission sind, ohne diese Formalität meine Wünsche und Anschauungen auseinandersetze". Das tat Weber im folgenden, und er verwies dabei auch auf die unter seiner Ägide zustande gekommene Doktorarbeit von Richard Gans (1880–1954): "Über die Induktion in rotierenden Leitern" [Gans 1902], denn Gans' Bezeichnungsweise entspräche Webers Vorstellungen. Diesen Brief Webers sandte Sommerfeld an Mehmke und Prandtl mit folgendem Zusatz:

> Ich möchte vorschlagen, Herrn Prandtl zur Seele der Kommission zu ernennen, was tatsächlich der historischen Entwicklung entspricht. Sollen wir Herrn Weber kooptieren, auch Jahnke und Grassmann, wie vorgeschlagen wurde? Die Initiative müßte von der Kommissionsseele ausgehen. Wie soll die Kommission weiterarbeiten? A. Sommerfeld.

Prandtl unterbreitete am 30.10.1903 Mehmke konkrete Vorschläge:

> Wenn zunächst Herr Sommerfeld vorschlägt, mich zur Seele der Kommission 'zu ernennen', so hoffe ich, daß ich mich dieser Ehre werde würdig erweisen können. Allerdings könnte—einstweilen wenigstens—die Kommission richtiger von sich sagen: drei Seelen wohnen, ach, in meiner Brust! Möge es mit der Zeit besser werden. Bezüglich des Arbeitsplanes habe ich mir folgendes ausgedacht: Zunächst soll jeder von uns dreien in einem Bericht niederlegen 1.) was er für das zu findende Einheitssystem als unverrückbare Grundlage fordert, 2.) was er als wünschens- und erstrebenswert hält. Dann würden zunächst wir Drei unsere Ausarbeitungen vergleichen und zusehen,

was zu machen ist. Mit dem neuen Elaborat würden wir dann vor die erweiterte Kommission treten. Zur Bildung derselben würde ich vorschlagen, daß jeder von uns zwei Herren hinzuwählt, die von den andern beiden bestätigt werden, so daß sich eine neungliedrige Kommission ergibt (Herr Weber würde sich dabei wohl unter Herrn Sommerfelds Mannen finden). Von den neuen Mitgliedern würde dann jeder die Akten zu lesen bekommen und seine Meinung abgeben. Da auf Stimmeneinheit kaum zu hoffen ist, soll in einer Schlußabstimmung eine Zweidrittels-Mehrheit entscheiden. Wo keine solche zu erreichen ist, soll auch die Ansicht der Minderheit in dem Bericht an die Mathem. Vereinigung aufgeführt werden.

Umgehend informierte Mehmke Sommerfeld über Prandtls Arbeitsplan (Brief vom 12.11.1903) und schlug seinerseits Hermann Grassmann junior (1857–1922) (Halle) und Emil Müller (1861–1927) (Techniche Hochschule Wien) vor. Mehmke begründete seine Wahl wie folgt: "Herr F. Klein sagte schon in Kassel, daß die Grassmannsche Richtung in der Kommission stärker vertreten sein sollte und er nannte selbst den Namen Grassmann der Jüngere. Herrn Müller schlage ich deshalb vor, weil er vielleicht der beste Kenner der Grassmannschen Methode ist, wie seine zahlreichen Arbeiten auf diesem Gebiet beweisen". In der Tat war Mehmke schon früh ein überzeugter Anhänger Graßmanns [Reich 1993a, 265, 272–274].

Mit seinen Vorschlägen stieß Mehmke auf den entschiedenen Widerstand Sommerfelds, denn, so schrieb Prandtl am 23.12.1903 an Mehmke: "Herr Sommerfeld wünscht sich keine Grassmannleute". Prandtl schlug deshalb August Föppl (München) und Karl Heun (1859–1929) (Karlsruhe) für die erweiterte Kommission vor. Mehmke lehnte jedoch, wie aus einem Brief Prandtls an Mehmke vom 28.2.1904 hervorgeht, Heun ab, weshalb ihn Prandtl zurückzog. Ebenso erklärte sich Sommerfeld mit dem Vorschlag von Hermann Graßmann junior nicht einverstanden, weshalb Mehmke diesen durch Jakob Lüroth (1844–1910) ersetzte. Sommerfeld wollte Max Abraham in die Kommission gewählt wissen.

Schließlich wandte sich Sebastian Finsterwalder (1862–1951), der zunächst ein Anhänger der Gibbsschen Methoden war, an Sommerfeld, nachdem ihm dieser offensichtlich die Vektorrechnung betreffende Unterlagen geschickt hatte. Finsterwalder unterstrich die Notwendigkeit einer Vereinheitlichung und erläuterte seine eigenen Methoden, indem er auf einen seiner Aufsätze verwies. Wahrscheinlich meinte er entweder seinen Artikel "Eine Grundaufgabe der Photogrammetrie und ihre Anwendung auf Ballonaufnahmen" oder aber "Bemerkungen zur Analogie zwischen Aufgaben der Ausgleichsrechnung und solchen der Statik" [Finsterwalder

1903, 1906]; in beiden Arbeiten verwendete er den Vektorkalkül in Gibbsscher bzw. Wilsonscher Manier. Dazu bemerkte er:

Man kann zweifelhaft sein, ob das Rechnen mit Dyaden nützlich ist und ob man nicht auf die alten Hamilton'schen und Graßmann'schen Dinge zurückkommen soll. Ich halte das letztere für verfehlt. Hamilton und Graßmann sind mit ihren Symbolen nicht durchgedrungen, weil sie einem Phantom nachjagten und den Boden der Wirklichkeit verließen. Die Einführung der Vektoren ist eine harte Notwendigkeit geworden. Man kann nicht auf die Dauer in Vektoren denken und in Kolonnen schreiben. Jetzt handelt es sich nicht mehr um ein philosophisches Problem, sondern um ein praktisches Handwerkszeug, das auch für jene erreichbar ist, die den Umweg über die Quaternionen und Extensivgruppen scheuen. Die Bedürfnisse der Mechanik sind entscheidend für die Einführung der Vektoren und deren Symbole; der Zeitpunkt für eine Vereinheitlichung ist da. (Brief vom 18.11.1903).

Sommerfeld schickte diesen Brief an Mehmke, daraufhin schrieb Mehmke am 10.1.1904 an Finsterwalder: er erläuterte ausführlich die Vorteile und Entwicklungsmöglichkeiten des Graßmannschen Kalküls und wies auf die vielen Anhänger im Auslande hin, nämlich Alfred Whitehead (1861−1947) und Homersham Cox (1821−1897) in England, Edward Hyde (1843−1930) in Amerika und Giuseppe Peano (1858−1932), Filippo Castellano und Cesare Burali-Forti (1861−1931) in Italien. Peano, Hyde und Whitehead waren Autoren einschlägiger Lehrbücher [Peano 1888; Hyde 1890; Whitehead 1908]. Was die Methoden anbelangte, so betonte Mehmke vor allem die zentrale Bedeutung des Begriffs des Bivektors. Mehmke kam zu dem Schluß, daß er eine Methode, wie sie 1902 Lüroth angewandt habe, bevorzugen würde [Lüroth 1902].

Inzwischen erwartete Prandtl mit Spannung die Gutachten; er ahnte wohl schon die bevorstehende Konfrontation. In einem Brief an Mehmke vom 13.2.1904 versuchte Prandtl, diesen zu einem Kompromiß zu bewegen: "Ich hatte gehofft, auf Grund von Ihrem und Herrn Sommerfelds Bericht, einen Vorschlag, der eine mittlere Linie einhält und von allen angenommen werden könnte, noch vor den Ferien vorzulegen, so daß wir uns in den Ferien schlüssig werden und nachher vor die erweiterte Kommission treten könnten ... Da die Kommission den Auftrag hat sich zu einigen, so wird schließlich jede Seite ein Zugeständnis machen müssen. Wenn Sie etwa mit Ihrem Bericht die Absicht verfolgten, die Physiker zur Graßmann-Methode zu bekehren, so würde ich dies nach dem Bisherigen

für ganz aussichtslos halten." Prandtl sah im folgenden eine Unverein-
barkeit der beiden Richtungen voraus, er konnte sich auch ein
Nebeneinander vorstellen. Zu diesem Zeitpunkt lag Prandtl Sommerfelds
Gutachten bereits vor, denn er berichtete Mehmke, daß es nur $4\frac{1}{2}$ Brief-
seiten umfasse. Leider befindet sich dieses Gutachten nicht im Mehmke-
Nachlaß. Mehmkes Gutachten ließ noch einen Monat auf sich warten, in
seinem Nachlaß befindet sich lediglich ein Entwurf eines "Vorläufigen
Gutachtens," datiert vom 14.2.1904. Mehmke plädierte ausschließlich für
die Graßmannsche, d.h. für die von ihm sog. deutsch-italienische Richtung:
"Trotzdem die deutsch-italienische Richtung in ihren Begriffsbildungen
die umfassendere ist, braucht sie weniger Zeichen... Die Bezeichnungsweise der deutsch-italienischen Schule ist deshalb einfacher als zum
Beispiel die Bezeichnungsweisen von Gibbs. Sie ist überhaupt die denkbar
einfachste. Von der Erwägung ausgehend, daß das Beste und Einfachste
für unsere studierende Jugend gerade gut genug ist, bin ich deshalb der
Meinung, daß die Lehrer an den deutschen Hochschulen, welche Vektoranalysis im Unterricht benützen wollen, sich der deutsch-italienischen
Richtung, welche auch in anderen Ländern immer mehr Anhänger findet,
anschließen sollten." Die englische Richtung, das ist die Quaternionen-
theorie, und die amerikanische Richtung (Gibbs) lehnte Mehmke strikt ab.
Der wesentliche Unterschied zwischen der amerikanischen und der
deutsch-italienischen Richtung bestand nach Mehmke darin, daß die
amerikanische Richtung die von den Physikern als notwendig erkannte
Unterscheidung von Polar- und Axialvektoren ebensowenig kennt wie den
Übergang von Vektoren zu Bivektoren und umgekehrt.

Prandtl antworte daraufhin am 28.2.1904:

> Nach ihrem Bericht scheint mir einstweilen die Hauptsache die Frage zu
> sein, ob das Produkt zweier Vektoren ein Bivektor oder ein Vektor sein soll,
> wenn dies geklärt ist, dann ist es leicht, über die Bezeichnungsweise einig zu
> werden. Rein logisch betrachtet hat unzweifelhaft der Bivektor den Vorzug;
> was aber nicht ausschließt, daß aus praktischen Gründen das Vektorprodukt
> vorgezogen werden kann. Denn dem Physiker kommt der Unterschied zwi-
> schen beiden einzig und allein bei Symmetriebetrachtungen vor... .

2.1. Mehmkes Aufsatz "Vergleich" und dessen Rezeption

Inzwischen war Mehmkes ausführliche Darstellung "Vergleich zwischen
der Vektoranalysis amerikanischer und derjenigen deutsch–italienischer
Richtung" in Druck gegangen [Mehmke 1904]. Mehmke legte ein
29–Punkte-Programm vor, das in einer Gegenüberstellung der beiden

Bezeichnungsweisen gipfelte [Mehmke 1904, 227]:

	deutsch-italienische Schule	Gibbs
$\vert a$	Ergänzung des Vektors, ein Bivektor	fehlt
ab	äußeres Produkt, ein Bivektor	fehlt
$a\vert b$	inneres Produkt	$a \cdot b$
$\vert ab$	Vektorprodukt, Ergänzung des Bivektors	$a \times b$
abc	äußeres Produkt, Rauminhalt des Trivektors	$a \cdot b \times c$
$ab\vert c$	ein Vektor in der Ebene des Bivektors ab, senkrecht zum Vektor c	$(a \times b) \times c$
$a\vert bc$	ein Bivektor, parallel zum Vektor a und senkrecht zum Bivektor bc	fehlt
$ab\vert cd$	inneres Produkt der Bivektoren ab und cd	$(a \times b) \cdot (c \times d)$
ab^2	Quadrat des Flächeninhaltes des Bivektors ab	$(a \times b)^2$
$ab \cdot cd$	ein Vektor in der Schnittlinie der Ebenen der beiden Bivektoren ab und cd.	$(a \times b) \times (c \times d)$

Mehmke schickte Fahnen- bzw. Separatabzüge an Max Abraham (Brief Mehmkes vom 13.4.1904), Sebastian Finsterwalder, August Föppl, Richard Gans, Carl Runge (1856–1927), Arnold Sommerfeld und wahrscheinlich noch weiteren Personen. Hier seien folgende im Mehmke-Nachlaß befindlichen Reaktionen erwähnt (von Abraham ist kein Antwortschreiben vorhanden):

Finsterwalder, Brief vom 21.3.1904 an Mehmke:

Ich habe mich inzwischen überzeugt, dass die Einführung der von Ihnen vorgeschlagenen Bezeichnung in die Geometrie der Vektorfelder keine Bedenken hat . . . . Von größter Wichtigkeit scheint mir die Frage, was soll in Zukunft geschehen? Ich zweifle, dass es gelingen wird, die amerikanische Schule mit Gründen mathematischer Logik zu verdrängen . . . . Den schliesslichen Sieg wird zweifellos jene Bezeichnung davontragen, in welcher die bedeutendsten Arbeiten publiciert werden. Helmholtz, Boltzmann und Hertz haben die Vektorbezeichnung verschmäht, wozu sie schliesslich durch die ablehnende Haltung der Mathematiker ganz berechtigt waren, daher datiert die geringe Würdigung der deutschen Vektoranalysis. Möge das bald anders werden.

Mehmke schloß aus diesem Brief, daß es ihm gelungen war, Finsterwalder auf seine Seite zu ziehen. Er machte sich daher Hoffnungen, die Verhältnisse in der Vektorkommission zu seinen Gunsten entscheiden zu können. So schrieb er am 26.3.1904 an Lüroth: "So ist es nicht unmöglich, daß wir in der erwähnten Kommission die Mehrheit bekommen werden".

August Föppl, Brief vom 22.3.1904 an Mehmke:

"...Ich verkenne übrigens keineswegs die Bedeutung der Gründe, die Sie
für Ihre Bezeichnungen anführen, namentlich auch, daß es für manche
Anwendungen recht nützlich sein wird, zwischen dem Bivector und seiner
Ergänzung zu unterscheiden. Für die Mechanik, die mich dabei allein
interessiert, halte ich allerdings die Unterscheidung nicht für nothwendig
und—im Interesse der Einfachheit der Darstellung—daher auch nicht für
wünschenswerth. Ein Kräftepaar wird z.B. nach meiner Meinung am besten
durch seinen Momentanvektor gekennzeichnet. Eine ganz andere Frage ist
es aber, ob nicht Jemand, der so denkt, wie ich, bereit ein sollte, eine ihm
zunächst etwas unbequemere Darstellung anzunehmen, um dadurch
berechtigte Interessen von Forschern auf anderen Gebieten entgegen zu
kommen. Diese Frage bin ich, unter der Voraussetzung, daß sich hierdurch
eine Einigung aller Autoren—oder doch der überwiegenden Mehrzahl—herbeiführen läßt, zu bejahen gern bereit. Ich fürchte freilich, daß es
nicht ganz leicht sein wird, eine solche Einigung wirklich zu erzielen.

Richard Gans, Brief vom 9.4.1904 an Mehmke:

Nehmen Sie bitte meinen herzlichsten Dank für den mir zugesandten Fah-
nenabzug und für den Brief, in dem Sie mir die Anwendung der deutsch-
italienischen Vektoranalysis auf Vectorfelder darlegten. Ich muß sagen, daß
ich damals in Nürtingen noch nicht völlig von dem Vorteil der deutschen
Richtung überzeugt war, jetzt bin ich es aber völlig.

Carl Runge, Brief vom 20.4.1904:

Ich will keine neue Richtung begründen, ich will nur constatiren, dass
deutsche Physiker wie Abraham und Gans die Vectoren in andrer Weise
unterscheiden als der deutsche Mathematiker Mehmke. Denn wenn Abraham und Gans glauben, du machtest dieselben Unterscheidungen wie sie, so
haben sie dein schwarzes Gemüt nicht verstanden. Ich habe zwar weder mit
Abraham noch mit Gans darüber gesprochen; aber ich bin fest überzeugt,
dass sich beide viel lieber an den Beinen würden aufhängen lassen, als dass
sie das Product zweier axialen Vectoren für einen polaren Vector erklärten.
Und das müssten sie doch, wenn das was du Vector nennst mit dem polaren
Vector identisch ist, und das was du Bivector nennst mit dem axialen Vector.

Arnold Sommerfeld, Brief vom 28.4.1904. Dieser Brief soll hier in voller
Länge wiedergegeben werden, da er sicher neben einer Stellungnahme zu
Mehmkes Aufsatz auch das Wesentliche enthält, was in Sommerfelds nicht
mehr erhaltenem Bericht stand.

Endlich habe ich Ihre Abhandlung zur Vektorrechnung gründlich studiren
können. Vor allem dieses: Es wäre töricht, nicht zugeben zu wollen, dass die
Einführung des Bivektors und des Trivektors sachlich begründet ist, daß sie
den Aufbau der Vektoranalysis abrundet und vermöge des herrschenden

Dualismus vereinfacht. In einem vollständigen Lehrgebäude der Vektoranalysis wird man also diese Begriffe nicht entbehren wollen und dürfen.

Was die Bezeichnung "Trivektor" betrifft, so sollte man vielleicht lieber "Raumskalar" sagen, wobei es lediglich an der dreidimensionalen Beschränkung liegt, dass dieser Trivektor ein Skalar wird. (Nebenbei bemerkt, scheint mir der Vorwurf in §8 nicht ganz gerechtfertigt. Ihr System der Vektoranalysis ist ebenso dreidimensional, wie das englische und bedarf für 4 Dimensionen einer Weiterbildung, wie schon aus dem Trivektor hervorgeht, der in 4 Dimensionen eine gerichtete Größe wird. Allerdings wird diese Weiterbildung bei der consequenten Grassmann'schen Begriffsbildung leichter sein, wie bei der englischen, die aus den unmittelbaren dreidimensionalen Anwendungen in Mechanik und Physik hervorgegangen ist.

Die principielle Unterscheidung von äusserem Produkt und Vektorprodukt ist mir sehr sympathisch. Übrigens ist in der mathematischen Physik Maxwell'scher Richtung lediglich das Wort Vektorprodukt üblich (Maxwell, Heaviside, H. A. Lorentz, *Math. Encykl.*) u. zw. in dem von Ihnen gut geheissenen Sinn. Andrerseits wird das Produkt $a|b$ stets skalares Produkt genannt. Die Bezeichnungen collidieren also mit der allgemeinen Terminologie deutscher Richtung nicht.

Die Ableitung der Vertauschungsregel, die Sie schreiben $\mathfrak{ABC} = \mathfrak{BCA}$ und die ich schreiben würde $(\mathfrak{A}[\mathfrak{BC}]) = (\mathfrak{B}[\mathfrak{CA}]) = \ldots$ geschieht natürlich überall durch Betrachtung des Parallelopipeds des Trivektors, indem man sich klar macht, daß $(\mathfrak{A}[\mathfrak{BC}])$ nichts anderes als jenen Rauminhalt bedeutet. Daß dieser Rauminhalt sich bei der Graßmann'schen Einführung als etwas Notwendiges aus dem Aufbau der Theorie darbietet, während er in der englischen Vektorrechnung durch einen Kunstgriff hereinkommt, ist ein großer Vorzug Ihres Standpunktes.

Ich will nun meinen persönlichen Standpunkt zu der Bezeichnungsfrage angeben. Derselbe ist, wie ich von vornherein zugebe, etwas voreingenommen deshalb, weil ich für die Bedürfnisse der Encyklop. Bezeichnungen verabredet habe (mit den massgebendsten Physikern), die ich nicht wieder fallen lassen kann, da die Lorentzschen Artikel bereits in diesen Bezeichnungen gedruckt sind. Außerdem wird mein Standpunkt stark durch die Bedürfnisse der Elektrodynamik beeinflußt, in der die Vektorrechnung weitaus die reichhaltigste Anwendung findet. Hätte Maxwell* u. nach ihm Heaviside und Lorentz nicht mit Vektoren gerechnet, so würde heute ein halb Dutzend mathematischer Specialisten sich mit Vektorrechnung befassen, aber es würde das allgemeine Interesse u. Bedürfnis fehlen. (Beachten Sie, dass auch Föppl u. die technische Mechanik erst durch den Umweg über die Electrizität u. Heaviside zu den Vektoren gekommen sind).

In der Elektrizitätslehre hat nun der Bivektor keinen rechten Platz. Zwar ist die magnetische Feldstärke $\mathfrak{H}$ sehr wahrscheinlich vom Charakter eines

---

*Ich wundere mich, dass Sie immer von der "amerikanischen" Richtung sprechen, während doch Maxwell u. seine Schule (Heaviside) unverhältnismässig mehr in Vektoren gemacht haben. Gibbs hat nur das eine Lehrbuch geschrieben, aber sonst keine Vektoren angewendet.

Bivektors. Wir können sie aber nicht als äußeres Produkt zweier Vektoren allgemein darstellen. Das womit wir zu rechnen haben ist immer die Ergänzung des Bivektors, z.b. wenn wir die mechanische Kraft $\mathfrak{F}$ des magnetischen Feldes auf eine mit der Geschwindigkeit $\mathfrak{v}$ bewegte elektrische Einheitsladung bestimmen wollen. Da ist (Biot-Savart)

$\mathfrak{F} = [\mathfrak{v}\,\mathfrak{H}]$ nach meiner Bezeichnung des Vektorproduktes

$\mathfrak{F} = |\mathfrak{v}|\mathfrak{H}$ nach Ihrer Bezeichnung, wenn Sie $\mathfrak{H}$ als Bivektor auffassen

oder $\mathfrak{F} = |\mathfrak{v}\,\mathfrak{H}$ wenn Sie unter $\mathfrak{H}$ bereits die Ergänzung des Bivektors verstehn.

Es kommt ferner oft das Moment dieser Kraft $\mathfrak{F}$ vor, die ich schreiben würde $\mathfrak{F} = [\mathfrak{r}[\mathfrak{v}\,\mathfrak{H}]]$, während Sie (unter $\mathfrak{H}$ die Ergänzung des Bivektors verstanden) schreiben würden $\mathfrak{F} = \mathfrak{r}|\mathfrak{v}\,\mathfrak{H}$, was natürlich einfacher ist; es scheint mir aber discutirbar, ob nicht die Zusammenfassung durch Klammern dem Verständnis entgegenkommt u. die Übersicht erleichtert.

Wenn ich nicht weiß, ob eine Größe Vektor oder Bivektor ist, wird Ihre Bezeichnung zweideutig. Sie können dies theoretisch als Vorteil, praktisch aber als Nachteil bezeichnen.

Was die Differential-Vektor-Rechnung betrifft, so werde ich mich gegen die Zeichen $|\nabla, \nabla|, \nabla \times$ energisch sträuben, wenn diese statt grad, rot, div eingeführt werden sollen. Wir würden dadurch die Benutzung dieser Operationen wesentlich erschweren. Ihre Meinung ist sicher auch die, dass man die Zeichen grad, rot, div beibehalten, aber sich ihrer Entstehung aus dem $\nabla|$ und $|\nabla$ bewusst sein soll; das kann ich billigen. Ebenso möchte ich aber sagen: Man soll sich des Zusammenhanges von vektoriellem u. skalaren Produkt mit Bivektor und Trivektor bewusst sein. Trotzdem kann man Bezeichnungen dafür verabreden, welche eingebürgerter sind, wie die Grassmann'schen und, indem sie die doppelte Operation der Bildung des Bivektors und der Ergänzung in Eins zusammenfassen auch vielleicht etwas einfacher. Als solche empfehle ich (mit Rücksicht auf den Gebrauch der Encykl.) $[\mathfrak{A}\,\mathfrak{B}]$ für Vektorprod; $(\mathfrak{A}\,\mathfrak{B})$ für skalares Produkt.

Können wir uns auf diesem Wege einigen?

Mit bestem Dank u. Gruß Ihr A. Sommerfeld."

Die Reaktion Prandtls auf Mehmkes "Vergleich" erfolgte nicht nur mittels eines Briefes, es bahnte sich eine rege Korrespondenz an, zahlreiche Briefe eilten hin und her. Prandtl veröffentlichte noch in demselben Jahr seine Meinung unter dem Titel "Über die physikalische Richtung in der Vektoranalysis" [Prandtl 1904b]. Schon am 1.4.1904 hatte Prandtl Mehmke mitgeteilt, "daß ich nicht bekehrt bin." In seiner Publikation benannte Prandtl gleich am Anfang die von Mehmke als "amerikanisch" bezeichnete Richtung um in "physikalische" und die von Mehmke als "deutsch–italienisch" bezeichnete in "geometrische". Und Prandtl glaubte nun: "Dabei ist aber zwischen den beiden Richtungen eine Verständigung so gut wie undenkbar" [Prandtl 1904b, 437]. Im folgenden rechnete er fast

Punkt für Punkt mit Mehmke ab. Schließlich kam er zu dem Schluß, "daß mein Streben darnach geht, Klarheit zu schaffen, und reinliche Scheidung. Da es sich nicht bloß um eine Schreibweise, sondern um diametral verschiedene Denkweisen handelt, wäre ein Versuch zur gewaltsamen Einigung einfach aussichtslos. Was dringend nottut, wofür ich auch in Kassel eingetreten bin, ist eine einheitliche Symbolik für den Unterricht, besonders in Mechanik und Physik. Daß sich hierfür das physikalische System als das einfachste und zweckmäßigste erweist, weil es aus dem Stoff selbst hervorgeht, darüber bin ich nicht zweifelhaft" [Prandtl 1904b, 445]. In Mehmkes Nachlaß befindet sich ein nicht datierter, drei Seiten umfassender Entwurf Prandtls unter dem Titel "Vektorbezeichnungen für die physikalische Richtung", der aus drei Leitsätzen, Vorschlägen, Ausführungen und einem Fragebogen besteht.

Mehmke verfügte, dank seiner guten Beziehungen zum Herausgeber des *Jahresberichtes der Deutschen Mathematiker-Vereinigung* August Gutzmer (1859–1924), über Prandtls Aufsatz, bevor dieser veröffentlicht wurde. In einem Brief vom 6.7.1904 äußerte Mehmke Gutzmer gegenüber seine Meinung: "Der Artikel von Prandtl ist so voller Widersprüche, Irrtümer und gewagter Behauptungen, daß eine gründliche Widerlegung nottut." Mehmke kündigte zwar eine Antwort, die er Gutzmer zu veröffentlichen bat, an, aber er fand nicht die dafür nötige Zeit. Schon vorher hatte Mehmke bei Prandtl selbst seine Kritik angemeldet, in einem zwei Seiten umfassenden Brief vom 29.6.1904 hatte er versucht, die Irrtümer und Widersprüche aufzudecken. Eine Kopie dieses Briefes schickte Mehmke auch an Sommerfeld, dem er noch folgendes handschriftlich hinzufügte: "Zu Ihrer Orientierung erlaube ich mir, Ihnen diese Abschrift eines von mir heute an Kollegen Prandtl geschickten Briefes zu zu stellen. Allmählich schwindet bei mir die Hoffnung, mit Prandtl zu einer Verständigung zu kommen, immer mehr. Vielleicht ist es überhaupt noch zu früh zu einer Vereinbarung, da die Vektoranalysis noch viel zu geringe Verbreitung hat und die meisten vom Vorhandensein verschiedener Systeme nichts wissen, also noch keine Gelegenheit gehabt haben, sich ein Urteil zu bilden." [3]

3. DAS ERGEBNIS

Weder Mehmke noch Sommerfeld waren Mitglieder der Vektorkommission geworden in der Absicht, Kompromisse zu suchen. Jeder beharrte letztendlich genau auf seinem System. Sommerfeld akzeptierte zwar, wie sein Brief vom 28.4.1904 zeigt, manche von Mehmkes Argumenten, dennoch hielt er weiter an den für die Encyklopädie gemachten Verabredungen fest. Und Mehmke kam Sommerfeld wahrlich keinen Millimeter entgegen. Prandtl bekundete zwar anfangs Kompromißbereitschaft, aber, bedingt durch die starre Haltung seiner beiden Mitkommissionsmitglieder, ging er mit seiner physikalischen Richtung sogar noch einen dritten Weg. Dieser lag zwar naturgemäß Sommerfelds Vorstellungen näher, dennoch waren Prandtls Vorschläge durchaus nicht mit denjenigen Sommerfelds identisch. An der Lage der Dinge noch etwas zu ändern, war schließlich keiner der drei Hauptakteure bereit. Felix Klein beschrieb in seiner "Elementarmathematik" die Situation noch etwas beschönigt wie folgt: "Wir haben auf der Naturforscherversammlung zu Kassel (1903) eine Kommission zu diesem Zwecke [4] eingesetzt; ihre Mitglieder konnten sich aber nicht einmal untereinander völlig einigen; da aber jeder doch den guten Willen hatte, von seinem ursprünglichen Standpunkte dem andern einen Schritt entgegenzukommen, war der einzige Erfolg der, daß ungefähr drei neue Bezeichnungsweisen entstanden!" [Klein 1924–1925, I, 71].

Schließlich schrieb Mehmke in treffender Weise folgendes am 21.6.1904 an Sommerfeld: "Ich komme immer mehr zu der Überzeugung, daß nicht die Bezeichnungen es sind, die uns (die Mitglieder der Kommission) noch trennen, sondern die Begriffe. Haben wir uns einmal über diese geeinigt, so werden uns die Bezeichnungen keine Schwierigkeiten mehr machen". Genau dieses war das unlösbare Problem.

Als zeitlich letztes, direkt mit dem Thema in Zusammenhang stehendes Dokument aus dem Mehmke-Nachlaß ist noch ein Schreiben von Sommerfeld an Prandtl und an Mehmke vom 7.12.1904 zu nennen. Sommerfeld schlug nun als cooptierende Mitglieder Heinrich Weber und Sebastian Finsterwalder vor. Diesem Schreiben legte Sommerfeld die Bezeichnungen der Deutschen Physikalischen Gesellschaft [Vorschläge 1903] und die inzwischen veröffentlichten Bezeichnungen der *Encyklopädie* [Sommerfeld 1904] bei. Auch betonte er, daß sowohl der Elektrotechnische Verein als auch der Ingenieurverein über diese Bezeichnungen beraten würden. Im Falle des Ingenieurvereins muß allerdings hinzugefügt werden, daß Sommerfeld diesem seit 1903 als Vorstand angehörte [Eckert u.a. 1984, 24].

Ein positives Ergebnis der Vektorkommission war eigentlich schon im Januar/Februar 1904, als die Berichte von Mehmke und Sommerfeld vorlagen, nicht mehr zu erwarten. Es ging den drei Hauptakteuren nicht eigentlich um einen Kompromiß, sondern darum, mit Hilfe der Stimmen weiterer wählbarer Kommissionsmitglieder eventuell doch noch eine Mehrheit für die eigene Partei zu erzielen. So schlug Mehmke nur Graßmannanhänger als Mitglieder vor, im Gegensatz dazu wollte Sommerfeld keine Graßmannianer. Schließlich führte dies zur Unterbrechung der Arbeit der Kommission. Dies geht aus einem Brief hervor, den Mehmke am 14.5.1907 an Giuseppe Peano schrieb: "Die von Ihnen genannte Kommission zur Vergebung der verschiedenen Bezeichnungen in der Vektoranalysis hat ihre Arbeit unterbrochen. Es bestand die Gefahr, daß die Anhänger der Heaviside-Gibbs'schen Richtung, welche Graßmann gar nicht kennen, einseitige und für den Fortschritt schädliche Beschlüsse erzielen würden." Wie die Mehrheiten im einzelnen aussahen, ist nicht bekannt, denn offensichtlich kam es gar nicht zu einer Abstimmung.

Wie weit sich Mehmke in seiner Auffassung von Felix Klein unterstützt wußte, läßt sich schwer beurteilen. Der erhaltene Briefwechsel Mehmke—Klein kann dafür leider nicht herangezogen werden, denn dieser endet bereits im Jahre 1900. Mehmkes Äußerung, daß nach Klein die Grassmannsche Richtung in der Kommission stärker vertreten sein sollte (s. S. 11), könnte wohl auch eine Übertreibung Mehmkes sein, obwohl das nicht seinem Charakter entspräche. Dafür, daß Klein Graßmannanhänger war, spricht vielleicht seine breite Darstellung der Inhalte von Graßmanns Ausdehnungslehre in seiner "Elementarmathematik" [Klein 1924–1925 II, 22–42]. Wie weit Klein bezüglich der Vektorrechnung Kompromisse zu schließen bereit gewesen wäre, ist unbekannt.

Im August 1904 fand der 3. Internationale Mathematiker-Kongreß in Heidelberg statt. Dort trafen sich, wie aus dem Mitgliederverzeichnis hervorgeht [Verhandlungen 1905, 11–22], die Kontrahenten, nämlich Burkhardt, Finsterwalder, Gans, Jahnke, Klein, Lüroth, Mehmke, Prandtl, Runge und Sommerfeld. Dazu gesellten sich noch die französischen Vektorspezialisten Emmanuel Carvallo (1856–1945) und Charles A. Laisant (1841–1920) sowie der Schotte Alexander MacFarlane (1851–1913) und der Amerikaner Edwin B. Wilson (1879–1964). Von einem Dialog wird in den Kongreßakten nichts berichtet.

Tatsächlich versuchte man kurze Zeit später auf internationaler Ebene, eine Vereinheitlichung der Vektorrechnung zu erreichen. 1908 wurde anläßlich des 4. Internationalen Mathematiker-Kongresses in Rom eine nunmehr internationale Vektorkommission ins Leben gerufen. Doch was schon national nicht gelingen wollte, scheiterte erst recht international. Nachdem diese Kommission anläßlich des 5. Kongresses in Cambridge 1912 keine Ergebnisse vorlegen konnte, vertagte man sich auf den nächsten Kongreß [Reich 1992, 219f., und 1989, 288ff.]. Der erste Weltkrieg setzte diesen Bemühungen schließlich ein Ende.

ANMERKUNGEN

[1] Im Unterschied zu Sommerfeld schrieb Lorentz Produkte mit ·, also ($\mathfrak{A} \cdot \mathfrak{B}$) und [$\mathfrak{A} \cdot \mathfrak{B}$].

[2] Dieser Brief befindet sich im Sommerfeld-Nachlaß (Deutsches Museum, München) unter der Signatur 28/A, 226.

[3] Der Zweck war die verbindliche Bezeichnungsweise in der Vektoranalysis.

LITERATURVERZEICHNIS

Abraham, M. 1901. Geometrische Grundbegriffe (abgeschlossen im Februar 1901). *Encyklopädie der mathematischen Wissenschaften*, Bd. IV, 3, S. 3–47. Leipzig: Teubner 1901–1908.

Benz, U. 1974. *Arnold Sommerfeld. Eine wissenschaftliche Biographie.* Dissertation Universität Stuttgart (maschinenschriftlich).

— 1975. *Arnold Sommerfeld. Lehrer und Forscher an der Schwelle zum Atomzeitalter 1868–1951.* Stuttgart: Wissenschaftliche Verlagsgesellschaft (= Große Naturforscher Bd. 28).

Bryan, G. H. 1903. Allgemeine Grundlegung der Thermodynamik (abgeschlossen im Januar 1903). *Encyklopädie der mathematischen Wissenschaften*, Bd. V, 1, S. 71–160. Leipzig: Teubner.

Burkhardt, H. 1901. Über Vectoranalysis. *Jahresbericht der Deutschen Mathematiker-Vereinigung* 5, 43–52.

Crowe, M. 1967. *A History of Vector Analysis. The Evolution of the Idea of a Vectorial System.* Notre Dame, London: University Press of Notre Dame.

Eckert, M., Pricha, W., Schubert, H., und Torkar, G. 1984. *Geheimrat Sommerfeld —theoretischer Physiker. Eine Dokumentation aus seinem Nachlaß.* München: Deutsches Museum.

Finsterwalder, S. 1903. Bemerkungen zur Analogie zwischen Aufgaben der Ausgleichsrechnung und solchen der Statik. *Sitzungsberichte der Akademie der Wissenschaften München, Mathematisch-Physikalische Klasse* 33, 683–689.

— 1906. Eine Grundaufgabe der Photogrammetrie und ihre Anwendung auf Ballonfahrten. *Abhandlungen der Mathematisch-Physikalischen Klasse der Königlich Bayerischen Akademie der Wissenschaften* 22, 223–260.

Gans, R. 1902. *Über die Induktion in rotierenden Leitern.* Dissertation Straßburg 1901. Leipzig: Teubner.

Hyde, E. W. 1890. *The Directional Calculus based upon the methods of Hermann Grassmann.* Boston: Ginn & Comp.

Jahnke, Eugen. 1904a. Elementare Herleitung der Formeln für die Reflexion und Brechung des Lichtes an der Grenze durchsichtiger isotroper Körper. *Archiv der Mathematik und Physik* 7 (3), 278–286.

— 1904b. Eine einfache Anwendung der Vektorrechnung auf die Theorie der veränderlichen Ströme. *Boltzmann-Festschrift*, S. 487−492. Leipzig: J. A. Barth.

— 1905. *Vorlesungen über die Vektorrechnung. Mit Anwendungen auf Geometrie, Mechanik und mathematische Physik.* Leipzig: Teubner.

Klein, F. 1901. Ueber die Encyklopädie der mathematischen Wissenschaften. In *Verhandlungen der Gesellschaft Deutscher Naturforscher und Ärzte, Tagung in Aachen 16.2−22.9.1900*, Leipzig: F. C. W. Vogel, S. 161−168.

— 1924−25. *Elementarmathematik vom höheren Standpunkte aus*, 2 Bde., 3. Auf. Berlin: Springer-Verlag.

Lorentz, H. A. 1903a. Maxwells elektromagnetische Theorie (abgeschlossen im Juni 1903). *Encyklopädie der mathematischen Wissenschaften* Bd., V, 2, S. 63−144. Leipzig: Teubner.

— 1903b. Weiterbildung der Maxwellschen Theorie. Elektronentheorie (abgeschlossen im Dezember 1903). *Encyklopädie der mathematischen Wissenschaften* Bd. V, 2, S. 145−280. Leipzig: Teubner.

— 1909. Theorie magneto-optischer Phänomene (abgeschlossen im März 1909). *Encyklopädie der mathematischen Wissenschaften* Bd. V, 3, S. 199−281. Leipzig: Teubner.

Lüroth, J. 1902. Zwei Beispiele für die Ableitung der wahren aus der scheinbaren Gestalt eines Körpers. In *Festschrift der Universität Freiburg zum 50−jährigen Regierungsjubiläum Seiner kgl. Hoheit des Großherzogs Friedrich von Baden 24.4.1902*, S. 179−205. Universitätsbuchdruckerei.

Mehmke, R. 1904. Vergleich zwischen der Vektoranalysis amerikanischer Richtung und derjenigen deutsch-italienischer Richtung. *Jahresbericht der Deutschen Mathematiker-Vereinigung* **13**, 217−228.

von Meyenn, Karl. 1993. Sommerfeld als Begründer einer Schule der Theoretischen Physik. In *Naturwissenschaft und Technik in der Geschichte. 25 Jahre Lehrstuhl für Geschichte der Naturwissenschaft und Technik am Historischen Institut der Universität Stuttgart*, H. Albrecht, Hrsg., Stuttgart: GNT-Verlag. S. 241−261.

Peano, G. 1888. *Calcolo geometrico secondo l'Ausdehnungslehre di H. Grassmann preceduto dalle operazioni della logica deduttiva.* Turin: Fratelli Bocca.

Prandtl, L. 1904a. Über eine einheitliche Bezeichnungsweise der Vektorenrechnung im technischen und physikalischen Unterricht. *Jahresbericht der Deutschen Mathematiker- Vereinigung* **13**, 36−40.

— 1904b. Über die physikalische Richtung in der Vektoranalysis. *Jahresbericht der Deutschen Mathematiker-Vereinigung* **13**, 436−449.

Reich, K. 1989. Das Eindringen des Vektorkalküls in die Differentialgeometrie. *Archive for History of Exact Sciences* **40**, 275−303.

— 1992. Who needs Vectors? Discussion of Calculus in History. In *Learn from the Masters. Proceedings of the Kristiansand Conference on the History of Mathematics, August 1988*, O. Bekken u.a., Hrsg., Pennsylvania State University. S. 214−228.

— 1993a. Der Mathematiker Rudolf Mehmke: Bausteine zu Leben und Werk. In *Naturwissenschaft und Technik in der Geschichte*. *25 Jahre Lehrstuhl für Geschichte der Naturwissenschaft und der Technik am Historischen Institut der Universität Stuttgart*, H. Albrecht, Hrsg. Stuttgart: GNT-Verlag. S. 263–285.

— 1993b. *Vom absoluten Differentialkalkül zur Relativitätstheorie*: *Die Entwicklung des Tensorkalküls*. Basel und Boston: Birkhäuser.

Sommerfeld, A. 1901. Theoretisches über die Beugung der Röntgenstrahlung. *Zeitschrift für Mathematik und Physik* **46**, 11–97 (= *Gesammelte Schriften* IV, Braunschweig: Vieweg, 1968. S. 240–326.

— 1904. Bezeichnung und Benennung der elektromagnetischen Grössen. *Encyklopädie der mathematischen Wissenschaften*, V: *Physikalische Zeitschrift* 1904, 467–470 (= *Gesammelte Schriften* I, S. 671–674. Braunschweig: Vieweg, 1968).

— 1903–1921. Vorrede zum fünften Bande. *Encyklopädie der mathematischen Wissenschaften*, Bd. V, 1, Leipzig: Teubner. S. V–VI.

Timerding, H. E. 1902. Geometrische Grundlegung der Mechanik eines starren Körpers (abgeschlossen im Februar 1902). *Encyklopädie der mathematischen Wissenschaften*, Bd. IV, 1, Leipzig: Teubner. S. 125–189.

Tobies, Renate. 1986–1987. Zu Veränderungen im deutschen mathematischen Zeitschriftenwesen um die Wende vom 19. zum 20. Jahrhundert. Teil I *NTM 23*, 19–33; Teil II *NTM* **24**, 31–49.

Verhandlungen der Gesellschaft Deutscher Naturforscher und Ärzte, Kassel 20.–26.9.1903. 1904. A. Wangerin, Hrsg. Leipzig: F. C. W. Vogel.

Verhandlungen des III. internationalen Mathematiker Kongresses in Heidelberg 1904. 1905. A. Krazer, Hrsg. Leipzig: B. G. Teubner.

Vorschläge des wissenschaftlichen Ausschusses der Deutschen Physikalischen Gesellschaft für eine einheitliche Benennung, Definition und Regel in der Physik. 1903. *Berichte der Deutschen Physikalischen Gesellschaft* (zusammen mit den Verhandlungen) 1903, 68–71.

Whitehead, A. N. 1908. *A Treatise on Universal Algebra with applications*, Vol. I. Cambridge, UK: Cambridge University Press.

MATERIALIEN AUS DEM MEHMKE-NACHLAß (UNIVERSITÄTSARCHIV STUTTGART), CHRONOLOGISCH GEORDNET:

Sommerfeld, Encyklopädiepapier 1901. 8 S., ohne Signatur.

H. Weber an Sommerfeld, 25.10.1903, Sign. SN 6 II.79.

Prandtl an Mehmke, 30.10.1903. Sign. SN 6 II.78.

Mehmke an Sommerfeld, 12.11.1903, ohne Signatur.

Finsterwalder an Sommerfeld, 18.11.1903, Sign. SN 6 II.59.

Prandtl an Mehmke, 23.12.1903, Sign. SN 6 II.63.

Mehmke an Finsterwalder, 10.1.1904, Sign. SN 6 II.64.

Prandtl an Mehmke, 13.2.1904, Sign. SN 6 II.73.

Mehmkes Entwurf, vorläufiges Gutachten, 14.2.1904, ohne Signatur.

Finsterwalder an Mehmke, 21.3.1904, Sign. SN 6 II.51.

Föppl an Mehmke, 22.3.1904, Sign. SN 6 II.50.

Mehmke an Lüroth, 26.3.1904, Sign. SN 6 II.49.

Mehmke an Abraham, 13.4.1904, Sign. SN 6 II.31.

Sommerfeld an Mehmke, 28.4.1904, Sign. SN 6 II.57.

Mehmke an Sommerfeld, 3.4.1904, Sign. SN 6 II.42.

Gans an Mehmke, 9.4.1904, Sign. SN 6 II.37.

Mehmke an Sommerfeld, 10.4.1904, Sign. SN 6 II.35.

Runge an Mehmke, 20.4.1904, Sign. SN 6 II.52.

Mehmke an Sommerfeld, 20.6.1904, Sign. SN 6 II.55.

Mehmke an Sommerfeld, 21.6.1904, Sign. SN 6 II.69.

Mehmke an Gutzmer, 6.7.1904, Sign. SN 6 II.23.

Sommerfeld an Prandtl und Mehmke, 7.12.1904, Sign. SN 6 II.93.

Prandtl. Entwurf. Vektorbezeichnungen für die physikalische Richtung, ohne Signatur.

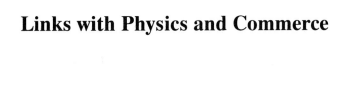

Links with Physics and Commerce

Experimental Physics at the University of Leuven during the 18th Century

Paul Bockstaele

Graetboslaan 9, B-3050 Oud Heverlee, Belgium

In the first half of the 18th century, attempts were made by the Faculty of Arts at the University of Leuven to supplement the pure theoretical teaching of physics with experiments. The *Schola Experimentalis*, started in 1755, was, however, not an unqualified success, due to the defective organization and the persistent lack of funds to purchase instruments. In order to remedy this unsatisfactory situation, an attempt was made in 1759 to establish a chair of experimental physics and to appoint the English priest John Needham to this post. The plan, however, was dismissed. Definite improvement of the situation came only in 1771 through the appointment of Professor Johannes Thijsbaert as Director of the School of Experimental Physics.

Philosophy, in the old University of Leuven, was taught from the 15th century in four Pedagogies, or Colleges, with the colorful names of Porcus (Pig), Lilium (Lily), Falco (Falcon), and Castrum (Castle). The students lived there with the teachers and followed lectures there. In every College there were four professors, two primarii and two secundarii. The duration of the studies in the Faculty of Arts was two years, and the whole curriculum could be followed in each of the four Colleges. Only for certain courses did all the philosophy students come together in the *Vicus Artium*, the building of the Faculty of Arts, known in short as the *Vicus*.

Instruction by the Faculty of Arts comprised logic, physics or natural philosophy, metaphysics, and ethics. The second year was mainly devoted to the study of physics. Until the 17th century, the approach was predominantly philosophical. For every topic, the opinions of the philosophers were presented and the arguments for and against were discussed. Alongside this purely speculative study of nature, there developed towards the end of the 17th century the teaching of physics "by means of experiments."

345

In Leuven, before the middle of the 18th century, there is no reference at all in the official curriculum of the Faculty of Arts to the use of experiments. However, from surviving student notebooks, it appears that simple physics experiments were already being described and commented on in the 17th century, e.g., Torricelli's experiment with the column of mercury, the working of the siphon and the thermometer, and the pneumatic experiment of the "Magdeburg hemispheres." Whether they were also carried out in front of the students cannot be known with certainty from the course notes that have come down to us [1].

From the beginning of the 18th century, changes were regularly introduced in the curriculum of the Faculty of Arts at Leuven. Occasional comments in the *Acta Facultatis* point more than once to the wish for a more experiment-oriented teaching of physics. In 1736, on the occasion of modifications in the curriculum, the Pig College was beginning systematically to acquire instruments. Before the middle of the 18th century, the other Colleges also owned a number of instruments, often donated by professors. The problem of extending the collection was mainly of a financial nature. The scarce resources of the Faculty of Arts were not sufficient to acquire often expensive instruments. The only remaining way out was to call upon the support of the government. In 1743 the Faculty asked the government in Brussels for a subsidy to buy instruments and to repair the buildings, especially the *Vicus*. The answer was a refusal, based on the bad financial situation as a result of the War of the Austrian Succession. There is an allusion to this in the letter that Count Friederich von Harrach, interim Governor General, wrote to the Faculty of Arts on March 14, 1743, when he left The Netherlands [2]:

> Une chose que je regrette en partant, est de n'avoir pas eu le le temps d'y etablir une physique experimentale, qui contribueroit beaucoup a la faire fleurir, et que je Vous prie de ne point perdre de vue: en travaillant aux moyens les plus promps pour y en avoir une Leçon.

Harrach's advice was not forgotten, and the Faculty kept hoping that it could realize its plans regarding the teaching of experimental physics.

A new attempt was made in 1751. The occasion was the announcement of the sale in Brussels of a collection of physics instruments from the estate of a certain Bauwens. The Faculty set up a commission to investigate the matter. On December 4 it proposed to the Faculty a set of rules regarding the creation of a *Camera experimentalis*. In this proposal, 10 articles established how experimental physics was to be taught within the

Faculty of Arts. The first articles concerned the formation of a fund whereby the purchase of instruments could be financed. To this end would be devoted the money that the students paid for the use of the libraries of the Colleges. However, this money would not be sufficient, and so article 9 provided that recourse should be made to the generosity of rulers and states. Article 5 proposed that the *Camera experimentalis* be installed for the time being in the *Vicus*. Cabinets to keep the instruments in should be installed at the cost of the Faculty. Article 6 specified that demonstrations be given once a week, 2 hr in the morning and 2 hr in the afternoon, and this from mid-March to mid-August. They would be carried out by four professors, one from each College [3]. On December 7 the Faculty council designated two professors who were to examine the instruments for sale in Brussels, and see what could be acquired. They reported to the Faculty on December 21. In the end, nothing was purchased. Six months later the money intended for the purchase was redeposited in the Faculty's account. After that, there was no more mention of the *Camera experimentalis* in the *Acta* of the Arts Faculty.

Even after this second failure, the Faculty of Arts did not lose heart. Two years after establishing the regulations for the *Camera experimentalis*, it turned once again to the government with the request to make funds available to set up a *Schola experimentalis*. And again it met rejection.

With the arrival in 1744 of Prince Charles of Lorraine as governor, a quiet but radical reform of the existing structures began to take place in the Austrian Netherlands. The University of Leuven could not remain unaffected, especially after various bodies in the University, including the Faculty of Arts, repeatedly called upon the help of the government. On May 11, 1753, Charles of Lorraine sent "à ceux de l'Université de Louvain" a list of ordinances aiming to eradicate a number of abuses. About a year later, on July 18, 1754, the Austrian government in Brussels appointed Count Patrice de Neny [4] as Royal Commissioner for matters concerning the University. His task was to oversee everything to do with the management, the discipline, and the studies of the University. On the same day Charles of Lorraine signed a decree obliging the University, and in particular the Faculty of Arts, to make use in the future of experiments in the teaching of physics [5].

> Tandis que nous donnons les attentions les plus suivies à tout ce qui peut intéresser l'honneur et les avantages de l'Université, nous apprenons avec regret que la physique s'enseigne dans vos écoles sans le secours des expériences, ce que nous voulons bien attribuer à un défaut de fonds pour

l'achat des machines et instruments nécessaires; il est important d'y pourvoir, tant pour la réputation de vos écoles, que pour l'instruction de la jeunesse.

For the financial implications connected with introducing experiments in the teaching of physics, the following solution was prescribed:

> Comme il nous revient que, dans chaque collège de philosophie, il se trouve grand nombre de livres de théologie, de droit, de médecine et autres, qui n'ont point de connexion immédiate avec la philosophie, Nous voulons que, dans le terme de trois mois, tous les livres qui sont dans les bibliothèques des quatre collèges de philosophie et qui ne traitent pas directement de la philosophie, soient vendus publiquement au plus offrant, et que l'argent qui en proviendra soit employé d'abord à l'achat de machines et d'instruments servant aux expériences physiques.

In addition, the contributions that the students made for the use of the libraries were henceforth to be used to acquire machines and instruments, and this until there were enough. Only then were philosophy books allowed to be bought, but no others. It was expected that the Faculty of Arts would start its lectures in experimental physics the following year.

Neny at once devoted himself to carrying out the decree and organizing the lectures. The Faculty of Arts was ready to cooperate, but the four Colleges had great objections to the sale of their so-called "useless" books. The professors stubbornly defended their libraries and, after negotiations with the Royal Commissioner, the sale was decided against.

On February 23, 1755, a delegate of the Faculty of Arts presented Neny with a *Projet pour les Demonstrations de la Physique Experimentale* [6]. It announced the beginning of lectures for the following month of March. They would take place in the *Vicus* for all students in the second year of philosophy. Every year four professors, one from each College, would be appointed to give the lectures. The instruments in the four Colleges were to be brought together, and, as soon as the necessary means were available, any lacks were to be made up. For demonstrations that the professors could not carry out, because of a lack of instruments or because they did not know how to operate them, recourse was to be made to an expert. The Faculty pointed out further that a start could be made only when the necessary funds to repair the instruments and to pay the four professors and the expert were found. The *Projet* concluded with the observation that the *Vicus* was totally unsuitable for the demonstrations and that the building could not easily be modified because it was in a poor state of repair. Despite the fact that the building was ideally located, it would have to be completely demolished, which would have cost at least 35,000 florins.

Another solution was to leave the building as it was, and to build a *Schola experimentalis* somewhere else, which would certainly cost 20,000 florins.

At the request of Neny, a list was drawn up of the instruments in the possession of the four Colleges and a plan made for the course in experimental physics. This program, probably worked out by professor J. P. Sauvage [7], shows how ill prepared the professors were for their new task: it is nothing more than a reduced copy of the table of contents of the first four books of Jean Antoine Nollet's *Leçons de Physique Expérimentale* (Paris, 1743–1748), supplemented by three experiments borrowed from Willem Jacob 's Gravesande's *Physices Elementa Mathematica, Experimentis Confirmata* (Leyden, 1720–1721). At the end it is stated that some experiments in electricity, optics, and magnetism would also be demonstrated. These are the subjects that Nollet deals with in Parts 5 und 6 of his *Leçons*.

In the list of instruments compiled by the Faculty, the note *non est in statu* appears repeatedly. Apparently various pieces of apparatus were in bad condition and unusable. Consequently, Neny sent the Brussels instrument-maker Henri de Seumoy [8] to Leuven to see what could be repaired. Seumoy reported "que la plupart étoient actuellement hors d'état de servir, mais qu'il pourroit les retablir avec très-peu de peine et de dépense, et qu'ils étoient d'ailleurs de la main de bons maîtres." He attached to his report a list of instruments that he would quickly be able to supply, so that they could be used in the courses due to start in March 1755.

In a *Memoire pour l'Etablissement des Leçons de Physique Experimentale dans l'Université de Louvain* [9], submitted to Neny, the Faculty of Arts again took up the most important points from the *Projet* of February 23. It was hoped that the lectures would begin by mid-March. They would be given in the *Vicus*, for want of anywhere better. Experience would teach whether it was a suitable place. There was no more talk of an expert from outside the University. The program drawn up by Professor Sauvage could be carried out in the current year unless the outlay it presumed formed an insurmountable obstacle. For the University had absolutely no reserves to finance such an undertaking.

Eventually, on March 11, 1755, Neny was able to submit the final report of the discussions with the Faculty of Arts concerning the establishment of a lecture in experimental physics to the Governor General. He appended to it the *Mémoire* of the Faculty, together with the list of instruments in possession of the Colleges and the list made up by Seumoy. In the latter list the price was noted for every instrument; the total amount was 776 florins. In his report Neny observed that the Faculty at that moment could not afford such an expenditure and that without this apparatus it would be impossible to carry out the demonstrations in experimental physics in Leuven. Consequently, in order to help the Faculty from the start, he pleaded for the purchase to be made through Her Majesty's finances.

The definitive regulations for the School of Experimental Physics, drawn up by Neny and approved by Charles of Lorraine, were handed to the Faculty of Arts on March 17 [10]. It contained the following provisions:

1. Every year, for three months beginning around March 15, demonstrations in experimental physics would be given.
2. The demonstrations would take place on Tuesdays and Fridays, from 9 till 11 A.M.
3. For the moment, the demonstrations would be held in the *Vicus*, where all the second year students would assemble.
4. Every year, each College would appoint one of its professors; these four professors were together charged with carrying out the experiments.
5. To encourage them to fulfil their task as well as possible, they would receive every year an allowance of 42 florins.
6. The apparatus and instruments that were then housed in the four Colleges were to be transferred to the *Vicus* for common use. In addition, the Faculty would receive, at the expense of Her Majesty's finances, some more instruments, the list of which was attached. It is identical to Seumoy's list mentioned above.

Because it was necessary to have funds available to gradually extend the collection of instruments, a yearly contribution would be requested of all the students of philosophy. From this fund the professors would also be paid.

In the course of 1755 the instruments promised were delivered by Seumoy, who also repaired several of them. In the following years he also sold various physics instruments to the Faculty of Arts.

With these regulations a rational basis for teaching experimental physics was established for the first time in Leuven. Nevertheless, the whole organization still had serious shortcomings, including the continuing insufficiency of funds and the fact that the lectures were entrusted to four professors, without any special competence in what they had to teach and changing every year. In the end, no one was responsible either for looking after the instruments or for organizing the experiments. From a list of the lecturers in experimental physics from 1755 to 1762, it emerges that in those 8 years no fewer than 19 professors were in charge of the demonstrations, but that none of them gave the lectures two years in succession, the only exception being professor Sauvage, who was on the list from 1755 to 1759. He seems to have been the only one to have had any real interest in experimental physics. For the others, teaching this was probably an unwelcome chore. In such cirumstances even the good will and dedication of the lecturers could not make the enterprise succeed. Neny also saw that the School of Experimental Physics was not immediately a success. He was aware of the numerous difficulties resulting from the continual change of lecturers in the Arts Faculty, their lack of preparation and knowledge of experimental physics, and the appalling state and inadequate arrangement of the buildings. As a consequence of all this, the course in experimental physics was neither as sound nor as useful as might have been hoped for. Neny was increasingly convinced that the only way out of this unsatisfactory situation was to establish a sufficiently remunerated chair of experimental physics and to entrust it to someone who was fully competent in physics.

One chance to realize this seemed to offer itself in the beginning of 1759. At that time the English priest John Needham [11], who already had a name as an expert and had been a member of the Royal Society since 1747, was staying in Brussels. Adrian van Rossum (1705–1789), who was professor in the Faculty of Medicine, discussed this with Neny and pointed to the possibility "d'attacher à l'université un Phisicien illustre." Without the Faculty of Arts being informed or consulted, contact was initiated with Needham. The proposal was for him to look after the teaching of experimental physics in Leuven and, in addition, to be responsible for forming a natural history collection. Needham was prepared to accept this commission, but under certain conditions. In a letter dated February 28, 1759, and written entirely in the third person, he outlined how he hoped to realize this project [12]:

> A la tête de tous les cabinets de l'Histoire naturelle, et de la Philosophie experimentale, ils se trouve ordinairement deux personnes; la premiere en qualité d'Intendant, comme à Paris, dont la fonction est de diriger le progres, procurer les pieces necessaires, les distribuer selon leurs classes, tenir les correspondances litteraires, et instruire son associé, ou subdelegué; la seconde personne à lui sujette doit avoir sa place en qualité de demonstrateur, et son office est, de faire les experiences, de donner les leçons publiques apres avoir reçu prealablement ses instructions journalieres de son Principal.

As remuneration for the work, Needham proposed

> de partager la pension de l'université attachée à sa place entre lui, et

quelque jeune homme à talens son éleve, et demonstrateur, dont la nomination dependra du gouvernement, et pour supleer à ce defaut d'appointement,
il ose esperer que le gouvernement le dedommagera par quelque benefice
ecclesiastique, dont la jouissance lui restera apres avoir rempli sa tache, à
savoir de completter l'assemblage de toutes les pieces absolument neces-
saires pour la physique experimentale, de mettre la collection d'histoire
naturelle en bon main, de bien detailler la methode de la rendre complete,
qui doit être l'ouvrage du tems, et de former parfaitement son eleve, ou son
demonstrateur; apres quoi, et avec l'agrement du gouvernement, il propose
de lui ceder sa place avec ses profits en entier.

At the end of his letter Needham asked for his proposals to be
sympathetically considered and for a decision to be made as soon as
possible, because he was free for the moment and would have to turn
down the offer of a job as a tutor that he had been promised for the
following spring. Neny hurried things along and discussed everything with
the minister plenipotentiary Cobenzl, who went along, in principle, with
the suggestion. In any case Neny wrote on March 2 to van Rossum, whom
he had given the task of discussing with Needham the concrete execution
of the proposal. Neny once again stressed the need to further deal with the
matter wholly in secret.

On the same day Neny instructed professor F. A. Graven, primarius of
Lily College, to show Needham the instruments that were in the *Vicus*.
After inspecting the collection, Needham's judgment was devastating. In
his report of March 7 [13] he wrote to Neny:

I have seen, and examinded, as you desired, all the Physical instruments,
disposed for that purpose in one view, and ranged in a certain order by
Monsieur Sauvage's directions in the school of arts. I must tell you freely my
opinion, you may easily imagin the consequences; a certain collection of
instruments, mostly constructed in wood, where metal is absolutely necessary
for the entire success of an experiment, selected without knowledge, tast,
order, or principles, ill modelled many of them by aukward work-men from
abbé Nolet (sic), who is himself an author more brillant, than scientifical,
chosen here and there from scattered lessons indeliberately, more, because
they strike the senses, than the understanding, some of them merely of
secondary nature, pretty, if you please, but merely accessory, while the less
expensive essentials are wanting: what can I do with this chaos of unwieldy,
and intractable instruments, unless the government will please to suppose,
that nothing in the experimental way exists here at Louvain, and advance me
at lest the sum of three, or four thousand florins; then I will endeavour to
bring into their several places, all these at present useless instruments with
some alterations, and thus at lest economise the publick money by rendering
useful, what is at present useless.

Needham further reported that he had already begun preparing the first lecture. It is also apparent from his letter that he found a student-assistant, the young and ambitious Nelis [14], a confidant of Neny's. Referring to the conversation he had had with professor Sauvage, who had shown him the instruments, Needham returned at the end of his letter to the difficult and unsatisfactory situation of experimental physics at Leuven.

Meanwhile, Van Rossum tried to solve a number of practical problems such as finding a suitable room for the experiments and housing the natural history collection. On March 12 he wrote about this to Neny [15]. He firmly believed in a good outcome, "persuadé de bonnes intentions du gouvernement, qui ne manquera point de saisir l'occasion que le ciel nous offre dans Mr. Needham, peut être l'unique qui se presentera de nos jours, pour mettre en execution un projet si profitable." The whole affair gained momentum, and two days later, on March the 14th, in a letter to Neny [16], Van Rossum noted in a postscript: "J'ai actuellement entre les mains les premiers cahiers du traité de la physique experimentale au quel M. Needham travaille. j'admire sa methode, justesse etc. le publique en jugera, dessinateur, graveur, imprimeur nous avons tout a la main, *modo inspires in faciem ejus spiraculum vitae.*"

The proposal, worked out by Van Rossum and Needham, was reworked by Neny into a detailed *Mémoire* to Charles of Lorraine on March 17 [17]. On March 21, Charles sent it with his recommendations to Vienna, where Count von Kaunitz gave a report of it to Empress Maria Theresia. In this report of April 14 [18], there is here and there the suggestion of a criticism of Neny. Kaunitz wrote, for example, aiming at Neny, "que ce n'est que pour trouver une Niche au Prêtre Needham, qu'on a songé aux nouveaux Etablissemens, qu'on propose". In the end, Kaunitz advised in the negative, and on April 18 Maria Theresia informed Charles of Lorraine that the proposed plan could not go ahead. Again it was the poor financial situation that would be given as reason for the refusal. Disillusioned, Neny informed Needham on May 2 "que l'Impératice n'a pas jugé à propos d'agréer l'Etablissement d'une chaire de Physique experimentale et d'un Cabinet d'Histoire naturelle à Louvain."

Meanwhile, the Faculty of Arts kept trying, given the available means, to make something of the lectures in experimental physics. Reports were made to Neny concerning the first two years of their existence, 1755 and 1756. This seems not to have happened in the following four years. An overview of the experiments carried out was first sent again in 1761 and then in 1762. Neny reacted to this last report, made by professor Thijsbaert [19], dean of the Faculty, with the wish to be informed which professors had given the lectures in experimental physics since 1755 and whether they had been paid for them, as provided in the regulations of 1755. Thijsbaert replied on January 25, 1763 and took the opportunity to refer to the unsatisfactory state of the collection of instruments. He reported that, insofar as was possible, the apparatus was taken care of and maintained, but that the collection left a lot to be desired. He made this comment [20]:

> Le plus grand mal de tous, est, qu'on a à Louvain tres peu d'ouvriers qui puissent raccommoder un instrument, au quel il vient à manquer quelque chose: car, outre que les ouvriers n'en connoissent pas la structure, il leur manque souvent de la delicatesse, et sur tout ils n'ont pas la precision, et l'extreme exactitude, que de tels ouvrages exigent: ce qui est cause qu'en se servant de differens instruments, on doit souvent se contenter des à-peu-pres; et que d'autres sont hors d'etat de servir, faute de n'avoir personne pour les retablir.

Thijsbaert requested that Neny give his sympathetic consideration to this difficult and unfavorable situation.

Still concerned about how everything was going in the Arts Faculty, Neny, in a detailed letter of November 22, 1764 [21], proposed a number of important reforms for teaching in the Faculty. Some concerned the teaching of mathematics and the natural sciences. The basics of astronomy, botany, anatomy, physiology, and geography were to be treated more thoroughly than was the case. The ecclesiastical calendar must henceforth be taught, alongside the principles of gnomonics. After everything had been discussed in detail by the Faculty, the reforms proposed were for the most part accepted. To implement them, provisional regulations, which were submitted to Neny on December 15, were drawn up. As for physics, the professors admitted "qu'on y omettoit quantité de choses aussi utiles qu'agreable, que les divisions de nos Traités n'étoient point faites avec toute l'exactitude, ni en même tems avec toute l'étendu, qu'exigent les nouvelles decouvertes de la Physique Experimentale" [22]. The assurance that in the new curriculum a greater place would be made for physics was given, but it was also observed that this would inevitably entail greater expenses than the Faculty was able to bear.

It was always the same difficulties that stood in the way of a sound teaching of experimental physics. Something drastic had to be done about both the financial situation and the lamentable state of the lecture rooms. Neny set himself to the task once again. As a result of his intervention, the income from the vacant chair of French was from 1765 devoted to the *Schola experimentalis*. About a year later, on March 1, 1766, he sent a *Mémoire sur la Restauration de l'Ecole générale de la Faculté des Arts à*

Louvain to Charles of Lorraine. Plans were made for extensions to and renovations of *Vicus* College, and in August 1766 work began. The income from the vacant chair of Mathematics was also diverted for the reconstruction of *Vicus* College and, after 1769, for the refurbishment. Eventually the lectures in experimental physics could be given in more suitable surroundings. Apart from the improved material conditions, the situation remained the same. The collection of instruments was still far from adequate and the lecturers were insufficiently prepared. Most of all it was the absence of someone in charge of the lectures and responsible for the buying and maintaining the instruments that was at the root of the poor state of affairs. Neny referred to all this in a letter of July 19, 1771, directing "à M. M. les Doyen et autres de la Faculté des arts de Louvain." He proposed appointing a Director as head of the *Schola experimentalis* who, alongside the tasks of purchasing, maintaining, and repairing the instruments, would alone and to the exclusion of anyone else have the keys of the cabinets and would be charged with making the necessary instruments available to the professors and, if necesary, with showing them how to use them. Neny awaited the advice of the Faculty concerning this suggestion. The dean began his reply by explaining how it was that various instruments were in poor repair or unusable. Aware of its shortcomings, the Faculty agreed with the appointment of a Director and made a few suggestions in connection with his statute. The affair was finally settled in the decree of November 5, 1771, in which the following stipulations appear [23]:

1. The responsibility for purchasing, maintaining, and repairing the instruments shall henceforth be entrusted to one of the professors of philosophy with the title of Director of the School of Arts.
2. Her Majesty, knowing the talent, dedication, and application of Mr. Thijsbaert, Professor Primarius of Philosophy at Castle College, appoints him as Director of the School of Arts.
3. The Director, to the exclusion of every other person, will have control over the keys to the cabinets in which the instruments are kept; he will make the necessary instruments available to those professors who wish to carry out experiments at prearranged time.

The decree further stipulated that the Director was in charge of the choice of experiments and that he was responsible for the care of the School of Arts, for the tidiness of the lecture rooms, for the maintenance of and the repairs of the building. Moreover, he was to make an annual report to the Faculty and to the Royal Commissioner of the University concerning the use of the funds allocated to the School of Arts.

Joannes Franciscus Thijsbaert, since 1759 a professor at Castle College, did an excellent job. For the purchase and maintenance of the instruments he had available the unused income from the vacant chairs of French and Mathematics. He was also able to extract funds from the Faculty, and after 1775 there was more income available from the vacant chair of Latin. Immediately after his appointment he began to set the collection of instruments in order. He was also greatly concerned about extending the collection. In order to purchase good and reliable instruments, he got in touch with foreign instrument makers. For orders in London he called on the Portuguese scholar John Hyacinth de Magellan, who advised and assisted him. In a *Mémoire pour l'Ecole des Arts* of April 12, 1778 [24], he listed a number of instruments that he had bought with the income from the chair of Latin:

> avec ces deniers le Cabinet experimental s'est enrichi de plusieurs instru-mens de la premiere classe, faits par les meilleurs ouvriers de Londres: tels que, microscope solaire, deux telescopes, l'un achromatique, l'autre de Newton, deux grands miroirs, l'un concave, l'autre convexe, montés chacun fort joliment. barometre de Ramsden. un appareil complet pour les experi-ences sur l'air fixe. un fort beau Niveau, avec division de limbe, boussolle, et lunette achromatique. un beau Theodolite. un Quadrant, ou plutot Octant, d'Hadley. un Quart de Cercle astronomique de 30 guinées: et finalement la plus belle piece du Cabinet *l'orrerij* de 85 guinées.

These instruments were probably ordered in 1775, and in 1776 they were already in Leuven. This emerges from what Magellan wrote to Thijsbaert from London on August 9, 1776 [25]:

> Si vous etes contant de vos orreries, je le suis aussi bien que vous le soyez: Car c'est mon plus grand plaisir que de rendre service à qui l'entend et en connoit la valeur. Vous n'avez rien à faire avec Adams pour le payement de ces instrum. C'est moi qui les ai payés, comme toujours, avant de les envoyer.

Again in 1783 Thijsbaert bought a number of instruments in London via Magellan, including an astronomical clock and an Adwood's machine for demonstrating the laws of gravity.

It was not part of Thijsbaert's task to give the lectures in experimental physics himself. His letter of appointment specified only that the profes-sors should discuss with him the choice of the experiments. As before, the lectures were given by four professors, one from each College. Neverthe-less, Thijsbaert's influence on the teaching of experimental physics in Leuven was very great. Besides a *Geometria Elementaria et Practica* (1774), he published a handbook on optics and perspective for the students [26]. On top of that he designed several series of engravings to serve as illustrative material for the physics lectures. They were sold to the students, and the proceeds went to the School of Experimental Physics. Alongside the engravings for optics and perspective from his handbook and a series on astronomy, there also appeared the series *De Motu, De Gravitate, Mechanica, De Hydrostatica, De Aere, De Lumine,* and *De Electricitate.*

The most famous of the professors to give the lectures on experimental physics in the last decades of the 18th century in Leuven was Jan Pieter Minckelers [27]. In 1771, not yet 23 years old, he was appointed professor of philosophy in Falcon College. Together with Thijsbaert and Karel van Bochaute, who taught chemistry in the Faculty of Medicine, he himself undertook some scientific research. In search of a gas that was lighter than air to fill balloons, he discovered in 1783 a simple and economical process for making coal gas [Minckelers 1783, Jaspers and Roegiers 1983]. In 1785 he was the first to use this gas to illuminate his lecture room.

After the death of the Empress Maria Theresia in 1780, her son Joseph II got the opportunity to put his political ideas into practice. In 1781 he traveled through all of his possessions in the Southern Netherlands and, on June 21, came to Leuven, where he was received by the representatives of the University. He visited not only the theater of anatomy and the botanical garden, but also the *Vicus*, where the School of Experimental Physics was housed. With the intention of bringing the University wholly under royal authority, in line with the Vienna model, he began to take a series of measures. Some acquired rights and privileges of the University of Leuven were abolished; this led to a break between the University and the government, and ended in the decree of July 17, 1788 transferring the whole University, with the exception of the Faculty of Theology, to Brussels. Under the supervision of Thijsbaert and assisted by Minckelers, the physics collection was transferred to the former Jesuits' College in Brussels. Thijsbaert drew up a new curriculum for philosophy according to the Vienna model. The duration of studies was increased from two to three years, and every course was entrusted to one professor. On October 11 Minckelers was promoted to full professor in physics. Thijsbaert was confirmed in his function as Director of the physics laboratory.

The reformed University was not granted a long life. The Brabant Revolution at the end of 1789 drove out the Austrian armies, and in January 1790 the independence of the Etats-Belgique-Unies was declared. The possessions of Leuven were brought back from Brussels, and on

March 1 the reestablished University solemnly opened its doors. Professors who had cooperated in the transfer of the Faculties to Brussels lost their academic functions. These included Thijsbaert and Minckelers.

The reestablished University was, however, not granted peace and quiet for study and teaching. At the end of 1790, Austrian troops reconquered the Southern Netherlands. About two years later, in November 1792, it was the French who occupied Leuven, only to abandon it once again in March 1793 to the Austrians. In June 1794 followed a second French occupation, and on October 1 of the following year the Southern Netherlands were officially incorporated in France.

The end of the University of Leuven was fast approaching. During the last turbulent years few students enrolled, whereas many professors left or were banned. In a decree of October 25, 1797 the centuries-old University was suppressed and its possessions requisitioned. The books, instruments, and collections that were in the *Vicus* also underwent the same fate.

Meanwhile, in mid-1797, an *ecole centrale* had already started in Brussels. Jan Baptist van Mons (1765–1842) (see [Debiève 1985]) was appointed professor of experimental physics and chemistry. In order to equip a laboratory for physics and chemistry, the confiscated collection in Leuven was raided. By November 11, 1797, the departmental administration had already decided that "Le citoyen Van Mons, professeur de chimie et de physique expérimentale, se rendra incessamment à Louvain et y choisira dans les dépôts provenant de l'université supprimée les instrumens et ustensiles de chimie et de physique dont il peut avoir besoin pour son cours. Ces objets seront transportés à Bruxelles" [De Vreught 1938, 55].

Over the course of 1803 and 1804 the *ecoles centrales* were abolished and replaced by lycea. The lyceum in Brussels was opened in mid-1803. The suggestion was made to the director that the machines and instruments left behind in the lecture rooms of the *ecole centrale* be taken and used to set up a physics laboratory. After inspecting them, the director declared that they were in a bad state of repair and were of no use. They were stored in two rooms, where they could still be seen in 1814. The slow disintegration of this collection that had been built up with so much difficulty had thus begun.

NOTES

[1] Concerning the teaching of physics at the University of Leuven during the 17th and 18th centuries, see Vanpaemel [1986, 1988].

[2] Copy in the archives of the Episcopal Seminary in Ghent, Fonds Van de Velde, No. 315b.

[3] For more details on the *Camera experimentalis* and for the full text of the regulations, see Van Tiggelen [1988].

[4] Patrice François, count de Neny (Brussels, 1716–Brussels, 1784), graduate in canon and civil law of Leuven, secretary and member of the Privy Council in 1744, and chief president from 1758 until 1783 (for more biographical details see Carton de Wiart [1943]).

[5] General Archives of the Realm, Brussels (henceforth shortened as G.A.R.), Conseil Privé autrichien, 1068A.

[6] G.A.R., Conseil Privé autrichien, 1078A, ff. 365–366.

[7] Joannes Petrus Sauvage, born in Luxemburg on July 13, 1725, graduated at Leuven in February 1754 as *juris utriusque licentiatus*. On June 13 of the same year, he was appointed professor of philosophy at Falcon College. He died in Leuven on June 25, 1771. (see Vanpaemel [1987a]).

[8] For information on Henri de Seumoy (Brussels, 1720–Brussels, 1798) see Vanpaemel [1990].

[9] G.A.R., Conseil Privé autrichien, 1078A, ff. 315–316.

[10] G.A.R., Ancienne Université de Louvain, No. 771.

[11] John Turberville Needham (London, 1713–Brussels, 1781) was ordained a priest at Cambrai on May 31, 1738. From 1736 to 1740 he taught rhetoric at the English College of Douai. In 1740 he became director of the school for Catholic youth at Twyford, near Winchester. In 1744 he went to Lisbon to teach philosophy in the English College. From 1746 on, he supported himself by accompanying young English Catholic noblemen on the grand tour, until he settled in Brussels in 1768 as director of what was to become the Royal Academy of Brussels (for more biographical details, see Westbrook [1974]).

[12] Autograph in G.A.R., Conseil Privé autrichien, 1078B, ff. 160–161.

[13] Autograph in G.A.R., Conseil Privé autrichien, 1078B, ff. 158–159.

[14] Corneille François de Nelis (Mechelen, 1734–Florence, 1798) graduated at Leuven in 1760 as *theologiae licentiatus*. From 1758 to 1768 he was librarian of the University. Philosopher, historian, and scholar, he was one of the founding members of a Literary Society, which in 1772 became the Imperial Academy. In 1785 he was consecrated a bishop of Antwerp (see Price [1970]).

[15] Autograph in G.A.R., Conseil Privé autrichien, 1078B, ff. 156–157.

[16] Autograph in G.A.R., Conseil Privé autrichien, 1078B, ff. 162–163.

[17] G.A.R., Chancellerie des Pays-Bas autrichiens, No. 442 D69/3, Lit. L.

[18] G.A.R., Chancellerie des Pays-Bas autrichiens, No. 442 D69, Lit. L.

[19] Joannes Franciscus Thijsbaert (Waasmunster, 1736–Harelbeke, 1825), *magister artium* in 1756, professor of philosophy at Castle College in 1759, excluded from the University in 1790 (see Vanpaemel [1984, 1987b]).

[20] G.A.R., Conseil Privé autrichien, 1078B, ff. 327–328.

[21] G.A.R., Ancienne Université de Louvain, No. 794.

[22] Letter from Professor P. Wuyts, dean of the Faculty, to Neny, G.A.R., Conseil Privé autrichien, 1076A.

[23] G.A.R., Ancienne Université de Louvain, No. 771.

[24] G.A.R., Conseil Privé autrichien, 1076A.

[25] Autograph in the University Library of Liège, MS. 2617 D, ff. 193–194.

[26] Published, without the author's name, as *Elementa Opticae et Perspectivae*. Lovanii, E Typographia Academica, 1775.

[27] Born at Maastricht in 1748, where he died on July 4, 1824 (for more biographical details, see Jaspers [1983]).

REFERENCES

Carton de Wiart, H. 1943. *Nény et la vie belge au 18ème siècle*. Brussels: Office de Publicité.

Debiève, M. 1985. Un pharmacien révolutionnaire: Jean-Baptiste Van Mons (1765–1842). In *Figures de professeurs de pharmacie à l'Université de Louvain*, pp. 7–93. Louvain-la-Neuve: Cabay.

De Vreught, J. 1938. L'enseignement secondaire à Bruxelles sous le Régime français: L'ecole centrale—Le Lycée. *Annales de la Société d'Archéologie de Bruxelles* **42**, 5–134.

Jaspers, P. A. T. M. 1983. *J. P. Minckelers, 1748–1824*. Maastricht/Leuven: Stichting Historische Reeks/Universitaire Pers.

—, and Roegiers, J. 1983. Le Mémoire sur l'air inflammable de J. P. Minckelers (1748–1824). *Lias* **10**, 217–251.

Minckelers, J. P. 1783. *Mémoire sur l'air inflammable tiré de différentes substances*. Louvain: (publisher not mentioned).

Price, W. J. 1970. Nelis, Cornelis-Franciscus. In *Nationaal Biografisch Woordenboek*, Vol. 4, Col. 609–625. Brussels: Koninklijke Academiën van België.

Vanpaemel, G. 1984. Jan Frans Thijsbaert (1736–1825) en de School voor Experimentele Fysika te Leuven. *Tijdschrift voor de Geschiedenis der Geneeskunde, Natuurwetenschappen en Techniek* **7**, 172–182.

— 1986. Echo's van een wetenschappelijke revolutie. De mechanistische natuurwetenschap aan de Leuvense Artesfaculteit (1650–1797). *Verhandelingen van de Koninklijke Academie voor Wetenschappen, Letteren en Schone Kunsten van België, Klasse der Wetenschappen*, No. 173.

— 1987a. Sauvage, Joannes Petrus. In *Nationaal Biografisch Woordenboek*, Vol. 12, Col. 656–658. Brussels: Koninklijke Academiën van België.

— 1987b. Thijsbaert, Joannes Franciscus. In *Nationaal Biografisch Woordenboek*, Vol. 12, Col. 726-730. Brussels: Koninklijke Academiën van België.

— 1988. Experimental Physics and the natural science curriculum in eighteenth century Louvain. *History of Universities* **7**, 175–196.

— 1990. Henri de Seumoy. In *Nationaal Biografisch Woordenboek*, Vol. 13, Col. 747–749. Brussels: Koninklijke Academiën van België.

Van Tiggelen, B. 1988. Du règlement de la 'Camera experimentalis' à la 'Schola experimentalis': Une page méconnue de l'histoire de la faculté des arts de Louvain au XVIIIe siècle. *Lias* **15**, 129–143.

Westbrook, R. H. 1974. Needham, John Turberville. In *Dictionary of Scientific Biography*, C. C. Gillispie, Ed., Vol. 10, pp. 9–11. New York: Charles Scribner's.

Zur Gründungsgeschichte der Polytechnischen Gesellschaft zu Leipzig, 1825–1827

Hans Wußing

Braunschweiger Straße 39, D-04157 Leipzig, Germany

In the Leipzig city archives there exist many documents on the history of the "Polytechnische Gesellschaft" (founded in 1825), hitherto fairly unknown, giving important historical evidence of the interaction between the developing sciences and the rise of industry during the first decades of the 19th century. This will be examplified here by two subjects: 1. Documents are presented illustrating the history of the foundation of this pioneering Leipzig group (illuminating its scientific and commercial strategies). 2. Immediately after the start of the first German long-distance railroad between Leipzig and Dresden in 1835, the electromagnetic telegraph signalling system constructed by Gauß and Weber in Göttingen was discussed during some meetings of the Polytechnic Society, with regard to both scientific and economic problems.

I

Sachsen gehört seit dem 16. Jahrhundert zu den am stärksten wirtschaftlich entwickelten Gebieten Deutschlands, mit allen sich auf dieser Grundlage herausbildenden kulturellen Leistungen sowie der Institutionalisierung von Wissenschaft, Technik und Verwaltung. 1698 wurde—um Beispiele heranzuziehen [1]—in Leipzig die erste Staatsbank Deutschlands gegründet. Unter Kurfürst Friedrich August I. ("August der Starke"), seit 1697 zugleich König von Polen, wurde in Freiberg die montanwissenschaftliche Ausbildung und Forschung systematisch gefördert; 1765 wurde dort die Bergakademie als erste derartige Hochschule gegründet, die mit A. G. Werner eine internationale Spitzenstellung erreichte. 1710 kam es zur Gründung der Königlichen Porzellanmanufaktur, wie überhaupt zahlreiche Manufakturgründungen einen wesentlichen Bestandteil merkantilistischer Wirtschaftspolitik in Sachsen darstellen. Schon im 18. Jahrhundert besaß Sachsen eine hochentwickelte, auf Wasserkraft beruhende Textilindustrie. Von 1732 bis 1754 wurde im Verlag von J. H. Zedler in Halle/Leipzig das hochberühmte 68bändige "Große Vollständige Universallexikon Aller Wissenschaften und Künste" gedruckt. 1764 wurde in Leipzig eine "Ökonomische Gesellschaft" mit dem Ziel gegründet, landwirtschaftliche und gewerbliche Produktion durch Einbeziehung technischer Verbesserungen zu heben. In dichter Folge entstanden wissenschaftliche Gesellschaften: 1774 die Fürstlich Jablonowskische Gesellschaft in Leipzig, 1779 die Oberlausitzer Gesellschaft der Wissenschaften in Görlitz, 1784 die Philologische Gesellschaft in Leipzig, 1787 die Naturforschende Gesellschaft in Leipzig. 1778 wurde das erste Lehrerseminar für Volksschullehrer in Dresden-Friedrichstadt ins Leben gerufen.

Obwohl das mit Frankreich bzw. Napoleon verbündete Sachsen (Königreich seit 1806) durch Festlegungen des Wiener Kongresses von 1815 mehr als die Hälfte (58%) seines Staatsgebietes und fast die Hälfte der Bevölkerung (42%) verloren hatte, besaß Sachsen bei der am Ausgang des 18. Jahrhunderts einsetzenden sogenannten Industriellen Revolution eine besonders günstige Ausgangsposition. Trotz der bewußt konservativen Haltung der sächsischen Herrscher gelangte Sachsen schon zu Beginn des 19. Jahrhunderts in die Rolle eines Wegbereiters der Industriellen Revolution in Deutschland. Einige wenige Angaben mögen dies belegen: 1816 Gründung der Forstakademie Tharandt, 1816 Inbetriebnahme der ersten Gasanstalt auf dem europäischen Festland, 1818 mechanischer Baumwollwebstuhl in Zschopau, 1820 die erste Dampfmaschine im sächsischen Steinkohlenbergbau, 1822 wurde die "Gesellschaft deutscher Naturforscher und Ärzte" gegründet. Mit der Aufstellung einer Dampfmaschine in Chemnitz entwickelte sich diese Region zu einem bedeutenden Industriezentrum, 1823 folgt das erste Walzwerk, 1825 wurde in Leipzig der "Börsenverein des Deutschen Buchhandels" begründet, 1828 erfolgte die Gründung der Polytechnischen Schule in Dresden, aus der die spätere Technische Hochschule, die heutige Technische Universität hervorgehen wird. B. G. Teubner hatte 1811, Anton Philipp Reclam 1828 seine verlegerische Tätigkeit aufgenommen, beide in Leipzig. Und 1829 wurde der Sächsische "Industrie-Verein" vom sächsischen Hofe bestätigt, als übergreifende Organisationsform zur Belebung des Fabrikwesens, des Maschinenbaues, von Handel und Verkehr.

Die Reformperiode Sachsens begann 1830 mit der Einsetzung B. A. von Lindenaus zum Kabinettsminister; 1834 erhielt die Leipziger Universität endlich eine moderne liberale Verfassung. 1835 verkehrte das erste Dampfschiff auf der Elbe, ausgerüstet noch mit einer englischen Dampfmaschine; 1836 schon wurden zwei Elbdampfer in heimischer Produktion in Dresden-Übigau gebaut. Die entscheidende Person ist J. A. Schubert (1808–1870), der die erste Lokomotive Sachsens, die "Saxonia," konstruierte. Schon 1835 war der Bau der Leipzig-Dresdener Eisenbahn, der ersten deutschen Fernlinie, begonnen worden. Der erste Eisenbahntunnel Deutschlands wurde bei Oberau gebaut. 1836 wurde die Höhere Gewerbeschule in Chemnitz begründet. (Auch hierzu findet der Leser weitere Einzelheiten in [Naumann 1991].)

II

Ein Spezifikum der Industriellen Revolution ist der Grundgedanke polytechnischer Bildung und Ausbildung sowie die Institutionalisierung der Ingenieurausbildung in polytechnischen Schulen nach dem Vorbild der 1793 während der Französischen Revolution in Paris gegründeten Ecole Polytechnique. In diesen allgemeinen und spezifischen Zusammenhang seien die nachfolgenden Ausführungen zur Geschichte der Industrialisierung Sachsens eingeordnet.

Im Stadtarchiv der Stadt Leipzig befindet sich ein fast völlig unerschlossener außerordentlicher Archivbestand (Kap. 35, Nr. 2, Band 1 bis 28) zur Geschichte der "Polytechnischen Gesellschaft" in Leipzig, deren Gründungsgeschichte auf das Jahr 1825 zurückgeht. Die Erschließung [2] des äußerst aussagekräftigen Quellenmaterials wird—so viel kann nach ersten Detailstudien schon jetzt festgestellt werden—ein Grundelement sächsischer Geschichte des 19. Jahrhunderts—die rasche Industrialisierung—von einem neuen Blickwinkel her verdeutlichen: Der hohe Grad der Industrialisierung Sachsens schon zu Beginn des 19. Jahrhunderts sowie die Entwicklung der Gewerbe beruhen nicht unwesentlich auf dem Grundgedanken der polytechnischen Bewegung [3], die Fortschritte der modernen Naturwissenschaften am Beginn des 19. Jahrhunderts systematisch daraufhin zu mustern, inwieweit sie für Gewerbe, Industrie und Produktion nutzbar zu machen seien.

Diese Grundidee läßt sich verfolgen anhand der Protokolle der alle 14 Tage stattfindenden Begegnungen ihrer Mitglieder, am Ausbildungsprogramm der von der Polytechnischen Gesellschaft betreuten Gewerbeschule, anhand der ihr angeschlossenen Sammlung von Geräten und Modellen, anhand der Statuten, an der Reaktion des Sächsischen Hofes und des

Rates der Stadt Leipzig auf die Vorstöße, Bitten und Forderungen der Polytechnischen Gesellschaft. Insgesamt dürfte es sich, trotz erkennbarer Lücken und einiger Verluste im Bestand, um ein Archivmaterial handeln, das in seltener Deutlichkeit das enge Wechselverhältnis von Fortschritt in den Naturwissenschaften mit der raschen Industrialisierung während der Industriellen Revolution dokumentiert, nicht nur am Resultat erkennbar, sondern auch aus den Absichtserklärungen der seinerzeit agierenden Persönlichkeiten. Die Polytechnische Gesellschaft Leipzigs hat, so viel sei hier pauschalisierend zunächst festgestellt, in der Zeit bis etwa 1850/1860 eine bedeutende Rolle für Leipzig und sogar für ganz Sachsen gespielt, trat dann aber gegenüber der 1828 gegründeten Polytechnischen Schule bzw. der aus ihr hervorgegangenen Technichen Hochschule zu Dresden in den Hintergrund. Lediglich die Leipziger Gewerbeschule, Bestandteil der bis nach 1945 existierenden Polytechnischen Gesellschaft, konnte ihre Arbeit mit deutlich erkennbarer Wirkung in die Zeit des Deutschen Kaiserreiches und darüber hinaus fortsetzen.

III

Hier, in einem kurzen Beitrag, den ich meinem langjährigen Kollegen Chr. J. Scriba in Dankbarkeit überreichen möchte, kann natürlicherweise nur eine Kostprobe zur Geschichte der Polytechnischen Gesellschaft zu Leipzig geboten werden. Ich wähle dazu, sozusagen exemplarisch, zwei Aspekte aus: Erstens die Initialphase der Gesellschaft, in der 1825 erste Schritte zur Etablierung der Gesellschaft unternommen wurden, wobei die Absichtserklärungen eben den Grundgedanken der Polytechnischen Bewegung hervortreten lassen. Zweitens greife ich aus den umfangreichen Protokollen der regelmäßigen Sitzungen den Begiun der Diskussion um die Ausrüstung der Eisenbahnlinie Leipzig-Dresden, der ersten deutschen Fernstrecke, mit einer elektromagnetischen Telegraphenverbindung im Gefolge der Entdeckung von Gauß und Weber in Göttingen heraus.

IV

Aus der Gründungsgeschichte der Polytechnischen Gesellschaft soll hier das Protokoll der allerersten Begegnung einer Initiativgruppe zitiert werden [4]. Der Text wird transliteriert, mit allen orthographischen Besonderheiten, auch zeilengerecht. Wörter, die im Original gestrichen wurden, sind in geschweifte Klammern { } eingeschlossen.

[*Kap. 35, Nr. 2. Band 3, Blatt 1, Vs*]
Protocol.

Leipzig am 21^{en} Oct. 1825

Anwesende:
Herr Osterland, Mechanikus;
Herr [...pe]; Kameralist;
Herr Pohl, Prof. d. Oekonomie und Technologie.
Herr M. Quarch [?], Privatgelehrter,
Herr M. Rüming, Besitzer einer Kunsthandlung.
Herr Wolf, Kaufmann.

Es versammelten sich neben verzeichnete
Freunde der Gemeinnützigkeit, um den Ver
such zu machen, eine Geselschaft zu consti
tuiren, welche sich so wohl die Gewerbswißen
schaften, als auch nach Möglichkeit das Ge
werkswesen selbst gemeinschaftlich zu
beförden zur Aufgabe macht, um so einen in Leipzig längst
laut gewordenen Wunsche entgegen zu
kom̄en

Zunächst sprach der Prof. Pohl einige Worte als Einleitung.
    In der vorläufigen Verhandlung über
die Constituirung einer solchen Gesel
schaft, wurden folgende Puncte näher
verabredet und zur Richtung für die
eingeleitete Thätigkeit angenom̄en.
1.) Zweck. Die Gesellschaft nim̄t zwar das
    ganze Gewerbswesen in ihr [?] Bereich auf, wen̄
    es Bezug auf die technischen Gewerbe hat, macht
    sich aber die letzteren zu beförden zur Haupt
    sache.
    Sie glaubt diesen Zweck zu erreichen
    1) wenn sie die Gewerbswißenschaften in theo
       retischer Hinsicht fördert;
    2) wenn sie deren Anwendung auf
       die Gewerbe selbst näher ermißt [?];
    3) wenn sie sich besonders bemüht,
       die neuer Erfindungen bekannt zu machen,
       und nähere Prüfungen anzustellen.
    4) Zu diesem behufe {wird} die Geselschaft
       Samlungen an Büchern, Abbildungen
       {Sachen}, Stoffe und Waaren und [?] was den Zweck
       sonst fördern kann, anlegen.

2. Zeit der Zusammenkünfte. Die Versam̅lungen
sollen vor der Hand Mittwochs, Abends 6 Uhr,
in der Behausung des Prof. Pohl, eine Woche um
die andere stattfinden.
3. Die Vorträge sollen theils mündlich, theils
schriftlich gehalten werden, und wurde verab
redet, daß ein schriftlicher Vortrag nicht über
eine Viertelstunde dauern solle, es wäre de̅n
dazu eine besondere Veranlaßung.
Auszüge aus Büchern und Zeitschriften wer
den ebenfalls willkommen sein ud. es sollen
die Herren Mitglieder besonders dazu veranlaßt
werden. Man hofft so die zeitgemäße Vorgän
ge zu faßen [?] und vorläufig kennen zu lernen.

[*Kap. 35, Nr. 2, Band 3, Blatt 1, Rs*]
3.) In Ansehung der Geldmittel ward beschloßen,
daß jedes eintretende neue Mitglied zur Ge
selschaftskaße Einen Thaler Eintrittsgeld,
und jährlich zwei Thaler in zwei Terminen
zahlen soll. Diese jährlichen Beiträge sollen
jedoch erst mit dem neuen Jahre ihren An
fang nehmen. Demnach zahlt bis Weihnach
ten l.J. das Mitglied nur Einen Thaler.
  Das Direktorium soll jedoch befugt sein,
wenn das eintretende Mitglied sehr unbe
mittelt ist, demselben nicht nur das Eintritts
geld, sondern auch die jährlichen Beiträge ent
weder ganz oder zum Theil zu entlaßen.
  Endlich wurde beschloßen die nächste Versam̅lg
zum 2ten November zu halten, welche ebenfalls
zunächst dazu bestim̅t sein solte, die Ange
legenheiten der Geselschaft zu consolidiren.
In dieser Absicht fand man es berathen
unter seinen Bekannten noch solche
Männer zur Theilnahme an dieser Versam
lung zu veranlaßen, von welchen es
anzunehmen ist, daß sie das Intereße der
Geselschaft befördern werden. {Eine}
Eine noch nicht festgesetzte Anzahl der er
sten Mitglieder, sollen lebenslänglich
als Stifter der Geselschaft, stets Norm
gebend und entscheidend an der Spitze der
Geselschaft stehen und so nach als beständige
Deputirte oder Vorsteher der Geselschaft
gelten.
  Als Geschenke gingen ein
Pohls Beiträge der neuesten Geschichte
der Landwirthschaft 1^r und 2^r Bd. v. Verf.

Die gegenwärtige Sitzung [....] wurde mit
dem lebhaften Gefühle geschloßen, daß durch
sie eine Geselschaft ihren Anfang genom
men habe, die künftig wichtigen Einfluß
auf das Gewerbswesen überhaupt und das
vaterländische insbeson. {Einfluß} haben könte

Prof. Hans Friedrich Pohl,
Protocolant.

Das Dokument zeigt in seltener Klarheit die Absichten der Initiativgruppe. Der Professor Pohl spielte in den ersten Jahren eine herausragende Rolle in der Polytechnischen Gesellschaft, insbesondere bei den
Bemühungen, öffentliche Anerkennung für die Gesellschaft zu erhalten.

Es ist hier nicht der Platz, anhand der realen Geschichte der Polytechnischen Gesellschaft zu verfolgen, wie sich im einzelnen die Vorstellungen
der Initiativgruppe haben realisieren lassen. Das Grundanliegen jedoch
konnte verwirklicht werden.

Anfang 1827 hatte sich Pohl an den Sächsischen Hof gewandt mit der
Bitte um das königliche Privileg. Der Hof unterstützte im März in einem
Schreiben an den Stadtrat von Leipzig grundsätzlich das Anliegen der
Polytechnischen Gesellschaft, machte aber die endgültige Zusage zur
Erteilung der "landesherrlichen Bestätigung" von einigen Auflagen ab-
hängig. Ein endgültiges Statut [5] wurde in Leipzig 1829 gedruckt, vom
sächsischen Hofe "confirmiert," d.h. mit Bestätigung durch den sächsischen
König.

V

Ich greife aus der Fülle der Versammlungsprotokolle, die schätzungsweise
2000 engbeschriebene, großformatige Seiten umfassen, einen Verhand-
lungsgegenstand heraus, der einerseits zeigt, daß die Mitglieder der
Gesellschaft die aktuellen Entwicklungen sehr wohl zu verfolgen imstande
waren, und andererseits, daß sich Wille zum technischen Fortschritt mit
wissenschaftlicher Gründlichkeit paarte.

Die Verhandlungen der Gesellschaft vom Spätherbst 1835 beziehen sich
u.a. auf die Möglichkeit, die Eisenbahnlinie Leipzig–Dresden mit elektromagnetischen Telegraphen auszurüsten. Am 6. Mai 1835 hatte die
Leipzig–Dresdener-Compagnie die staatliche Konzession für den Bau
dieser Eisenbahnlinie erhalten; im Herbst war mit dem Bau der Eisen-
bahnbrücke bei Wurzen begonnen worden [6]. Wenige Monate später
schon debattierte die Polytechnische Gesellschaft darüber. Gauß und

Wilhelm Weber hatten im April 1833 in Göttingen einen voll funktionsfähigen elektromagnetischen Telegraphen eingerichtet, der eine Strecke von beinahe zwei Kilometern zwischen dem physikalischen Kabinett und der Sternwarte überbrückte. Erst 1845 zerstörte ein Blitzschlag die Leitung. Als Leitungsträger hatte man gefirnißten Eisendraht verwendet. Die Begeisterung über die Möglichkeiten eines vervollkommneten elektromagnetischen Telegraphen muß sehr groß gewesen sein; Gauß sprach schon 1835 davon, daß die elektromagnetische Telegraphie "zu einer Vollkommenheit und zu einem Maaßstabe gebracht werden könnte, vor der die Phantasie fast erschrickt."

Ebenfalls 1835 hatte der Münchener Professor C. A. Steinheil (1801–1870), Schüler von Gauß und im Kontakt mit ihm, eine Telegraphenlinie zwischen der eben in Betrieb genommenen Kurzeisenbahnlinie Fürth–Nürnberg einzurichten gesucht. Steinheil folgte der Anregung von Gauß, statt der teuren Leitungsdrähte die schon vorhandenen Schienen für die Hin- und Rückleitung des Stromes zu benutzen. Diese Idee erwies sich als Fehlschlag—man hatte einen Kurzschluß wegen des unerwartet hohen Leitungsvermögens des Erdbodens. Immerhin war nun klar, daß man eine isolierte Leitung benötigen würde.

Auf diesem Hintergrund diskutierte man in Leipzig. Der Initiator war Professor Weber; das Protokoll vermerkt nicht, um welchen der Brüder—Ernst Heinrich (Anatom und Physiologe) oder Eduard Friedrich (Anatom)—oder den Göttinger Wilhelm Eduard Weber (Physiker) es sich handelte [7] Aus dem weiteren Zusammenhang ergibt sich das, daß Ernst Heinrich Weber gemeint ist.

Der elektromagnetische Telegraph war Gegenstand (mindestens) zweier Versammlungen, am 27. November und am 11. Dezember 1835. Protokollauszug vom 27. November 1835:

[*Kap. 35, Nr. 2, Bd. 4, S. 48, Vs*]
In Abwesenheit des Directors
Herrn Prof. Erdmañ führte in
heutiger Versamlung der Vicedirector
Herr Stadtrath Porsche den Vor-
sitz und eröffnete, nachdem das
Protocoll von voriger Versamlung
vorgelesen und genehmigt worden
war die Sitzung.
Das Mitglied Herr Prof.
Weber hielt einen Vortrag
über die Möglichkeit der Mittheilung in weite Ferne nach Art

der Telegraphie durch Galvanismus mittelst eines von
einem Orte zum andern fortge
leiteten mit Seide umsponnenen
Kupferdrahtes. Er beschrieb

[*Kap. 35, Nr. 2, Bd. 4, S. 48, Rs*]

die von den Mathematikern
Gauß und Weber in Göttingen
erfundene Methode und machte
dieselbe den Anwesenden [....?]
durch Zeichnung anschaulich
und deutlich. Die Ausführ-
barkeit einer solchen Com̄uni-
cation selbst für die weitesten
Entfernungen kañ nach dieser
Darstellung nicht bezweifelt
werden und ist sogar vielleicht
schon nahe [?], indem sie mit
der ihrer Ausführung nahen
Eisenbahn zwischen Leipzig
und Dresden verbunden
werden soll. Die Vortheile,
welche diese Einrichtung von
der Telegraphenverbindung
haben würde, sind sehr be
deutende. Vor allem ist die
Mittheilung noch viel schneller
möglich als durch Telegraphen,

[*Kap. 35, Nr. 2, Bd. 4, S. 49, Vs*]

die selbst bei einer über den ganzen
Umfang der Erde geschehenden
Mittheilung {solcher Art} durch
Galvanismus der Zeitabschnitt,
biñen welcher dieß geschehe kaum
bemerkbar seyn würde. Sodañ [?]
findet auch {bei dies} hierbei {nicht}
wie {bei den Telegraphen} eine
Unterbrechung oder Umöglich [?]
keit der Mittheilung, bei trübem
Wetter oder bei Nacht, wie solches
bei den Telegraphen der Fall ist
nicht statt.

Die Versammlung vom 27. November 1835 beschäftigte sich noch mit
weiteren Themen. In der darauffolgenden Versammlung vom 11. Dezember 1835 ging man nochmals auf die "galvanische Telegraphenverbindung"

(optische Telegraphen waren weithin in Gebrauch, auch auf Fernlinien, z.B. zwischen Berlin und Koblenz) ein. Insbesondere äußerte sich Professor Erdmann (Chemiker an der Leipziger Universität) skeptisch, vor allem hinsichtlich der Kosten. Protokollauszug aus der Sitzung vom 11. Dezember 1835.

[*Kap. 35, Nr. 2, Bd. 4, S. 50, Vs*]

Nach Verlesung des Protocolls sprach der
Herr Director Prof. Erdmann über die
Schwierigkeiten welche die beabsichtigte
electro-galvanische Telegraphenverbindung,
nach der Theorie des Dr. Gauß in {Wittenberg} Göttingen
bei Anwendung auf die Leipzig—Dresden
Eisenbahn im Wege stehen, finden möchten
Er zählte 3 verschiedene Momente auf, erstens
die einfache Eisenbahngleis [?] 2) die mögliche
Ableitung 3) die Kosten, welche sich bis
Dresden von Leipzig aus, auf 60 Centner
Gewicht herausstellen würden. Der
Ableitung ließe sich durch Üiberspinnen
begegnen, aber auch dieses Mittel

[*Kap. 35, Nr. 2, Bd. 4, S. 50, Rs*]

möchte viel Kosten verursachen, da man
gewohnt ist, bespoñenen Drath theuer be-
zahlen zu müssen. Der Herr Director
glaubt, daß man zuerst einen Versuch
bis Wurzen anstellen werde.

Soweit die Durchsicht der weiteren Sitzungsprotokolle bisher ergeben hat, ist man auf dieses Thema der Signalgebung mittels elektromagnetischer Telegraphen im Eisenbahnwesen Sachsens zwar noch mehrfach zurückgekommen, hat aber dann die Sache fallengelassen. Dies wird—sozusagen im Quervergleich—auch bestätigt durch Erinnerungen von Zeitzeugen des frühen sächsischen Eisenbahnwesens. So berichten Ludwig Neumann (Damals Finanzrath) und Paul Ehrhardt (damals Bezirksmaschinenmeister) in der Zeitschrift "Civilingenieur," Bd. 36, 1890, mit "Erinnerungen an den Bau und die ersten Betriebsjahre der Leipzig—Dresdener Eisenbahn" über die Diskussion zum Signalwesen folgendermaßen:

Ein Vortrag des Dr. Wilhelm Crusius, stellvertretenden Vorsitzenden der Kompagnie, gehalten in der zweiten Generalversammlung derselben am 15. Juni 1836, empfahl 'die Errichtung eines galvanisch-magnetischen Telegraphen' und enthielt auch die Mitteilung der nach dem Gutachten des Professor Dr. Weber obwaltenden, 'ein solches Unternehmen als sehr sicher und zuverlässig' empfehlenden Umstände. Obwohl die Versammlung für eine Telegraphenanlage Leipzig–Wurzen 2000 Rth [8] genehmigte, konnte sich das Direktorium—infolge einer durch Magister Hülsse geleiteten Untersuchung, die eine etwa dreifach höhere Anschlagssumme ergab—doch nicht entschliessen, die Anlage auszuführen, worüber sich das Direktorium gegenüber der vierten Generalversammlung am 10. April 1838 folgendermassen rechtfertigte. 'Der unmittelbare Vortheil, welcher für die Gesellschaft aus der Benutzung der Telegraphen entspringen könnte, hat uns nicht bedeutend genug geschienen, um die Aufwendung eines so ansehnlichen Kapitals zu rechtfertigen, besonders da die finanzielle Lage des Unternehmens jede thunlichste Ersparniss doppelt zur Pflicht macht und die fragliche Anlage sich auch später noch ebensowohl ausführen lassen, wenn man das angemessen finden sollte.'

Sonach blieb es für die Eröffnung bei den Handfähnchen und den Lichtern der Handlaternen als urwüchsigste Signalmittel, ...

Die Signalisierung mittelst Flügel-Masten-Telegraphen wurde erst am 1. September 1840 eingeführt. [Neumann und Ehrhardt 1988, 33–35]

Da die in den "Erinnerungen ..." [Neumann und Ehrhardt 1988] wiedergegebenen Quellen verschollen oder verloren sind, lohnt sich eine Wiedergabe jener Passagen, in denen Neumann und Ehrhardt jener Feierlichkeiten gedenken, mit denen 1876 die Überführung der privaten Leipzig–Dresdener Eisenbahn in den Besitz des sächsischen Staates begleitet wurde. Sie schreiben 1890:

Wir können diese Erinnerungsblätter wohl nicht besser schliessen, als indem wir die Worte, welche der Staatsminister Dr. von Falkenstein in der denkwürdigen Landtagssitzung am 12. Mai 1876 zu Dresden, in welcher der Ankauf der Leipzig–Dresdener Eisenbahn durch den sächsischen Staat beschlossen wurde, gleichsam als Schwanengesang der aufgelösten ersten grossen Lokomotiv-Eisenbahngesellschaft Deutschlands nachrief, wiederholen:

'Es ist eine ernste Stunde, in der wir gewissermaassen Abschied nehmen von dem Direktorium der Leipzig–Dresdener Eisenbahn. Meine Herren! Ich bin vielleicht der Einzige oder wenigstens Einer der Wenigen, der so zu sagen der Geburtsstunde dieser Eisenbahn (als Königl. Kommissar) im Jahre 1835 beigewohnt hat. Ich entsinne mich noch sehr wohl der Zeit, da im Jahre 1835 die erste Generalversammlung von dem weltbekannten List und dem in unserem Lande ebenso bekannten Harkort geleitet wurde, wo diese beiden Männer, wenn auch von ganz verschiedenen Ansichten vielleicht ausgehend, aber doch in den Hauptpunkten zusammentreffend, dieses für die damalige Zeit—man kann wirklich sagen—unerhörte Werk besprachen und befürworteten und der Ausführung entgegenführten. Ich habe dann lange Jahre hindurch als Kommissar verschiedenen Sitzungen beigewohnt und mehr und mehr die Ueberzeugung gewonnen, mit welcher Solidität, mit welcher Umsicht das ganze Unternehmen geführt worden ist und wie sehr das Vaterland in der That diesem Direktorium damals Dank schuldig war, dass es den Mut hatte, in jener Zeit, vom Publikum mehr angefeindet als unterstützt, ein solches Unternehmen durchzuführen. Es hat sich das auch in der späteren Zeit bewährt; es ist bekanntlich die Verwaltung eine solche gewesen, die von allen Seiten nicht blos im Inlande, sondern auch im Auslande vollständige Anerkennung gefunden, ja sogar als Muster einer guten Eisenbahnverwaltung gegolten und anerkannt worden ist. Ich habe geglaubt, dass ich in diesem Augenblicke, wo nunmehr nach meiner Ueberzeugung feststehen dürfte, dass der Staat diese Bahn hoffentlich zum Segen des ganzen Vaterlandes übernehmen wird, also in diesem ernsten Augenblicke, dieses anerkennende Wort dem Direktorium der Leipzig–Dresdener Bahn gegenüber aussprechen darf.' [Neumann und Ehrhardt 1988, 48–49]

VI

Wie schon die Initiativgruppe der Polytechnischen Gesellschaft vorgeschlagen hatte, fanden die Versammlungen regelmäßig im Abstand von 14 Tagen statt. Die Themen waren überaus vielfältig. Im bunten Bild der Themen finden sich Produkte, die wir heute als Kunstgewerbe oder Spielereien einordnen würden, Erfindungen, die keine sich bewährende Produktion auslösen konnten trotz vielversprechender Prototypen, Diskussionen über wissenschaftliche Umwege und Irrwege, aber auch Erörterungen über zukunftsträchtige oder sogar wegweisende Erfindungen und Entdeckungen in Technik und Naturwissenschaften. Alle Bereiche der Naturwissenschaften waren einbezogen ebenso wie alle gewerblichen Zweige, von Heizung, Beleuchtung, Lederverarbeitung, Baumaterial bis hin zum Maschinenbau. Deutlich verschiebt sich von den 20er zu den 40er Jahren hin das Interesse vom Gewerbe zur Großindustrie, und deutlich rückt auch die Ausbildung der Arbeiterschaft in den Vordergrund. Die der Polytechnischen Gesellschaft angeschlossene Gewerbeschule wird seit der Mitte des Jahrhunderts zunehmend die Aktivitäten der Gesellschaft prägen.

ANMERKUNGEN

[1] Über diese und viele weitere Details der Entwicklung Sachsens, insbesondere auch auf kulturellem Gebiet, unterrichtet das Nachschlagewerk [Naumann 1991].

[2] Die Sächsische Landesregierung hat, in Verbindung mit der Sächsischen Akademie der Wissenschaften zu Leipzig, Fördermittel zur Aufarbeitung dieses

für die Geschichte Sachsens bedeutsamen Quellenmaterials bereitgestellt. Der Autor dankt den Mitarbeitern des Stadtarchivs für freundliche Unterstützung.

[3] Nach Auskunft von Herrn Dr. Priesner (München), der sich ausführlich mit der Geschichte des 1815 gegründeten Polytechnischen Vereins München befaßt hat, sind für München deutliche Anklänge an die europäische Aufklärung erkennbar. Vgl. dazu auch [Priesner 1989, 1983]. Ein Vergleich der Entwicklungen in München und Leipzig wäre interessant, ebenso ein Quervergleich mit der Geschichte der polytechnischen Schulen in Prag, Wien, Karlsruhe, München, Dresden, Stuttgart, Hannover, Kassel, Zürich, Lissabon, Kopenhagen, Riga und anderswo.

[4] Da die Protokolle (wie überhaupt der gesamte Bestand) in ständig wechselnden Handschriften verfaßt sind, teilweise bis in die Heftung der Blätter hineinreichen, sind die Texte gelegentlich schwer zu entziffern. Manche Eigennamen blieben unleserlich. Bei der Wiedergabe der Texte stehen in eckige Klammern eingeschlossene Punkte für unleserliche Buchstaben; ein in eckigen Klammern hinzugefügtes Fragezeichen deutet auf Unsicherheiten der Lesung hin.—Wegen des Umzuges des Stadtarchives (Wiederbenutzung ab 7. Juli 1994) war es nicht möglich, eine reproduktionsfähige Kopie des Originaltextes zu erhalten. Doch möchte ich der Leiterin des Archivs, Frau Dr. Berger, ausdrücklich danken, daß sie mir durch Anfertigung von Lesekopien einige Teile des Aktenbestandes auch während dieser Zwischenzeit zugänglich gemacht hat. [5] Die Bemühungen um die öffentliche Anerkennung der Polytechnischen Gesellschaft durch die Stadt Leipzig und die sächsische Regierung, verbunden mit einer Diskussion um die Statuten, finden sich im Kap. 35, Nr. 2, Bd. 1.

[6] Vgl. dazu wiederum [Naumann 1991].

[7] Gegenwärtig konnte noch kein Mitgliederverzeichnis der Polytechnischen Gesellschaft aufgefunden werden.

[8] = Thaler.

LITERATURVERZEICHNIS

Naumann, G. 1991. *Sächsische Geschichte in Daten*. Berlin und Leipzig: Koehler und Amelang.

Neumann, L., und Ehrhardt, P. 1988. *Erinnerungen an den Bau und die ersten Betriebsjahre der Leipzig–Dresdener Eisenbahn*. Leipzig: Zentralantiquariat der DDR.

Priesner, C. 1989. Förderung technischer Bildung außerhalb der Schulen. Polytechnische Vereine und technische Bildung. In *Technik und Kultur. Herausgegeben von der Georg-Agricola-Gesellschaft*, VDI Verlag, S. 235–259.

— 1983. Der Polytechnische Verein—seine Geschichte und sein Wirken für die Industrialisierung Bayerns im 19. Jahrhundert. *Journal of the Korean History of Science Society* 5, 59–80.

Index

Please note: All items are indexed according to their English spellings; thus *Casopeia* in German will be found here under "Cassiopeia." To avoid inconsistencies in the transcription of Arabic names, all of them are given in the usual English transcription. Commonly used words or objects are also alphabetized according to their English equivalents; thus *Fässer* is indexed here as "Barrels." Technical terms like *Einschneideverfahren* and *Meßtisch,* however, are alphabetized as they would be spelled in German. Arabic names like al-Kindī are indexed under *K*; names like von Hohenburg will be found under *H*.

A

Aachen, 321–322

Abacus, 307, 309, 312

Abelian groups, 115, 118, 120, 122

Abraham, Max (1875–1922), 321, 324, 328, 331–332, 341

Abū ʿAbd Allāh Muḥammad ibn Muʿādh (*see* Ibn Muʿādh)

Abū Saʿd al-ʿAlāʾ ibn Sahl (2nd half of 10th century), 209

Book on the Synthesis of Problems (K. Tarkīb al-masāʾil), 209

Abu l-ʿAbbās al-Faḍl b. Ḥātim al-Nairīzī (*see* al-Nairīzī)

Abu l-Ḥasan ʿAlī ibn Yaḥyā (died 888), 207–208

Abū l-Wafāʾ, (940–997/998), 288

Abū Jaʿfar al-Khāzin (died 961/971), 212

Abū Maʿshar (787–886), 208, 238–239, 253

Acceleration, 314–315

Achillini, Alessandro (1463–1512), 241

Acta eruditorum (see *Leipziger Akten*)

Adams, George (1750–1795), 356

Adelard of Bath (12th century), 173–175, 177–190, 192–194, 196–199, 201–203

Aegidius de Thebaldis of Parma (13th century), 239

Aerodynamics, 101–110, 128

Ahmad ibn Shākir (*see* Banū Mūsā)

Alberti, Leone B. (1401–1472), 51, 54

Alcinous (Alkinoos) (2nd century), 241

Aleksandrov, P. S. (1896–1982), 138, 140

Alfons X the Wise (1221–1284), 250

Algebra, 112, 120, 160–161, 280, 289, 292, 301, 307, 312–314, 316

Algebraically closed fields, 117–119, 121, 130

Algebraists (*aṣḥāb al-jabr*), 215

Algebraization, 315

Algorithms, 310

Alhazen (*see* Ibn al-Haitham)

ʿAlī ibn Abi l-Rijāl (11th century), 238–239

ʿAlī ibn Riḍwān (998–1061/1069), 238–239

"All-Peoples' Union for the Struggle for the Rebirth of Russia," 142, 144, 147

Alma Alta, 143

Alpetragius (*see* al-Biṭrūjī)

Altdorf, 149, 151, 155, 157–159, 161, 163

Academy, 153, 164, 166

Library, 154

University of, 154, 157, 163

Amburger, E., 88
American Mathematical Society, 131
American Philosophical Association, 113
Amico, Giovanni Battista (1512–1538), 241
Amsterdam
    Colloquium 1954, 116, 131
    International Congress of Mathematicians
        1954, 130
    University of, 112
Analysis, 264, 274
    18th century, 263
Anamorphoses, 3–4
    Mirror Anamorphoses, 4, 6
        Conic, 21, 23–24, 27
        Cylindrical, 17–25
    Perspectival, 4, 6–7, 9–16
    Simple, 14–15
Anaritius (see al-Nairīzī)
Andersen, Kirsti, 280
Angle trisection, 212–213
Angles (see also Horn angle), 315
Anti-Semitism, 94
Apian, Peter (1495–1552), 240
Apollonios (fl. ca. 200 B.C.), 162, 167
    Conics, 213
Application of areas, 213
Applications, practical, 269
Applied mathematics, 93–94, 97, 100–110,
    126
Approximate value (Näherungswert), 266
Aqātūn
    Book of Assumptions (Kitāb al-Mafrūḍāt),
        207–208
Arabic astronomy, 237–261
Arabic mathematics, 288–289, 295–297, 308
Arabic sources, 156
Archimedean mathematical corpus, 50–53,
    55
Archimedean solids, 54–55
Archimedes (ca. 287–212 B.C.), 55, 162, 213,
    314
    On the Measurement of the Circle, 162
    The Sand Reckoner, 162
    On the Sphere and the Cylinder, 162
Architects, 56–57
Argyros (see Isaak Argyros)
Aristarchos (ca. 310–230 B.C.), 167
Aristotelian theory of elements, 250
Aristotle (384–322 B.C.), 239, 241, 243, 254,
    302, 314, 317
Arithmetic, 160, 280, 307–309, 314, 316
    Fundamental operations of, 310, 314

Armillary spheres, 157
Artists' manuals, 49, 51
Association for Symbolic Logic, 113–114
Astrolabes, 152, 175
Astrological argument, 253
Astrology, 154, 157–158
Astronomical observations, 150
    Calculations, 162
Astronomical texts (Babylonian), 281–284
Astronomy, 151, 154, 160, 162–163, 165
Athens, 58
Atomic bomb, 95
Atzema, Eisso, 23
Augsburg, 153
Auria, J. (16th century), 165
Autolykos (fl. ca. 300 B.C.), 165, 167
Averroes (see Ibn Rushd)
Averroists, 256
Avicenna (see Ibn Sīnā)
Axioms, 280

B

Babylonians, 281–284, 292
    Cuneiform tablets, 280, 282
Bach, J. S. (1685–1750), 29–30, 45
Bacon, Roger (ca. 1219–ca. 1292), 315
al-Baghdādī, 238
Baldini, Ugo, 252
Balmer, H., 66
Banū Mūsā (fl. 850), 213
    Aḥmad ibn Shākir, 213
    On the Measurement of Plane and Curved
        Figures, 213
Barbaro, Daniele (1514–1570), 13–14
Barlaam (ca. 1290–ca. 1350), 167
Barrels, 162
Basel
    Euler-Archiv, 229
    University of, 224–226
al-Battānī (before 858–929), 237–238, 240,
    246, 248, 250, 252, 257, 259
Beatty, Samuel (1881–1970), 93–95, 97,
    124–126
Bede (672/673–735; see also Pseudo-Bede),
    288–289, 309
van Beethoven, L. (1770–1827), 38–42, 45
Behnke, Heinrich (1898–1979), 66, 74
Benz, Ulrich, 319
Berkeley, University of California, 105, 128

Berlin, 163, 167
  Akademie der Wissenschaften, 65–66,
    68–69, 75, 80, 88–89, 226
  Autographen-Sammlung Darmstädter, 85,
    90
  Soviet Embassy, 145–146
  University of, 75, 79, 89, 94, 122
Bernoulli
  Daniel I (1700–1782), 224, 226–228, 234
  Daniel II (1751–1834), 233
  Genealogy (family tree), 233
  Jakob I (1655–1705), 224–226, 233–234,
    265–266
    Ars conjectandi (1713), 225, 234, 265
    Opera, 234
  Jakob II (1759–1789), 224, 233
  Johann I (1667–1748), 223–226, 233–234
  Johann II (Jean) (1710–1790), 224, 226,
    233
  Johann III (1744–1807), 233 (passim)
  Niklaus I ("the younger") (1687–1759),
    223–236, 266
    Correspondence with Euler, 223–236
      (passim)
    Dissertatio de usu artis conjectandi in jure,
      225, 234
  Niklaus II (1695–1726), 224, 233–234
  Niklaus "the older" (1662–1716), 224, 233
Bers, Eli, 130
Besthorn, R. O., 173–180, 182, 185, 191–192,
  195, 197
Beth, E. W., 112, 129
  Beth's Theorem, 113–115
von Bethmann, Johann Philipp, 83
von Bethmann, Moritz, 83
Betting strategies, 273
Bettini, Mario (17th century), 27
Bhaskara (ca. 1250), 301–302
Bieberbach, Ludwig (1886–1982), 50, 139
Binomial coefficients, 267
  For large n, 267
Binomial distribution, 263–275
  Limit value theorem, 263–275
Binomials, 162
al-Biṭrūjī (fl. ca. 1190), 237–241, 243–246,
  248–249, 251–252, 254–256, 259
Bivector, 319, 325, 329–334
Blum, Reinhard, 76–77
Blumenberg, Hans, 253
Bölling, Reinhard, 76–77, 88

Bochaute, Karel van (1732–1793), 357
Bodleian Library (Oxford), 164, 167
Boethius (ca. 480–ca. 524), 161, 307
Bologna, Academy of, 226
Boltzmann, Ludwig (1844–1906), 87, 326, 331
Bolzano, Bernard (1781–1848), 302–303
Book of Assumptions (see Aqāṭūn and
  Thābit ibn Qurra)
Borchardt, Carl Wilhelm (1817–1880), 74–75,
  88
Borchardt, Rosa (1840–ca. 1910), 74–78
Born, I. (1724–1791), 30–33
Born, Max (1882–1970), 94
Bos, Henk, 23
Boston University, 113
Botany, 164
Bouguer, Pierre (1698–1758), 223, 227–228
  Theorema Bouguerianum, 228
Bradwardine, Thomas (ca. 1290/1300–1349),
  315
Brahe, Tycho (1546–1601), 150, 153–155, 166
Brahmagupta (598–after 665), 288, 301–302
  Brāhmasputasiddhānta (628), 288
Brauer, Richard, 94
Brentjes, S., 190
Breslau, 163, 167
Brightness (Helligkeit), 245
British Columbia, University of, 112
Britten, E. B. (1913–1976), 44–45
Bruno, Giordano (1548–1600), 254
Bryan, George Hartley (1864–1928), 321
Burali-Forti, Cesare (1861–1931), 329
Burckhardt, J. J., 66
Burgers, J. M., 103
Burkhardt, Heinrich (1861–1914), 326–327,
  337
Busard, H. L. L., 174, 178–179, 183, 194–197,
  200–201
Byzantine texts, 160

                    C

Calculi (counters), 307, 309, 311–312
Calculus, 263
Calendar reform, 150
Calendars, 151, 154, 157, 310
  Gregorian calendar, 157
Calonymos, Calo (first half of 16th century),
  240, 243
Cambridge, 337
Camerarius, Joachim (the older, 1500–1574),
  53, 164

Camerarius, Joachim (the younger, 1534–1598), 157, 163–164, 167
Campanus of Novara (died 1296), 315
Campbell, L. L., 127–128
Canada, 93–136
    Defense Research Board, 102
    National Aeronautical Establishment, 101
    National Research Council, 101, 105, 112
Canadian Broadcasting Corporation, 98
Canadian Mathematical Congress, 123–124, 131
Canadian Mathematical Society, 130
Canadian Summer Research Institute 1956, 120–121
Cardano, Girolamo (lat. Hieronymus) (1501–1576), 162, 279, 297–301
    Ars magna (1545), 297
Carleton College, 110
Carmody, Francis J., 243, 254–255
Cartography, 158
Carvallo, Emmanuel (1856–1945), 337
Cassel, 327
Cassiopeia, 153
Castellano, Filippo (1860–1919), 329
Categoricity, 117
Cathedral construction, 314
Catholic Church, 150
Caus, Salomon de (ca. 1576–1626), 15
Celsius, Anders (1701–1744), 284
Central Antireligious Museum, 142
Changing world view, 241
Chaplygin, S. A. (1869–1942), 143–144, 147
Charles, Prince of Lorraine (1712–1780), 347, 350, 353, 355
Chinese mathematics, 284–288
    Counting boards, 286
    Counting rods, 284
    Jiuzhang suanshu (Nine chapters), 285
    Systems of linear equations, 284–287
Chrysostom, Saint John (ca. 346–407), 164
Chuquet, Nicolas (died 1488), 287, 291–292
    Triparty (1484), 291
Ciphers, 307
Circles (see also Quadrature), 315
Clagett, Marshall, 50, 53, 203
Clairaut, Alexis Claude (1713–1765), 228, 233
Clausthal, 320–321
Clavius, Christoph (1538–1612), 237–238, 241, 248–259
Closed expressions, 271

CNRS (see Paris)
von Cobenzl, Karl Johann Philipp (1712–1770), 352
Cohen, Paul J., 93
Cohn, Emil Georg (1854–1944), 324
Coin-toss experiment, 265–266, 270–271
Coinage, 311
Cold War, 102
Coleman, A. J., 97, 110, 127, 129
Comets, 153–154, 156–157
Commentarii (Petersburger Kommentare), 228
Communism, 96–97
Commutative fields, 115
Completeness (see Model theory)
Compositio Mathematica, 130
Conic sections, 55, 57, 162
Construction of alphabets, 54
Continuity, 314
Continuum, 314–316
Convergence, 267
    Rapid, 268
    Slow, 267
Cook, M., 131
Cooper, Mary, 125, 131
Copernican system, 150–151, 155, 237
Copernicus, Nicolaus (1473–1543), 149–150, 154–156, 162, 165, 210, 237–238, 241–244, 248–250, 255, 259
    De Revolutionibus (1543), 151
Cornell University, 123
    Summer Institute in Logic 1957, 123, 130
Coss (see also German Coss), 312–313
Cossists, 161–162
Counting boards, 286, 309, 312
Cox, Homersham (1821–1897), 329
Coxeter, H. S. M., 94, 100, 124–125, 127
Cracow, Poland, 94, 97, 152–153, 156, 165
    Jagiellonian University, 94
Craig, William, 114, 129
Craig's Interpolation Theorem, 115
Cranfield, 101, 105, 128
    Royal College of Aeronautics, 101, 105
Crowe, Michael, 319
Crusius, Paul (ca. 1525–1572), 157
Cubic equations, 162
Curtze, M. (1837–1903), 174–176, 178–179, 183, 185, 192, 195, 199–201
Czuber, Emanuel, 263

D

d'Abano, Pietro (1250–1315), 240
d'Alembert, Jean le Rond (1717–1783), 233
Danti, Egnazio (1536–1586), 14–15
Danzig, 322
darb-Version (of Euclid's *Elements*), 191, 193
Darmstadt, 322
da Vinci (*see* Leonardo da Vinci)
De Young, G. R., 174, 182–183, 186,
    190–192, 194–195, 197–200
Decline of English mathematics, 264
Deferents, 252
Definitions, theory of, 114
Delft, 103
Della Francesca (*see* Piero della Francesca)
Delos, 58
Delta wing, 104
de Magellan, John Hyacinth (1722–1790),
    356
Demidov, S. S., 141
de Moivre, Abraham (1667–1754), 225, 234,
    263–275
    Doctrine of chances, 264, 266
    Limit value theorem (*see* Binomial distri-
        bution)
    Miscellanea, 273
    Private instruction, 271
    Supplementum, 273
de Montmort (Monmort), Pierre Rémond
    (1678–1719), 226, 234
    Essay d'analyse sur les jeux de hasard, 226,
        234
de Nelis, Corneille François (1734–1798),
    353, 359
de Neny, Patrice François (1716–1784),
    347–355, 359
Desargues, Girard (1591–1661), 12
Descartes, René (1596–1650), 301
    Géométrie, 279
    Regulae ad directionem ingenii, 279
Deutsche Mathematik, 50
de Seumoy, Henri (1720–1798), 349–350, 359
Deutsche Mathematiker-Vereinigung,
    326–327
Deutsche Physikalische Gesellschaft, 325,
    336
Diagrams, 116–117
Dictionary of Scientific Biography, 224
Dietrich, Sebastian (d. 1574), 152–153
Differential equations, 106, 128–129, 227

Diophantos (fl. ca. 250), 161, 167, 281,
    287–292, 300
    Arithmetic, 280
Dirichlet, P. G. L. (*see* Lejeune Dirichlet)
Distance measurement, 158
Djursholm, Sweden
    Institut Mittlag–Leffler, 67, 76, 78, 88
Dörge, K., 113
Doppelmayr, Johann Gabriel
    (ca. 1671–1750), 151–152
Doubling the cube, 160, 162
Dresden, 363, 366, 369, 371–374
    Technische Hochschule, 366
Dürer, Albrecht (1471–1528), 5–6, 49–59
    Schriftlicher Nachlaß, 50, 52
    Underweysung der messung (1525), 50–54
    Underweysung der messung (2nd ed. 1538),
        54
    Vier Bücher von menschlicher Proportion
        (1528), 53, 57–58
Dublin Institute for Advanced Studies, 126
Dubreuil, Jean (1602–1670), 16
Dudith, Andreas (1533–1589), 152–153,
    157–158, 161, 163, 167
Duff, George, 100, 108, 127, 130
Dugac, Pierre, 68–69, 71, 74
Duplication of the cube, 53, 57–58
Dutch Academy of Sciences, 114
Dyads, 329

E

e-function, 268
Earth, central position of, 150
Earthquakes, 108–109, 129
Eber, Paul (1511–1569), 155
Ebert, Hermann (1861–1913), 324
Eccentric, 243
Eccentricity, 239
Eckmann's Theorem, 119
Eckmann, Beno, 118–119
Eclipses, 242
Ecole Polytechnique, 365
Edmonton Summer School 1957, 123–124,
    131
Eggenberger, Johann, 263
Egidiengymnasium, 164
Egorov, D. F. (1869–1931), 138, 141
Egyptian arithmetic, 31
Einschneideverfahrens, 158
Einstein, Albert (1879–1955), 94, 126
    Evolution of Physics, 94

Einstein Institute (*see* Jerusalem)
Electromagnetic telegraph system (*see* Telegraph)
Elements of Geometry (K. fī l-Uṣūl al-handasīya), 207
Eliava, Sh., 146
Empirical confirmation, 269
Engineering, 107
England, 97
Enlargements (*see* Model theory)
Encyklopädie der mathematischen Wissenschaften, 319–326, 333–334, 336
Envelope (of spheres), 57
Epicycles, 155, 239
Equant, 237
Equations
  Algebraic, 307
  Chinese solutions of, 284–287
  Cubic, 297–298
  Geometric solutions of, 279, 292, 295–296, 298
  Higher degree, 292
  Imaginary solutions, 298–299, 301
  Indeterminate, 161
  Linear, 279, 284–287, 291
  Quadratic, 279, 288, 292, 298–299, 312
  Systems of, 284–287, 291
Equivalence
  of mathematical expressions, 263, 271
  of Newtonian/Leibnizian versions of calculus, 263, 271
Erdmann, Otto (1804–1869), 372
Erlangen, 154, 157, 164
  University library, 149, 154
Essex College, Windsor, Ontario, 122
Eton, 164
Euclid (fl. ca. 300 B.C.), 52, 58, 165, 167, 173–175, 177–180, 182, 192–193, 196–199, 213, 240, 281, 287, 292–295, 307, 314–315
  ḏarb-version, 191, 193
  Data (K. al-Muʿṭayāt), 209, 211, 214
  Elements, 53, 160–161, 164, 173, 191, 307, 315
  Elements (K. fī l-Uṣūl al-handasīya), 207, 211, 215
  Fifth postulate, 207
Euclidean solids, 54–55
Eudoxos (ca. 408–ca. 355 B.C.), 160
Euler, F. W., 88
Euler, Leonhard (1707–1783) (*passim*)
  Correspondence with Niklaus I Bernoulli, 223–236
  De infinitis curvis ejusdem generis..., 228
  Euler edition, 223
  Euler-Kommision of the Swiss Academy of Sciences, 223
  Mechanica, 228
  Opera omnia (Euler-Ausgabe), 223, 232–233
Eutocius' Commentary on Archimedes, 53, 57

F

Falco, Jacob (1522–1594), 167
al-Farghānī (died after 861), 237–240, 257
Farnborough
  Royal Aircraft Establishment, 101, 103, 107
Faulhaber, Johannes (1580–1635), 160
Fellmann, Emil A., 223
Fibonacci, Leonardo (*see* Leonardo of Pisa)
Field measurement, 158
Fields Medal, 93
Fields, 117–119, 121, 130
  of characteristic zero, 118–119, 130
Fields, J. C., 93
Figurate numbers, 161
Film Board of Canada, 99
Finaeus, Orontius (1494–1555), 165
Finger numbers, 307, 309
Finsterwalder, Sebastian (1862–1951), 328–329, 331, 336–337, 340–341
First mover, 244
First-order logic (*see* Logic)
"*Fischblase*," 52
Florensky, P. A. (1882–1937), 137, 141–144, 147
  The Pillar and Foundation of Truth, 141
Fluid dynamics, 101–102, 105, 107, 109–110, 129
Föppl, August (1854–1924), 321, 326, 331–333, 341
Folkerts, Menso, 320
Fontaine, Alexis de la (1704–1771), 223, 227–228, 235
Forcing (*see also* Robinson forcing), 116
Fortifications, 158
Foundations of mathematics, 110, 121
Fowell, L. R., 104, 127
Fracastoro, Girolamo (ca. 1478–1553), 165, 241, 255
Fraction, 310

Fraction bar, 310
Fractions, powers of, 315
Fraenkel, A., 122
Franz, W., 113
Frauenberg, 149
Freiberg, Bergakademie, 363
French mathematics, 272
Friedrich August I (1670–1733), 363
Friedrich II (Frédéric, King of Prussia)
    (1712–1786), 233
Fürst, Josef, 99
Function concept, 264, 271, 314
Functions, theory of, 227
Fundamental operations of arithmetic, 310,
    314

G

Galilei, Galileo (1564–1642), 150, 246, 259
Game theory, 226
Games of chance, 266, 270
Gans, Richard (1880–1954), 327, 331–332,
    337, 341
Gauß, Carl Friedrich (1777–1855), 78, 88,
    363, 366, 369–372
Gaurico, Pomponio (ca. 1480–ca. 1530), 51
Geber (*see* Jābir ibn Aflah)
Gebhardt, Hans, 320
Gematria, 29, 34–35, 38–39, 43–45
Geminos (fl. ca. 70 B.C.), 164, 167
Geocentric system, 150, 155
Geodesy, 151, 158, 160, 163
Geometric methods, 264
Geometric proofs, 281, 292, 295–296, 298
Geometrical algebra, 215
Geometry, 127, 160, 162, 314, 316
    Constructive, 51–52, 58
    Descriptive, 57
    Euclidean, 51, 58–59
    Practical, 49, 51, 307
Gephirandus, Thomas (ca. 1600), 167
Gerard of Cremona (ca. 1114–1187),
    173–205, 240, 246, 312
Gerardy, Th., 70, 88
German Coss, 312–313
Gesellschaft deutscher Naturforscher und
    Ärzte, 321, 326, 336, 364
Ghubār numbers, 308–309
Gibbs, Josiah Willard (1839–1903), 321, 326,
    328–330, 333, 337
Gidulyanov, P. V., 142–144, 147
Gilmore, Paul C., 110, 112–113, 129

Gingerich, Owen, 155
Gnedin, Ye., 145
Göttingen, 320–321, 363, 366, 370, 372
Goldbach, Christian (1690–1764), 225, 234
    Goldbach conjecture, 234
Golden section, 314
Goldstein, Bernard R., 243
Goll, August, 69–70, 77
Gothic architecture, 51
Gouzenko affair, 96, 126
Gouzenko, Igor, 126
Grassmann, Hermann (1809–1877), 319, 321,
    325–330, 333–334, 336–337
Grassmann, Hermann, Jr. (1857–1922), 328
Grattan-Guinness, Ivor, 67
Graven, Franciscus Antonius (?–1805), 352
Gravesande, Willem Jacob, 349
Green, David, 99
Gregorian calendar, 157
Grosseteste, Robert (ca. 1168–1253), 314
Groups, 118, 279
Guidobaldo del Monte (1545–1607), 15
Gundisalvi, Dominicus (12th century), 240
Gutzmer, August (1859–1924), 335, 341

H

Hadamard theory, 106
al-Ḥajjāj b. Yūsuf b. Maṭar, 173, 182–183,
    186, 190–194, 198–200, 202
Haifa
    Technion = Israel Institute of
        Technology, 110, 121–122
Halley, Edmond (1656–1743), 225
Ham, James, 125
Hamel, Georg (1877–1954), 326
Hamilton, William Rowan (1805–1865), 324,
    326, 329
Hammel, Laurent (d. 1849), 83
Hamngren, Hans, 25
von Harrach, Friedrich August Gervais
    (1698–1749), 346
Hart, J. F., 101, 127
Heath, T. L. (1861–1940), 195–197
Heaviside, Oliver (1850–1925), 324, 333, 337
Heiberg, J. L., (1854–1928), 173–180,
    182–183, 185, 191–192, 194–195, 197,
    200–202
Heidelberg, 337
Heliocentric theory, 149–150, 155–156
von Helmholtz, Hermann (1821–1894), 65,
    79–81, 84–86, 88–90, 331

Henkin, Leon, 115–116
Herbrand functions, 121
Hermann of Carinthia (first half of 12th
   century), 186, 199–201, 239, 253
Hermann, Jakob (1678–1733), 226
Hermite, Charles (1821–1901), 71, 74
Hero of Alexandria (first century A.D.), 53,
   175, 191–192, 195, 199
Heron (first century A.D.), 53
Hertz, Heinrich (1857–1894), 65, 79–80,
   84–87, 331
Hesse, O., 72
Heun, Karl (1859–1929), 328
Heussenstamm-Stiftung, 83–84, 90
Heydrich, R. (1904–1942), 147
Heyting, Arend, 112, 114
Higginbottom, C. E., 131
Hilbert's Irreducibility Theorem, 112–113
   Nullstellensatz, 117
   17th Problem, 120
Hilbert-Ackermann, *Principles of
   Mathematical Logic*, 111, 124
Hiroshima, 95
Hitler, A. (1889–1945), 137, 143–147
Hofmann, Joseph E. (1900–1973), 50
von Hohenburg, Herwart (1553–1622), 156,
   160, 163, 166
Holbein, Hans (1497–1543), 7, 9
Holland, 112
Holtzmann, Wilhelm (Xylander)
   (1532–1576), 161
Hommel, Johannes (1518–1562), 155, 157
Homocentric spheres, 165
Homocentric world system, 239
Horn angle, 315
Hūlāgū (1217–1265), 210
Humanists, 151, 153, 163–165
Hutzler, Christoph, 167
Hyde, Edward (1843–1930), 329
Hyderabad/Deccan, 211, 216
Hydrodynamics, 128
Hyperbolic differential equations, 104
Hyperbolic logarithms, 267, 273
Hypotheses, 149, 155
Hypsicles (fl. ca. 175 B.C.), 167

I

IBM, 112
Ibn Ezra (ca. 1090–ca. 1164/1167), 238, 240
Ibn al-Haitham (965–ca. 1040), 208, 237–238,
   240

Ibn Mu'ādh (probably 11th century), 238,
   240
Ibn Rushd (1126–1198), 237–239, 241, 243,
   248, 249, 254–259
Ibn Sīnā, 183, 238, 240
   Kitāb al-Shifā', 183
Ibn Tibbon, Jacob ben Machir (ca.
   1236–1305), 238, 243
Ibn Tufail (died 1185), 239, 243, 254
Ibrāhīm ibn Sinān ibn Thābit ibn Qurra
   (908–946), 208–209
   *On Analysis and Synthesis (M. fī Ṭarīq at-
   Taḥlīl wa-l-Tarkīb fī l-masā'il al-
   handasīya)*, 208
Incompressible fluids, 109, 129
Indeterminate equations, 161
Indian mathematics, 288, 292
Indian number system, 308
Indivisibles, 316
"Industrial Party" affair, 142
Industrial Revolution, 364–365
Infeld, Halina, 94
Infeld, Leopold, 93–97, 126
   Evolution of Physics, 94
Infinite series, 225–226
Infinitesimal calculus (*see also* calculus), 263,
   271
Infinitesimal mathematics, 225
Infinity, 314–315, 317
Inner product, 326, 331
Institut für Geschichte der Naturwissen-
   schaften, Mathematik und
   Technik, Hamburg University, 49
Institute for Advanced Study (Princeton), 94
Institute for Defense Analysis (USA), 128
Instruments (*see also* Astrolabes, Armillary
   spheres, Jacobs staff, Quadrants, and
   Sundials), 152, 157
Integration symbol, 264, 271
Intension/remission of forms, 316
Intermediate Books (K. Mutawassiṭāt), 211, 215
   Redaction (taḥrīr), 210
International Congress of Mathematicians
   1954, 130
Intuitionism, 112
Iohanicius Babiloniensis, 201
Ireland, 125
Irrationals, theory of, 161
Isaak Argyros (d. ca. 1375), 167
Isḥāq b. Ḥunain (d. ca. 910), 183, 191–193,
   195, 202, 215
Isoperimetric figures, 315

Israel, 122, 125
Israel Institute of Technology (*see* Haifa)
Izvestiya, 139

J

Jābir ibn Aflaḥ al-Ishbīlī (died mid-13th century), 162, 237–241, 245–246, 254
al-jabr, 215
Jacob, Anatoli (ca. 1194–ca. 1256), 248–255
Jacob, Simon (d. 1564), 162
Jacobs staff, 158
Jacobus Cremonensis (fl. ca. 1450), 51
Jagiellonian University (*see* Cracow)
Jahnke, Eugen (1861–1921), 326–327
Jeffrey, Ralph, 130
Jeffreys, H., 129
Jehuda ben Mose (fl. mid-13th century), 239
Jerusalem, 122, 125, 131
    Einstein Institute, 122
    The Hebrew University, 122–123, 131
Jiuzhang suanshu (*see* Chinese mathematics)
Joachimsthal, 152
Johannes Hispalensis (12th century), 239–240
Johannes de Sacrobosco (ca. 1200–ca. 1256), 150, 239, 249, 255
Jordanus Nemorarius (ca. 1220), 161, 312
Joseph II, Emperor (1741–1790), 357
Journal für die reine und angewandte Mathematik, 75
Journal of Symbolic Logic, 116, 129
Junge, Gustav (1879–1959), 173–174, 176

K

Kamenev, L. B. (1883–1936), 140, 147
Kandelaki, D., 146
Kant, I. (1724–1804), 302, 304
Karlsruhe, 328
Kassel, 154, 326, 328, 336
von Kaunitz, Count, 353
Keisler, J., 117, 129–130
Kepler, Johannes (1571–1630), 150, 156
    Astronomia nova (1609), 156
    Mysterium cosmographicum (1596), 156
van Keulen, Ludolph (1540–1610), 167
KGB, 140–143
al-Khaiyāmī, 'Umar b. Ibrāhīm (1048–1131), 208
al-Khwārizmī, Mukḥ. b. Mūsā (fl. ca. 800–847), 215, 295–298, 300, 310, 312
    al-Kitāb al-mukhtasar . . ., 312

al-Kindī (died ca. 873),
    De Aspectibus, 168
Khinchin, A. Ya. (1894–1959), 138
Kingston, Ontario, 120–121
    Canadian Summer Research Institute 1956, 120–121
    Queens University, 120–121, 130
Kirov, S. M. (1888–1934), 146
Kitāb al-Mafrūḍāt (*see* Aqāṭūn and Thābit ibn Qurra)
Klamroth, M., 174
Klein, Felix (1849–1925), 319–322, 326–328, 336–337
Knoblauch, Johannes (1855–1915), 67–68, 74
Kochen, Simon, 113
Kochs, Peter M. and Anna G., 89
Koenigsberger, Leo (1837–1921), 65, 79–81, 84–87, 89, 90
Kokott, Wolfgang, 167
Kolman, E. (1892–1979), 139, 145
Kolmogorov, A. N. (1903–1987), 138
Korean War, 102
Kowalewsky, Fuffi (1878–1952), 76
Kowalewsky, Sonja (1850–1891), 69, 71, 76, 78
Kramp, Christian, 273
Kreisel, Georg, 131
Krivitsky, W., 146
Kronecker, Leopold (1823–1891), 75, 77
Krzhizhanovsky, G. M. (1872–1959), 140
Kunitzsch, P., 200–201

L

Lag theory, 241
Lagrange Interpolation Formula, 93
Lagrange, Joseph Louis (1736–1813), 233
Laisant, Charles Ange (1841–1920), 337
Laplace, Pierre Simon (1749–1827), 263, 271–274
    Limit value theorem, 263, 268, 271
    Théorie analytique des probabilités, 274
Latin schools, 153
Latitudes of forms, 316
Lattis, James, 255
Laurmann, John A., 105–107, 110–128
    Wing Theory, 110
Law of large numbers, 265
Lay, Juliane, 259
Lechler, Lorenz (ca. 1460–1519), 51
Leibniz, Gottfried Wilhelm (1646–1716), 224–225, 263–264, 271

Leibnizians, 264
Leiden, 162
Leipzig-Dresden railroad, 363, 366, 369–374
  Telegraph signal system, 363, 366, 369–374
Leipzig, 157, 164
  Polytechnic Society (Polytechnische
    Gesellschaft), 363–375
  Sächsische Akademie der Wissenschaften,
    374
  Scientific societies, 364
  Stadtarchiv, 365, 375
  University of, 364–365
Leipziger Akten (= *Acta eruditorum*),
  229–230, 232, 235
Lejeune Dirichlet, Peter Gustav (1805–1859),
  88
Lemmata, K. al-Ma'khūdhāt, 209
Leonardo da Vinci (1452–1519), 9, 54
Leonardo of Pisa (ca. 1170–ca. 1240), 287,
  289–291, 310
  Liber abbaci (1202; 2nd ed. 1228), 289–290,
    310
  Flos (1225), 290
Leuven, University of, 345–361
  Camera Experimentalis, 346–347, 359
  Schola Experimentalis, 345, 347, 349, 354
Levin, A. E., 139
Liebisch, Theodor (1842–1922), 320
Lighthill, M. J., 106
Lightstone, A. H., 110–112, 119–121, 131
Lindemann, Ferdinand (von) (1852–1939), 77
von Lindenau, Bernhard August
  (1779–1854), 364
Linear perspective, 54
Linearized theory, 106
Lines, 315
Lingelsheimius, G. M. (17th century), 166
Lipschitz, Rudolf (1832–1903), 79–81, 84–86,
  88–90
Liu Hui (3rd century), 285–286
ln *n*!, 267
Löwenheim-Skolem Theorem, 124
Lockheed Aircraft (Missile and Space Divi-
  sion), 128
Locomotives (*see* Railroads)
Loeffler, Hans, 83–84
Logarithms, 160, 267, 274, 314–315
  Hyperbolic, 267, 274
Logic (*see also* Mathematical logic, Symbolic
  logic), 110, 112, 121–124
  First-order logic, 115
Logistic, 160

Lomazzo, Paolo Giovanni (1538–1600), 14
London, 98
London, the Royal Society of, 225–226
Lorch, Richard Paul, 255
Lorentz, Hendrik Antoon (1853–1928),
  321–322, 324–326, 333, 338
Lower predicate calculus, 120
Lüroth, Jakob (1844–1910), 328–329, 331,
  337, 341
Luther, Martin (1483–1546), 150, 152
Luxemburg, W. A. J., 125
Luzin, N. N. (1883–1950), 137–148
Lyndon, R., 115, 129
Lyusternik, L. A. (1899–1981), 138

M

MacFarlane, Alexander (1851–1913), 337
Maestlin, Michael (1550–1631), 150, 154
Magic Squares, 160
Magini, Giovanni Antonio (1555–1617), 250
Maignan, Emmanuel (1601–1676), 16
Makowski, J. A., 129
al-Ma'mūn (caliph, reigned 813–833), 238
von Mangoldt, Hans (1854–1925), 322
Manitoba (Winnipeg), University of, 96
Mantino, Jacob (16th century), 239
Manuscripts, 154, 160–161, 163–164
  Augsburg MS Staats- und Stadtbibliothek
    8° Cod 1, 312
  Brugge MS Stadsbibliotheek 529, 183
  Escurial MS Derenbourg 907, 190, 194,
    199
  Istanbul
    MS Aya Sofya 4832, 220
    MS Köprülü 930/13 and 931/13, 210
    Topkapi Saray MS 3456/14, 210
  Leiden MS Univ. Libr. Arab. 399.1,
    173–205
  Leipzig MS Univ. Bibl. 1470, 313
  Leningrad MS Akademia Nauk, C 2145,
    198–199, 202
  London MS Brit. Libr., Burney 275, 202
  Madrid MS Bibl. Nac., 10010, 174
  Nürnberg MS Stadtbibliothek Cent 5, app
    56c, 313
  Oxford MS Bodl. Libr.
    Arch. Seld. A45, 217, 220
    Arch. Seld. A46, 220
    Lyell 52, 313
    Marsh 709, 217, 220
    Thurston 11, 199...

Paris MS Bibl. Nat.
   Arab. 2467, 217, 220
   Persan 169M, 190
   Suppl. lat. 49, 211
   Rabāṭ MS Ḥasanīya 53 and MS Ḥasanīya
      1101, 190, 194
   Teheran MS Malik 3586, 192
   Upsala MS Univ. Libr., O. Vet 20, 199
   Vatican MS Bibl. Apost.
      Reg. lat. 1268, 174
      Ross. 579, 198
   Vienna MS ÖNB 5277, 315
Maps, 158–159
Marāgha, 210
Maria Theresia, Empress (1717–1780), 353,
   357
Marolois, Samuel (1572–ca. 1626), 15
Marxism, 94
Mason's square, 31–32
Mass, 308, 311
Mathematical logic, 110–112, 123–124, 131
Mathematical physics, 101
Mathematical Quarterly (Riveon Lemate-
   matica), 122
Mathematical Reviews, 113, 121
Mathematics, 151, 154, 163
Maupertuis, Pierre-Louis Moreau de
   (1698–1759), 233
Maurolico, Francesco (1494–1575), 165
Max-Planck-Gesellschaft, 85, 89–90
Maximilian II, Holy Roman Emperor
   (1527–1576), 152
Maximus Planudes (1255–1310), 167,
   280–281
Maxwell, James Clerk (1831–1879), 321,
   323–325, 333
McCarthyism, 96
Mean value theorem, 316
Means (proportions), 161
Mechanical quadrature, 270, 272
Mechanical solutions, 160
Medea, 99
Mehmke, Rudolf (1857–1944), 319–341
Melanchthon, Philipp (1497–1560), 150,
   152–153, 164, 240
Menelaos (fl. ca. 100), 163, 165, 213
Menshov, D. E. (1892–1988), 138
Mensula Praetoriana, 158
Mercury, 242, 245–246
   phases of, 245–246
   transit of, 245
Merton College (Oxford), 164

Meßtisch, 158–159
Messung, 49, 58–59
Metamathematics (see also Model theory),
   115, 118–121
Mexico, 98
Mikhajlov, Gleb K., 223
Minckelers, Jan Pieter (1748–1824), 357–358
Minus, 281–284, 300, 307, 310
Mittag-Leffler, Gösta (1846–1927), 69,
   71–72, 74, 77–78, 88
Model theory, 110, 112–120, 128, 130–131
   Completeness, 115–121, 129–131
   Enlargements, 122
   Model companions, 130
   Persistence, 115–116
Models (of the planets), 156
Models, 106, 115
   Prime models, 130
Molodshii, V. N., 139, 145
Molotov-Ribbentrop pact, 145
Monge, Gaspard (1746–1818), 57
van Mons, J. B., 358
Montreal, Canada, 96
   University of, 115
Moscow
   Academy of Sciences of USSR, 138, 140
   Moscow Committee of the Party, 139
   Moscow School of the theory of functions,
      141
   Moscow University, 139, 141, 143
   State Electro-Technical Institute, 141
   Theological Academy, 141–142
Moses (ben Samuel ben Jehuda) ben Tibbon
   (13th century), 240
Motion,
   of the earth, 156
   of planets, 156
Mozart, W. A. (1756–1791), 30–38, 42, 45
   as Freemason, 30–32, 37
   Last three symphonies (1788), 35–37
   Die Zauberflöte (1791), 30–34, 45
   Sarastro, 31–34, 43, 45
Müller, Conrad (1878–1953), 321
Müller, Emil (1861–1927), 328
Müller, Hans, 79–80, 83
Müller, L. August, 80, 89
Müller, Peter-Wilhelm (1788–1881), 80,
   82–83, 89
Multiplication, 310
Multiplication tables, 300, 310
Munich, 154, 163, 167, 338
al-muqābala, 216

Murdoch, J. E., 174, 182, 185
Murs, Jean de (first half of the 14th century)
    De arte mensurandi, 53
al-Mutawakkil (caliph, 847–861), 208

N

n! (*see also* ln *n*!), 267
Napoleon I. (1769–1821), 364
al-Nairīzī (Anaritius, fl. ca. 897–922),
    173–176, 178, 183, 186, 199–201
NASA, 128
Naṣīr al-Dīn al-Ṭūsī (1201–1274), 182,
    198–199, 207–208, 210–211, 215
    *Redaction (taḥrīr) of the Intermediate Books
    (K. Mutawassiṭāt)*, 198, 207, 210,
    216–217
National Academy of Sciences (USA), 128
National Fascist Center, 137–148
Naturphilosophie, 250
Navier-Stokes equations, 106
Nazi "Brownshirts," 146
Nazi fascism, 145–146
Nazi-Soviet pact, 137
Needham, John Turberville (1713–1781), 345,
    351–353, 359
Nemorarius (*see* Jordanus Nemorarius)
Neuenschwander, Erwin, 66, 69
Neuhoff, Klaus, 80
von Neurath, K. F. (1873–1956), 146
Neusis, 213
New Brunswick, Canada, 110
New Haven (USA), 155
New York Times, 126
New York University, 101
Newton, Isaac (1643–1727), 225, 263–264,
    271, 274, 356
    Principia mathematica..., 225, 264
Newtonians, 264
Niagara-on-the-Lake (Canada), 98
Niceron, Jean-François (1613–1646), 3, 11,
    14–16, 23–26
Nicomachus (fl. ca. 100), 161, 307
Nifo, Agostino (1473–1538/1545), 241
Nizhny Novgorod, 142
NKVD, 146
Nollet, Jean Antoine (1700–1770), 349, 352
Nominalism, 248
Non-Euclidean geometry, 208
Nonius, Petrus (ca. 1502–1578), 161–162,
    165, 240
Nonuniform supersonic flow, 102–103

Normal distribution, 263–264, 272
    General case, 263, 270
    Symmetric case ($p = \frac{1}{2}$), 263
North-Holland Publishing Co., 117, 130
Novikov, P. S. (1901–1975), 138
Nürnberg, 51, 53, 149, 152–154, 157–158,
    163–164, 167
    German National Museum, 152
Nürtingen, 332
Number concept, 303, 308, 314
Number mysticism, 317
Number symbols, 307–308
Number systems
    Arabic, 308–310
    Base 10, 308
    Binary, 311
    Decimal, 311
    Duodecimal, 311
    Indian, 308–310
    Infinite, 315
    Positional, 308, 310
    Sexagesimal, 311
Number theory, 161, 207
Number words, 308
Numbers, 308
    As sets of units, 303
    Complex, 301
    Fictive, 299, 301, 304
    Indian, 288
    Irrational, 313, 315
    Negative, 279–306
    Rational, 315
    Subtractive, 279–281, 300–301
Numerical evaluation, 268
    of probability, 270
Numerical expansions, 270
Numerology, 29–45
Nuñez (*see* Nonius)

O

Observations, 155
von Oettingen, Arthur (1836–1920), 326
Ogonyok, 141
OGPU, 140, 142–145, 147
Ohio State University, 125
Olschki, Leonardo (1885–1962), 50
Omar Khayyam (*see* al-Khaiyāmī)
Operation words/symbols (*see also* Symbols),
    308, 310, 312
    Algebraic, 308
    Arithmetic, 310

Operations, 308
Optics texts, 164
Order and motion of the heavenly spheres, 241, 253
Oresme, Nicholas (ca. 1320–1382), 315–317
Orthogonal trajectories, 223, 227–228
Osiander, Andreas (1498–1552), 149, 156
Ostroukhov, 142, 144
Ottawa, Canada, 101, 110
Outer planets, 246
Outer product, 325–326, 331, 333
Oxford (*see also* Manuscripts), 163–164, 167

P

Pacioli, Luca (1445–1517), 51, 54–55, 59, 162
Panofsky, Erwin (1892–1968), 54
Pappos (fl. ca. 320), 54, 167
    Collection, 212–213
Parabolas, 162
Parallax, 157, 242
Parallel postulate, 314
Paris (*see also* Manuscripts), 155, 163, 167
    Académie des sciences, *Mémoires*, 228
    CNRS, 113
    Colloquium 1950, 113
Partial differential equations, 223–236
Partial differentiation, 233
Partitio numerorum, 226
"Party for the Rebirth of Russia," 142, 144
Patterson, G. N., 102, 127
Paumgartner, Hieronymus (16th century), 157
Peano, Giuseppe (1858–1932), 329, 337
Pearson, Karl, 263
Pennsylvania State University, 112
Pentagons, 162
Persistence (*see* Model theory)
Perspective, 4, 7, 11, 314
    Negative or decelerated, 6
Peter-Wilhelm-Müller-Stiftung, 65, 78–87, 89–90
Petrus de Regio (13th century), 239
Peucer, Caspar (1525–1602), 152–153
von Peuerbach, Georg (1423–1461), 237–259
Philoponos, Johannes (died second half of sixth century), 167
Philosophy, 314–317
Photogrammetrie, 328
Physics, 108, 112, 155, 314–317
    Experimental, 345–361

Piero della Francesca (ca. 1410–1492), 3, 7, 9–10, 54–55
Pirckheimer, Willibald (1470–1530), 51–53
Place value system, 310
Plan and elevation method, 56–57
Planck, Max (1858–1947), 80, 85, 87–90
Planudes (*see* Maximus Planudes)
Platonism, 156
Plato (427–347), 5, 53, 57, 248
    Timaeus, 248
Plato of Tivoli (first half of 12th century), 240
Platonov, S. F., 142, 144, 147
Plus, 307, 310
Pohl, Hans Friedrich, 367–369
Poland, 95–97, 126
    Polish Academy of Science, 97
Poleni, Giovanni (1683–1761), 226
Politbüro, 146
Polygonal numbers, 161
Polygons (regular), 52
Polyhedra
    Development of, 54–55
    Paper folding of, 55
    Regular (or Euclidean), 54–55
    Irregular (or Archimedean), 54–55
Polytechnic "movement," 365–366
Polytechnic schools (*see also* Ecole Polytechnique), 364, 375
Porcelain manufacture, 363
Pounder, I. R., 122–123, 131
Power series, 264, 274
Powers, 313
Praetorius, Johannes (1537–1616), 149–169
    Astronomy, 154–158
    Correspondence, 163
    Geodesy, 158–160
    Humanist, 163–165
    Lectures
        Hypotheses astronomicae, 155
        Mensula Praetoriana, 158
        Theoriae planetarum, 155
    Library of, 151, 156
    Life, 152–154
    Manuscripts, 154
    Mathematics, 160–163
Prague, 152
Prandtl, Ludwig (1875–1953), 319, 325–330, 334–337, 340–341
Pravda, 139
Precession, 156
Pre-Copernican astronomy, 242

Predicate calculus, 111, 115, 124
Prime Model Test, 118
Primum mobile, 249
Princeton University, 126
Priority, 265
Private instruction, 271
Probability
  Numerical determination of, 270
Probability theory, 226, 264
  Limit value theorem, 264
Proclus (410–485), 248
Product (*see* Inner product, Outer product,
  Scalar product, Vector product)
Prognostication, 154, 157
Progressions
  Arithmetic, 315
  Geometric, 315–316
Proportion, theory of, 160
Proportionals, geometric and mean, 160
Proportions, 314–315
  Irrational, 315
Propositional calculus, 124
Prutenic tables, 150
Pseudo-Archimedes, 207
  Book of Assumptions (Kitāb al-Mafrūḍāt),
    207
  *Elements of Geometry (K. fi l-Uṣūl al-
    handasīya)*, 207
Pseudo-Bede, 291
Pseudo-Euclid
  Optics, 53
Pseudo-Ṭūsī (*see* Ṭūsī)
Ptolemaic astronomy, 237
Ptolemaic system, 150, 155, 208
Ptolemy, Claudius (ca. 100–ca. 170), 156, 164,
  167–168, 237, 240–250, 254–255, 259,
  284
  Almagest, 164, 211, 237
  Optics, 168
Pühlheim, 158
Pyatakov, G. L. (1890–1937), 145–146
Pyenson, Lewis, 95
Pythagoreanism, 156
Pythagoreans, 161, 213

Q

Qazi, Nizam Ahmed, 125
Quadrant, 153 157
Quadrature of the circle, 315
Quadrupling the cube, 58
Quantifier elimination, 117

Quantum mechanics, 101
Quaternions, 326, 329–330
Queens University (*see* Kingston, Ontario)
Quercu, Simon a (Duchesne), 167

R

Radzivilovsky, 143
Raeder, J., 173
Railroads, 365, 369–374
Ramsden, Jesse (1735–1800), 356
Ramzin, L. K., 142
Rayleigh waves, 109
Réaumur, René Antoine Ferchault de R.
  (1683–1757), 284
Real numbers, 115–116
Real-closed fields, 117, 130
Reclam, Anton Philipp (1807–1896), 364
Regensburg, 153
Regiomontanus, Johannes (1436–1476), 152,
  154, 156, 162, 166, 237, 240–241,
  243–244, 246–249, 251–254, 257, 259,
  311–313, 317
  Regiomontanus-Walther collection, 51
Regular solids, 161–162
Reinhold, Erasmus (1511–1553), 150,
  152–153, 155
Relativity, theory of, 94
Relf, E. F., 105
Representations, analytic, 272
Retrograde motion, 244
Rheticus, Georg Joachim (1514–1574),
  149–150, 152, 155–156, 163, 165, 238,
  241–242
Richard of Wallingford (1292–1335), 255
Riemann method, 104
Ries, Adam (1492–1559), 300
Risner, Friedrich (d. 1580), 240
Riveon Lematematika, 122
Robert of Chester (ca. 1150), 186, 199–200,
  296, 312
Roberval, Gilles Personne (1602–1675), 24
Robinson forcing, 116
Robinson's Consistency Theorem, 115
Robinson, Abraham (1918–1974), 93–136
  Complete Theories (1956), 116–120,
    130–131
  On the Metamathematics of Algebra (1951),
    115, 119
Robinson, Gilbert, 94, 125–127
Robinson, Renée, 98–99, 123–125
Rockefeller Foundation, 94

Roehm, E. (1887–1934), 146
Roman number system, 310
Roman practical geometry, 307
Rome, 337
Rooney, P. G., 127
Roritzer, Matthäus (died ca. 1492–1495), 51–52
Ross, R. A., 99–100, 103, 108–109, 123, 129
van Rossum, Adriaan (1705–1789), 351–353
Rotor, 325
Royal Aircraft Establishment (*see* Farnborough)
Royal Canadian Air Force, 112
Royal Canadian Mounted Police, 96
Royal College of Aeronautics (*see* Cranfield)
Rudio, Alice, 66–67
Rudio, Ferdinand (1856–1929), 66–68, 70, 73–74
Runge, Carl (1856–1927), 322, 331–332, 337, 341
Rupprich, Hans, 50, 53,

S

Sacrobosco (*see* Johannes de Sacrobosco)
St. Egidien, Cloister of, 153
Saint John's College, Cambridge, 112
St. Petersburg, Academy of Sciences, 228, 234
Sauvage, Joannes Petrus (1725–1771), 349, 351–353, 359
Savile, Henry (1549–1622), 163–164, 167
Saxonius, Petrus (1591–1625), 151, 154, 159, 161
Scalar product, 324, 333–334
Scaliger, Joseph (1540–1609), 162, 167
Schikaneder, E. (1751–1812), 30–35
Schmuttermayer, Hans (end of the 15th century), 51
Schön, Erhard (1491–1542), 7–8
Schöner, Johannes (1477–1547), 240
Schubert, Johann Andreas (1808–1870), 365
Schuler, Wolfgang (d. 1575), 153
Schulz, H., 88
Schwarz, Hermann Amandus (1843–1921), 68–70
Schweinfurt, 165
    Municipal library, 149, 151, 154–155, 157
Schweizer Lexikon 91, 224, 234
Schwenter, Daniel (1585–1636), 159, 161
Scientific societies, 364
Scotus, Michael (died 1235), 243, 246

Screw, 57
Scriba, Christoph J., 3, 49, 93, 366
SD, 147
Sehnenvierecke, 154, 162
Seleucids, 282
Sergiev Posad, 142
    Moscow Theological Academy, 141–142
    Troitse-Sergieva Lavra, 142
Series, 267–268
    Expansions, 268, 271
    for ln *n*!, 267
Set theory, 121
Sets, 303
Sets, infinite, 315
Sexagesimal system, 163
Sezgin, F., 182
Shadows (of the sun), 157
Shakespeare, *Othello*, 99
Shelah, S., 129
Shelby, Lon R., 51–52
Shentalinski, V., 141–142
Shnirelman, L. G. (1905–1938), 138
Shuleiko, 143
Sidereal orbit, 244
al-Sijzī, 'Abd al-Jalīl (ca. 945–ca. 1020), 212
Sine tables, 311
Sine theorem, 254
Skolem normal form, 111
Slesinsky, R., 141
Smith, Sidney, 96, 123, 126, 131
Sommerfeld, Arnold (1868–1951), 319–341
Soviet Academy of Sciences (*see* Moscow)
Soviet Embassy in Berlin, 145–146
Soviet Union, 95–96, 126
Spanish Civil War, 145
Spheres, 162
Spherical geometry, 163
Spherical triangles, 163
Spherical trigonometry, 207
Spiral (Archimedean), 55–56
Sporus (2nd century A.D.), 53
Sputnik, 95
Square root, 288, 310
Squaring the circle, 162
Staigmüller, Hermann (1857–1908), 50–51
Stalin terror, 138
Stalin, I. V. (1879–1953), 137–138, 140, 145–147
Stalinism, 97, 126
Standard variation, 264
Standlinie, 158
Stanford University, 128

Statics, 208
Steam engines, 364–365
Steam ships, 365
Steck, Max (1907–1971), 50
St. Egidien, Cloister of, 153
Steinheil, Carl August (1801–1870), 370
Steketee, J. A., 97–98, 103–104
Stephan Batory, 97
Stephan Batory University (*see* Vilna)
Stevenson, A. F. C., 97, 101
Stevin, Simon (1548–1620), 15, 281, 292, 298, 300
Stiegler, S. M., 271, 274
Stifel, Michael (ca. 1487–1567), 160, 162, 281, 299, 301, 313
Stirling, James, 267
Stirling's formula, 267
Stokes' Law, 129
Stonemasons, 51, 56–57
Strassburg, 153
Strauss, Walter L. (1932–1988), 50
Studies in Logic and the Foundations of Mathematics, 117, 130
Stupa, Anton (died 1551), 239
Sturm, Johann (1507–1589), 153
Stuttgart, 320, 322
Substitutions, 271
Subtraction, 291, 303–304
Suez Crisis (1956), 125
Summation of infinite series
  Geometric methods, 316
Summation theorems for probability of disjoint events, 265
Summer Institute in Logic (*see* Cornell University)
Sun (shadows of), 157
Sundials, 152, 157
Supernova of 1572, 153
Supersonic aircraft, 101–103, 105–106, 128
Surveying, 158
Suslin, M. Ya. (1894–1919), 138
van Swieten, Baron W. (1734–1809), 37, 44
Swineshead, Richard (fl. 1340–1355), 316
Symbolic logic, 113
Symbolism, mathematical, 264, 268, 271
Symbols, 307–308, 310
  For integration, 264
  Fractions (division), 310
  Gobar symbols, 308–309
  Multiplication, 310
  Operational, 280, 299–300
  Square root, 310

Synge, John L., 94, 97, 125
Synodic period, 244
Syntactical transforms, 111, 119

T

Tables
  Multiplication/division, 160
  Prutenic, 150
  Trigonometric, 163
Taḥrīr, 198
Tallies (*Kerbholz*), 307, 311
Tangents, 315
Tarle, Ye. V., 142
Tarski, A., 116–117, 130
Tartaglia, Nicolo (ca. 1500–1557), 162
Tasso, Torquato (1544–1595), 241
Tausch des Lehrstuhls, 226
Technical Encyclopedia, 142
Technion (*see* Haifa)
Technische Hochschule, 364
Telegraph system, 363, 366, 369–374
Teleological argument, 247
Tensor, 324
Teubner, Benedikt Gotthelf (1784–1856), 364
Textile industry, 364
Thābit ibn Qurra (836–901), 182, 192–193, 197–200, 207–222, 237–238
  Book of Assumptions (Kitāb al-Mafrūḍāt), 207–222
  On the Verification of Problems in Algebra by means of Geometrical Proofs (Qawl fī tashīḥ masā'il al-jabr bi-l-barāhīn al-handasīya), 215
Thaer, Clemens (1883–1974), 182, 200
The Ensign, 96
The Hebrew University (*see* Jerusalem)
Theodosios of Bithynia (first century B.C.), 163–164, 167
Theon of Alexandria (4th century), 164, 197
Theon of Smyrna (fl. early 2nd century), 167
Theoretical physics, 95
Theory of equations, 301
Thermodynamics, 321
Thijsbaert, Joannes Franciscus (1736–1825), 345, 353–359
Thomson, W., 173
Tibbon, Moses ben (*see* Moses ben)
Timerding, Heinrich Emil (1873–1945), 321

Toronto
 Radio station, 99
 University of, 93–136
  Department of Applied Mathematics,
   94, 97, 100–110, 126
  Department of Mathematics, 93, 98, 126,
   131
  F.R.O.S., 125, 131
  Institute for Aerospace Studies, 102
  Institute of Aerophysics, 102
  The Varsity, 131
Torre, Giovanni Battista de la (fl. 1270), 241
Transfer principles, 119
Transfinite numbers, 121–122
Transformations, 264
Transits, 242
Trepidation theory, 156
Triangles, 162–163
Trigonometric functions, 163
Trigonometry, 152, 160, 163, 310–311, 316
Triklinios, 167
Tripling of the cube, 58
Trisection of the Rectilineal Angle, 212–213,
 217
Tschertte, Johann (ca. 1480–1522), 52
Tübingen, 164
Tukhachevsky, 145–146
Tummers, P. M. J. E., 174
Turmair, Johannes (Aventinus) (d. 1534), 309
al-Ṭūsī (*see* Naṣīr al-Dīn al-Ṭūsī);
 Pseudo-Ṭūsī (ca. 1300), 200, 238
Twisted column, 57

U

Uniform/difform motion, 316
United States, 95, 98, 110
Unity, 314, 317
Uryson, P. S. (1898–1924), 138
Utility, 49, 51, 57–58

V

Vacuum, 247
Vanishing point, 314
Variability of speed, 249
Variable distance of sun/moon, 256
Varignon, Pierre (1654–1722), 225, 234
Vaught, R. L., 130
Vaulezard, Jean-Louis (17th century), 3,
 17–18, 20–24, 26
Vector calculus, 319–341

Vector Commission, 319, 326–337
Vector product, 324–326, 331, 333–334, 338
Vectors, polar and axial, 321, 325, 330, 332
Venice, 155
Venus, 242
 Epicycle, 251
 Phases, 245–246, 259
 Transits, 245
Verapanish, Vichien, 125
Viète, François (1540–1603), 162, 165, 300
Vienna, 98–99, 163, 166, 320, 328
Vienna Congress (1815), 364
"Vierung" (square), 52
Vignola, Jacopo Barozzi da (1507–1573),
 14–15
Vilna, Lithuania
 Stephan Batory University, 122
Visiereinrichtung, 158
Visierrute, 162
Voigt, Woldemar, 321, 324–325
Volume of the sphere, 162

W

Wachsmuth, Richard, 86
Walcher, Heinrich, 71, 73, 88
Wallis, John (1616–1703), 302
 Arithmetic (1685), 302
Wapowski, Bernhard (16th century), 156
 Wapowski letter, 156
Warburg, Emil (1846–1931), 80, 85–90
Warsaw, Poland, 126
Wasserwaage, 158
Watson Research Center (*see* Yorktown
 Heights)
Watson, T. J., 112
Waves, 108
 Elastic waves, 108
Weber, Eduard Friedrich (1806–1871), 370
Weber, Ernst Heinrich (1795–1878), 370
Weber, Heinrich (1842–1913), 304, 324,
 327–328, 336, 340
Weber, Wilhelm Eduard (1804–1891), 363,
 366, 370–372
Weierstraß, Clara (1823–1896), 71, 76–78
Weierstraß, Elise (1826–1898), 67, 71–74
Weierstraß, Franz (1882–1898), 65, 67–78,
 88
Weierstraß, Karl (Carl) (1815–1897), 65–91,
 303–304

Weierstraß, Oskar, 68, 72–74
Weierstraß, Peter (1820–1901), 67, 73
Weight, 311
Werner, Abraham Gottlob (1749–1817), 363
Werner, Johannes (1468–1522), 52–53, 152, 156, 163
Whitehead, Alfred North (1861–1947), 329
Widmann, Johannes, 307
Wiechert, Emil (1861–1928), 324–325
Wien, Willy (1864–1928), 320, 322, 324–325
Wigner, Eugene, 124
Will, G. A. (1727–1798), 158
Wilson, Edwin Bidwell (1879–1964), 329, 337
Wind tunnels, 106
Windsor, Ontario, 122
Wing theory, 101, 105–108, 110
Winkelmessungen, 159
Witelo (Vitellio) (ca. 1230–1280), 241
Wittenberg, 149–153, 155, 164–165
  University of, 165
Wuyts, Petrus (1736–1788), 359

X

Xenophon (ca. 430–ca. 354 B.C.), 164
Xylander, *see* Holtzmann, W.

Y

Yaḥyā ibn Abī Manṣūr (died 832), 208
Yale University Library, 128, 155
Yezhov, N. (1895–1938), 146
Yorktown Heights, N.Y.
  Watson Research Center, 112
Young, Alec, 107
Yushkevich, A. P. (1906–1993), 140

Z

Zacuto, Abraham bar Samuel (ca. 1450–ca. 1522), 238, 240
Zakon, Elias, 110, 121–122, 131
Zamberti, Bartolomeo (1473–ca. 1539), 53
al-Zarqāllu (died 1100), 238, 240
Zedler, Johann Heinrich (1705–1763), 364
Zeitschrift für Mathematik und Physik, 322
Zermelo, Ernst (1871–1953), 326
Zero, 309–310
Zhirikhin, 142
Zhukovsky, N. Ye. (1847–1921), 141
Zinner, Ernst (1886–1970), 151, 154, 166
Zinoviev, G. E. (1883–1936), 140, 147
Zodiac, 157–158

ISBN 0-12-204055-4

90065

9 780122 040559

DATE DUE

Demco, Inc. 38-293